Topics in Applied Physics Volume 46

Topics in Applied Physics Founded by Helmut K. V. Lotsch

1 **Dye Lasers** 2nd Ed. Editor: F. P. Schäfer

2 **Laser Spectroscopy** of Atoms and Molecules. Editor: H. Walther

3 **Numerical and Asymptotic Techniques in Electromagnetics** Editor: R. Mittra

4 **Interactions on Metal Surfaces** Editor: R. Gomer

5 **Mössbauer Spectroscopy** Editor: U. Gonser

6 **Picture Processing and Digital Filtering** 2nd Edition. Editor: T. S. Huang

7 **Integrated Optics** 2nd Ed. Editor: T. Tamir

8 **Light Scattering in Solids** Editor: M. Cardona

9 **Laser Speckle** and Related Phenomena Editor: J. C. Dainty

10 **Transient Electromagnetic Fields** Editor: L. B. Felsen

11 **Digital Picture Analysis** Editor: A. Rosenfeld

12 **Turbulence** 2nd Ed. Editor: P. Bradshaw

13 **High-Resolution Laser Spectroscopy** Editor: K. Shimoda

14 **Laser Monitoring of the Atmosphere** Editor: E. D. Hinkley

15 **Radiationless Processes** in Molecules and Condensed Phases. Editor: F. K. Fong

16 **Nonlinear Infrared Generation** Editor: Y.-R. Shen

17 **Electroluminescence** Editor: J. I. Pankove

18 **Ultrashort Light Pulses** Picosecond Techniques and Applications Editor: S. L. Shapiro

19 **Optical and Infrared Detectors** 2nd Ed. Editor: R. J. Keyes

20 **Holographic Recording Materials** Editor: H. M. Smith

21 **Solid Electrolytes** Editor: S. Geller

22 **X-Ray Optics.** Applications to Solids Editor: H.-J. Queisser

23 **Optical Data Processing.** Applications Editor: D. Casasent

24 **Acoustic Surface Waves** Editor: A. A. Oliner

25 **Laser Beam Propagation in the Atmosphere** Editor: J. W. Strohbehn

26 **Photoemission in Solids I.** General Principles Editors: M. Cardona and L. Ley

27 **Photoemission in Solids II.** Case Studies Editors: L. Ley and M. Cardona

28 **Hydrogen in Metals I.** Basic Properties Editors: G. Alefeld and J. Völkl

29 **Hydrogen in Metals II** Application-Oriented Properties Editors: G. Alefeld and J. Völkl

30 **Excimer Lasers** Editor: Ch. K. Rhodes

31 **Solar Energy Conversion.** Solid-State Physics Aspects. Editor: B. O. Seraphin

32 **Image Reconstruction from Projections** Implementation and Applications Editor: G. T. Herman

33 **Electrets** Editor: G. M. Sessler

34 **Nonlinear Methods of Spectral Analysis** Editor: S. Haykin

35 **Uranium Enrichment** Editor: S. Villani

36 **Amorphous Semiconductors** Editor: M. H. Brodsky

37 **Thermally Stimulated Relaxation in Solids** Editor: P. Bräunlich

38 **Charge-Coupled Devices** Editor: D. F. Barbe

39 **Semiconductor Devices** for Optical Communication Editor: H. Kressel

40 **Display Devices** Editor: J. I. Pankove

41 **Computer Application in Optical Research** Editor: B. R. Frieden

42 **Two-Dimensional Digital Signal Processing I.** Linear Filters Editor: T. S. Huang

43 **Two-Dimensional Digital Signal Processing II.** Transforms and Median Filters. Editor: T. S. Huang

44 **Turbulent Reacting Flows** Editors: P. A. Libby and F. A. Williams

45 **Hydrodynamic Instabilities and the Transition to Turbulence** Editors: H. L. Swinney and J. P. Gollub

46 **Glassy Metals I** Editors: H.-J. Güntherodt and H. Beck

47 **Sputtering by Particle Bombardment I** Editor: R. Behrisch

48 **Optical Information Processing** Fundamentals. Editor: S. H. Lee

49 **Laser Spectroscopy of Solids** Editors: W. M. Yen and P. M. Selzer

Glassy Metals I

Ionic Structure, Electronic Transport, and Crystallization

Edited by H.-J. Güntherodt and H. Beck

With Contributions by
H. Beck J. L. Black P. J. Cote P. Duwez T. Egami
H.-J. Güntherodt J. Hafner U. Herold
W. L. Johnson U. Köster A. P. Malozemoff
L. Meisel J. Wong

With 119 Figures

Springer-Verlag Berlin Heidelberg GmbH 1981

Professor Dr. *Hans-Joachim Güntherodt*

Institut für Physik, Universität Basel, Klingenbergstraße 82,
CH-4056 Basel, Switzerland

Professor Dr. *Hans Beck*

Institut de Physique, Université de Neuchâtel, Rue A.-L. Brequet,
CH-2000 Neuchâtel, Switzerland

ISBN 978-3-662-30898-1 ISBN 978-3-540-38477-9 (eBook)
DOI 10.1007/978-3-540-38477-9

Library of Congress Cataloging in Publication Data. Main entry under title: Glassy metal. (Topics in applied physics; v. 46,) Bibliography: v. 1, p. Includes index. 1. Metallic glasses. I. Beck, Hans, 1939– II. Güntherodt, Hans-J., 1939–. TN693.M4G57 620.1′4 80-28575

Originally published by Springer-Verlag Berlin Heidelberg New York in 1981
Softcover reprint of the hardcover 1st edition 1981

2153/3130-543210

Preface

The field of glassy metals or metallic glasses has seen enormous development during recent years. For the uninitiated the notion of a "metallic glass" may seem rather unusual, since one would primarily associate "glass" with transparent and insulating window material. However, the term glass is nowadays almost unanimously used for an amorphous substance which is obtained by cooling the corresponding melt. Besides glassy metals there are other types of amorphous metals produced, for instance, by evaporation or sputtering techniques. In contrast to most dielectric glasses, conducting materials have to be quenched very rapidly in order to overcome crystallization in the cooling process. Thus the development of such quenching techniques was the necessary condition for an intensive and systematic investigation of glassy metals. The rapid progress and the broad interest in this field can be seen by the fact that the third international conference on rapidly quenched metals in Brighton, 1978, gathered 235 participants from eighteen nations.

Metallic glasses are new substances with exciting properties which are of interest not only for basic solid-state physics but also for metallurgy, surface chemistry, and technology. For the physicist interested in pure research they offer a unique opportunity to study bulk disordered metallic systems in solid form and to find connections with corresponding facts in the liquid phase. The theoretical description of ionic and electronic structure of amorphous materials is, of course, an outstanding problem which is, far from being adequately solved, in full development at the present time. Moreover metallic glasses have properties which are quite unexpected for solid metals and make them promising candidates also for technical applications. In order to list but a few facts, the following properties can be found in metallic glasses: high mechanical ductility and yield strength, high magnetic permeability, low coercive forces, unusual corrosion resistance, and temperature-independent electrical conductivity. Metallic glasses for magnetic shielding purposes are already on the market and applications for transformer materials and tape recorder heads seem to be underway.

The present volume is the first of two devoted to the physical properties of metallic glasses. Experimentalists and theoreticians active in the field present recent results and actual research problems in this domain of condensed matter physics. After an introduction which is meant to create a general and unifying framework for the various topics, P. Duwez gives a personal historical introduction into the development of the subject of metallic glasses, which began, about twenty years ago, in his laboratory. The topics covered in this first volume are the following:

- Special methods of structure investigation: EDXD (T. Egami) and EXAFS (J. Wong)
- Special aspects of ionic dynamics, namely low-energy excitations – which have their counterpart in insulating glasses (J. L. Black) – and Brillouin scattering (A. P. Malozemoff)
- A pseudopotential approach to a first principle treatment of structure and stability of simple metallic glasses (J. Hafner)
- Electrical transport properties: normal electrical conductivity (L. V. Meisel and P. J. Cote) and superconductivity (W. L. Johnson)
- An overview of the crystallization processes which occur when metallic glasses are heated up (U. Köster and U. Herold).

The second volume should include the following further topics: structure determination by model building and by X-ray and neutron scattering techniques, defects, mechanical and elastic properties, magnetism electron spectroscopy and the theory of electronic structure, corrosion resistance, and application.

The editors acknowledge stimulating discussions, active help, and valuable cooperation of their collaborators and colleagues working in the field of metallic glasses. Thanks are also due to all of the contributors for keeping the deadlines and for their cooperation.

Basel and Neuchâtel, January 1981 *H.-J. Güntherodt H. Beck*

Contents

Contributors

Beck, Hans

Institut de Physique, Université de Neuchâtel, Rue A.-L. Breguet 1, CH-2000 Neuchâtel, Switzerland

Black, James L.

Physics Department, Brandeis University, Waltham, MA 02154, USA

Cote, Paul J.

Physical Science Section, Benet Weapons Laboratory, Watervliet Arsenal, Watervliet, NY 12189, USA

Duwez, Pol E.

California Institut of Technology, Pasadena, CA 91125, USA

Egami, Takeshi

Department of Material Science and Engineering, University of Pennsylvania, Philadelphia, PA 19104, USA and Max-Planck-Institut für Metallforschung, Institut für Physik, D-7000 Stuttgart 80, Fed. Rep. of Germany

Güntherodt, Hans-Joachim

Institut für Physik, Universität Basel, Klingelbergstraße 82 CH-4056 Basel, Switzerland

Hafner, Jürgen

Institut für Theoretische Physik I, Technische Universität Wien, Karlsplatz 13, A-1040 Wien, Austria

Herold, Ursula

Institut für Werkstoffe, Ruhr-Universität Bochum, Postfach 102148, D-4630 Bochum, Fed. Rep. of Germany

Johnson, William L.

Department of Material Science, California Institute of Technology, Pasadena, CA 91125, USA

Köster, Uwe
Institut für Werkstoffe, Ruhr-Universität Bochum, Postfach 102148, D-4630 Bochum, Fed. Rep. of Germany

Malozemoff, Alexis P.
IBM Thomas J. Watson Research Center, P.O. Box 218, Yorktown Heights, NY 10598, USA

Meisel, Lawrence V.
Physical Science Section, Benet Weapons Laboratory, Watervliet Arsenal, Watervliet, NY 12189, USA

Wong, Joe
General Electric Corporate Research and Development, 1 River Road, P.O. Box 8, Schenectady, NY 12301, USA

1. Introduction

H. Beck and H.-J. Güntherodt

With 4 Figures

1.1 Introductory Remarks

Metallic glasses (MGs) are the subject of an increasing research effort, spurred by both science and technology. The research in this field helps our understanding of noncrystalline matter in general. The interpretation of the properties of MGs imposes a particular challenge since the understanding of crystalline solids has, in the past, generally been based upon their crystal periodicity. A fascinating theoretical concept based on translational invariance has been developed to deal with the lattice dynamics and electronic structure of crystalline matter. No such general theory has yet been developed for the disordered state. This volume and [1.1] deal with the experimental results and theoretical approaches for a special class of disordered solid metals. Amorphous metals can be prepared by a variety of methods: 1) Evaporation of metals in vacuum and condensation of their vapor on a cooled substrate. 2) Sputtering, by which the atoms are removed from the source under bombardment with energetic inert gas atoms. 3) Chemical or electroless deposition, a method in which ions in aqueous solution are deposited onto substrates by chemical reactions. 4) Electrodeposition, where the chemical reaction requires the presence of an external potential. 5) Rapid quenching from the liquid state. Amorphous alloys prepared by the latter method are the so-called metallic glasses or glassy metals. We concentrate in this book on amorphous metals made by melt quenching. Results on other kinds of amorphous metals are taken into consideration only where data on glassy samples are not available or for the sake of comparison.

1.1.1 Preparation

The basic principle of obtaining MGs by rapid quenching from the melt is that the liquid must be converted very rapidly from a droplet or jet into a thin layer in contact with a highly thermally conductive metal to produce thin splats or ribbons. The cooling rates achieved by some rapid quenching devices are around $10^6\,\mathrm{Ks}^{-1}$. Nowadays there are two main techniques used.

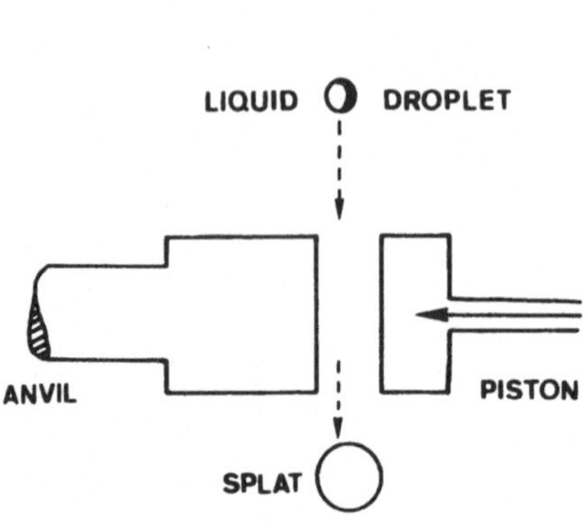

Fig. 1.1. Splat-cooling technique

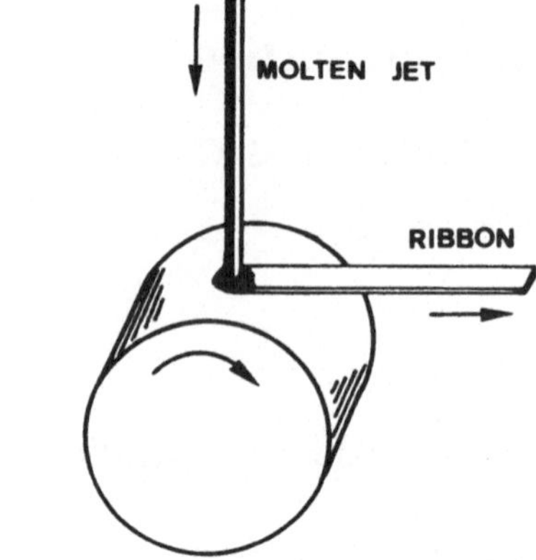

Fig. 1.2. Melt-spinning technique

a) Splat Cooling

For instance the piston and anvil method (Fig. 1.1), in which a liquid alloy droplet is squeezed between a rapidly moving piston and a fixed anvil. Also two rapidly moving pistons can be used. The obtained splat is between 15 and 30 mm in diameter and 20–80 μm thick. Such rapid quenching devices are suitable for basic research and for ultrahigh vacuum conditions. The splat cooling devices seem to give somewhat slower cooling rates than the technique described next.

b) Melt Spinning

In this method a rapidly spinning copper or steel wheel is used to conduct the heat away rapidly and continuously from the melt. In its simplest form (Fig. 1.2) this process is suitable to produce metallic glasses of relatively easy glass formers, e.g., $Fe_{40}Ni_{40}B_{20}$ and $Cu_{70}Zr_{30}$, at speeds up to 2 km per min in air. The ribbons have typical dimensions of 1–3 mm width and 20–60 μm thickness. More details and other modifications are covered by all the recent review papers. This is the reason why we did not include a chapter on methods of making metallic glasses in this book. However, there are important details, not yet discussed in the literature, to go beyond the simplest version. Those further steps are wide sheets up to 10 or 15 cm, the use of vacuum, and the handling of mass production items. Progress seems to be being made in industry. The methods of direct solidification of ribbons, foils, sheets, and wires from the melt represent a remarkable technological advance. This single step process is one of the reason for the growing industrial interest in this field and may well be used also for crystalline metals.

Other techniques to obtain metallic glasses are still under development. These are laser glazing [1.2], electric field emission of ions from the melt [1.3], and electric arc furnace quenching [1.4].

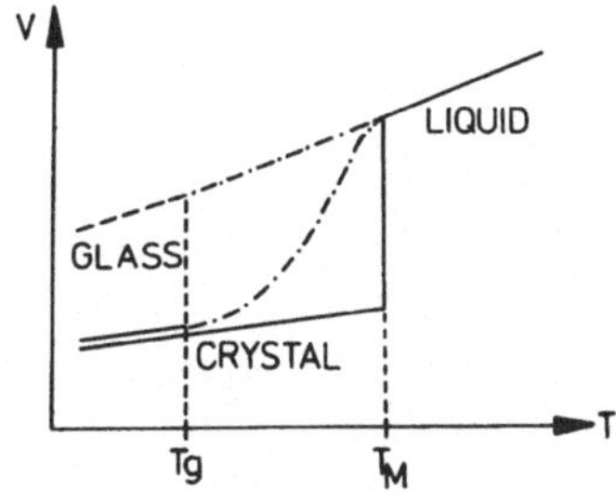

Fig. 1.3. Volume versus temperature

At present, the following families of metallic glasses are known: T–N (e.g., $Fe_{80}B_{20}$, $Pd_{80}Si_{20}$), N–N (e.g., $Mg_{70}Zn_{30}$, $Ca_{70}Mg_{30}$), T_L–T_E (e.g., $Ni_{60}Nb_{40}$, $Fe_{24}Zr_{76}$), RE–N (e.g., $La_{70}Al_{30}$), RE–MN (e.g., $La_{80}Au_{20}$), RE–T_L (e.g., $Gd_{70}Co_{30}$, $Gd_{70}Ni_{30}$), and U–T (e.g., $U_{70}Fe_{30}$, $U_{70}Cr_{30}$), where T is transition metal, N is polyvalent metal, T_L is late transition metal, T_E is early transition metal, RE is rare earth metal, MN is monovalent noble metal, and U is uranium.

The question of glass forming ability has been of great interest. Every experimental group has its own recipe on how to find alloy concentrations forming stable metallic glasses. Basically one consults the phase diagrams of the alloys. Nowadays, there are so many new metallic glasses known that one can start to ask whether there is any underlying general principle or not.

1.1.2 Characterization

Metallic glasses show strong similarities to ordinary glasses and liquid metals. Figure 1.3 shows the volume versus temperature for insulating (dashed) and metallic (solid line) materials in the liquid, crystalline, and glassy states. The indicated temperatures T_g and T_M are the temperatures of the glass transition and melting point, respectively. In contrast to ordinary glasses, the volume of metals is nearly the same in the glassy and crystalline states. Metallic glasses often show a reversible glass-liquid transition at the glass temperature T_g, which is manifested in the specific heat or viscosity. The schematic behavior of these properties is shown in Fig. 1.4. The abrupt increase in specific heat accompanies the sharp, reversible decrease in viscosity at T_g. The observed reversible changes in viscosity and specific heat suggest that metallic glasses, like glasses of other types, can revert to the undercooled liquid state without crystallization and that their atomic arrangements are closely related to those present in corresponding liquid alloys. At a temperature T_c only slightly higher than T_g the metallic glasses undergo recrystallization.

Since 1975, the field of metallic glasses has grown very rapidly, but it still seems to be possible to list all the existing reviews, books, and conference proceedings. Only three appeared before 1975; Since then their number has

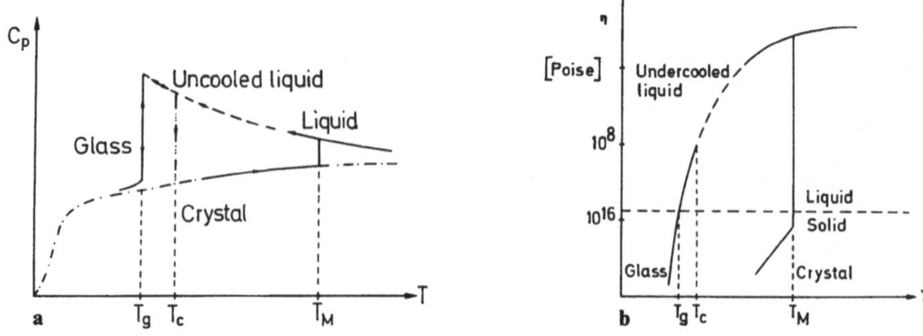

Fig. 1.4a, b. Specific heat (**a**) and viscosity (**b**) versus temperature

grown enormously. The bibliography gives an overview. The further development of the field will be seen at the large international conferences in Grenoble (LAM 4, July 7–11, 1980) and in Sendai (RQ IV, August 24–28, 1981).

1.2 Experiments

1.2.1 Static Structure

It is amazing in view of the quite successful study of liquid metal alloys, aqueous solutions, and ionic liquids how few details of scattering experiments from metallic glasses are evaluated. Until quite recently, the only information about the atomic scale structure of metallic glasses, in particular from the T–N group, had been derived mainly from single wavelength x-ray scattering experiments yielding the total structure factors or interference functions. Such structural information is easily obtained in the course of checking the glassy state of the prepared samples by x-rays. The status of our knowledge on the basis of this level was reviewed by several authors (see [1.5–8]). This is the main reason why we included the chapter on x-ray and neutron scattering experiments in [1.1]. In order to get more profound information on the structure, careful quantitative scattering experiments are called for, which seem at present to be underway. Unambiguous experimental structure determination is of great importance for comparison with statistical models which have been built. Usually from the total structure factor the radial distribution function (RDF) is calculated by Fourier inversion. To draw conclusions from a radial distribution function, it is necessary to start with a statistical model, from which a theoretical RDF can then be computed for comparison with the experiments.

A metallic glass in its simplest form is a binary alloy *AB*. Therefore, in general the diffraction experiments should determine the partial structure factors $S_{AA}(q)$, $S_{AB}(q)$, and $S_{BB}(q)$. Consequently, a set of three different experiments with different scattering factors and high accuracy is necessary.

The availability of melt-spun ribbons of metallic glasses facilitates the application of neutron scattering, for which large specimens are required. The three partial structure factors can be determined by the following techniques: isotope substitution in n scattering experiments and three different scattering experiments using neutrons, x-rays, and electrons on the one hand, and unpolarized neutrons, polarized neutrons, and x-rays on the other hand. In a few cases, information from the total structure factor can be extracted by using the Bathia-Thornton formalism in terms of $S_{CC}(q)$, $S_{NN}(q)$, and $S_{NC}(q)$, the correlation-correlation, the number-number, and the number-correlation functions by applying the so-called "zero alloy" method. The appropriate concentration of two elements having positive and negative scattering lengths can yield a zero scattering length for the alloy. Other, less direct methods make use 1) of the large anomalous dispersion terms in the x-ray scattering factor for wavelengths near the absorption edge of the scattering element to reveal the partial structure factors by using three different wavelengths and 2) of chemical substitution, in order to get the partials.

New experimental techniques such as pulsed neutron scattering methods, small angle scattering, Mössbauer effect, EXAFS (extended x-ray absorption fine structure), and EDXD (energy dispersive x-ray diffraction) have been introduced to the field of metallic glasses. Pulsed neutron scattering methods can provide a high resolution radial distribution function in the high wave vector region, which can show structural details of short range order not detected in previous diffraction studies. Small angle scattering seems to be sensitive to small clusters and defects. EXAFS has the unique capability of probing the near neighbor environment in multicomponent systems where the EXAFS from each element can be studied. The study of the Mössbauer effect is another tool to probe the local environment (see Chap. 4). The main application of the EDXD technique lies in real time study of the kinetics of structural relaxation by annealing and in the study of compositional short range order of metallic glasses, as described in Chap. 3.

Much current attention is directed towards the analogue of "defects" in metallic glasses. Such imperfections [1.9] may be defined as deviations in the structure of the metallic glass from that of the fully relaxed (ideal) metallic glass at 0 K. Direct or indirect evidence for the existence and nature of imperfections in metallic glasses comes from a variety of observations: the uniform chemical attack of etchants, the density, positron annihilation, $\gamma-\gamma$ angular correlation, viscous flow, diffusion, small angle neutron scattering, and crystal growth. In addition this field has been studied by computer simulation of defects.

1.2.2 Ionic Dynamics

Once the static structure is characterized in terms of local coordinations, short range order, and pair correlation functions, time-dependent processes can be investigated. One class of experiments detects the (collective) motion of the ions around some equilibrium position, corresponding to standard "lattice dynamics" in crystalline solids. Being macroscopically isotropic the response of MGs to mechanical forces is characterized by two elastic constants. The compressibility is in general not much different from crystalline alloys of the same composition. However, the shear modulus of many MGs is up to 20–30% smaller than in the crystalline phase. Besides ultrasonic techniques Brillouin scattering (Chap. 5) has also been used to measure sound velocities via the intermediate of the detection of surface waves. Recently, inelastic neutron scattering experiments have been performed on Cu–Zr, Ca–Mg, and Mg–Zn, aiming at the dynamical structure factor $S(q,\omega)$ and vibrational densities of states. Concerning collective phononlike excitations, glassy metals seem to lie between liquid metals, where longitudinal collective modes have been seen up to relatively high wave vectors, e.g., in Rb, and polycrystals for which the various phonon branches can be seen, though averaged over all spatial directions. The present neutron results for $S(q,\omega)$ do show inelastic tails besides a quasielastic peak, but a definite q variation could not yet been identified.

While these experiments in the realm of "lattice dynamics" of metallic glasses can probably be understood on the basis of interionic forces as in crystals, provided that the effect of disorder is properly taken into account, there are other results concerning thermal, elastic, and transport properties of the ion system which seem to be very specific for glasses. Measurements of sound velocity and attenuation, specific heat, and thermal conductivity of some MGs have revealed the same low temperature peculiarities as in insulating glasses. For example, the specific heat shows a term linear in T at very low T and the thermal conductivity decreases about as T^2, which are both quite different from crystalline matter. These observations, summarized in Chap. 8, are specially remarkable, because they demonstrate a basic similarity of two types of disordered materials, namely insulating and conducting glasses. This points to the fact that the origin of these anomalies lies in the disorder itself, rather than in a specific type of bonding between the atoms. On the other hand, some special results, like the numerical values of some relaxation times, are quite different in the two types of glasses, which shows the definite influence of the presence of mobile conduction electrons.

A disordered structure also allows for dynamical processes which are far less important in a regular lattice at not too high temperatures: motion and relaxation of "defects" and diffusion of particles through open channels in the random structure. From a mechanical point of view any small region of a disordered material should be in a force-free configuration when the system is in equilibrium. However, thermal motion or external strains can induce stresses. A regular system relaxes in general by going back to the *original* stress-free configuration, but an amorphous array of particles can also relax towards a *new*, slightly different state. Thus measurements of frequency and temperature dependence of, for example, the attenuation of elastic vibrations have shown peaks which can be interpreted as being produced by thermally activated, stress-induced relaxation processes.

On a much larger scale such mechanisms are certainly important in the region of the glass temperature where the glass is "ready" to leave the metastable state into which it was frozen during the quenching process, and in the various steps of recrystallization, as described in Chap. 10.

1.2.3 Electronic Structure and Transport

An important aim of modern solid-state physics is to obtain information on the electronic band structure. The most direct experimental techniques are electron, x-ray, and optical spectroscopy in terms of photoemission (UPS: **u**ltraviolet **p**hotoemission **s**pectroscopy, XPS: **x**-ray **p**hotoemission **s**pectroscopy) experiments on valence bands, XPS core level spectroscopy, AES (**A**uger **e**lectron **s**pectroscopy), SXS (**s**oft **x**-ray **s**pectroscopy), and optical reflectivity. However, the crystalline state of metals and alloys was mainly considered in the experiments for the following reasons: Experiments on noncrystalline metals and alloys can be carried out using the liquid or the amorphous state. The liquid state causes experimental difficulties in electron spectroscopy experiments due to the high vapor pressure. In the past, amorphous metals were generally not available, for example, at room temperature, where such experiments are mainly performed. Only recently the metallic glasses have opened a new field for systematic studies of the electronic structure of noncrystalline metals. The problems which can be addressed by these studies are the following: The electronic band structure of noncrystalline metals and alloys allows one to study the influence of the lack of crystal periodicity on the electronic states. Moreover, the density of states is the key for the explanation of many physical properties such as superconductivity, magnetism, compound formation, etc. The relationship between the band structure and the glass forming ability is of great interest. Only very few results on electron spectroscopy, SXS, and optical reflectivity are known. In order to cover the subject in a more comprehensive way and to review the recent progress this field is described in [1.1].

Our information on the electronic transport properties comes from the measurements of the electrical resistivity, the thermopower, and the Hall coefficient. There is a substantial quantity of resistivity data, as reviewed and discussed in Chap. 7. The experimentally observed magnitude, temperature dependence, and composition dependence of the electrical resistivity of metallic glasses at high temperatures ($T \geqq T_D$; where T_D is Debye temperature) are comparable to liquid metal data. The observation of a small positive, zero, or negative temperature coefficient of the electrical resistivity depends on the concentrations and can be changed continuously on alloying. The negative temperature coefficient of resistivity, observed in both liquid and amorphous states, played a key role in the theoretical description. However, there is little published information on the thermoelectric power of metallic glasses. The observed linear variation of the thermoelectric power over a large temperature range seems in accord with the diffraction model used to explain the resistivity. The study of the thermopower is particularly valuable to test theories of the electrical resistivity since it is given by the energy derivative of the electrical resistivity. The Hall coefficient of simple metal glasses indicates nearly free electron behavior. The positive Hall coefficient of liquid and glassy alloys containing transition or rare earth metals is still an unexplained fact. Nevertheless, data on the thermopower and the Hall coefficient might become very important in order to test new ideas on the electrical resistivity in the high resistivity regime, where the Mooij correlation between the resistivity and its temperature coefficient is observed. The proposed phonon-assisted hopping [1.10] which might occur even before localization should show up in the temperature and concentration dependence of the thermopower and the Hall coefficient. Those who are interested in more details on the Hall coefficient are referred to a recent review covering the Hall effect in amorphous alloys [1.11].

A number of MGs are superconducting, though the transition temperatures do not exceed 10 K. This might just reflect what a simple mixing of metals shows in superconductivity. An enhancement of superconductivity above 10 K seems to go back to the well-known influence of crystal structure and the lattice. As far as open questions and controversies are concerned the superconductivity of MG seems the most advanced field. The point is even reached where applications become feasible (Chap. 9).

1.2.4 Magnetic and Mechanical Properties

These two subjects are sharply divided into two fields. The first one deals with the application-oriented properties of the technologically significant alloys. Those MGs containing generally 75 to 85 at. % transition metals (Fe, Co, Ni) based on group T–N are ferromagnetic with Curie temperatures well above room temperature. They are extremely soft magnetic materials with very high initial and maximum permeabilities and very low coercive fields. In addition, they are resistant to plastic deformation and hence do not loose their high permeability on handling as do crystalline high permeability alloys. It is well established that these materials have power losses at 50–60 Hz that are much lower than those of the best commercial Fe–Si alloys. Recent research activities have yielded a substantial quantity of magnetic data on saturation magnetization, Curie temperature, temperature and field dependence of magnetization, low field properties, alloying behavior, magnetic anisotropy, magnetostriction, domain structure, magnetization processes, and magnetic microstructure (see, e.g., [1.12]).

Furthermore, metallic glasses exhibit a remarkable combination of mechanical properties. They are often very strong and stiff. The strongest at present known, $Fe_{80}B_{20}$, is stronger than the best carbon fiber and almost as stiff. Unlike other high-strength filaments, MGs are capable of some plastic deformation, though in tension the plastic elongation to fracture is generally less than 1%. By rolling, however, strains of over 30% can be attained in some glasses. This deformability leads to a high fracture toughness. MG ribbons are therefore much less sensitive to surface defects than are ordinary oxide glass fibers. These new materials are not brittle like silicate glasses, even at very high stress levels. In addition, metallic glasses have high hardness and good wear resistance. Another exceptional property of MGs is their high bend fatigue strength. These application-oriented properties will be described in the chapter on applications in [1.1].

The second field is more concerned with the basic properties of ferro-, para-, and diamagnetism in MGs containing T, RE, and simple metals, their dilute alloy, and spin glass behavior. These studies provide a good opportunity to explore further details of magnetism in the noncrystalline state. For the T–N group the questions which are addressed are always associated with ferromagnetism [1.13]. These are forming of magnetic moments, their relation to local environments, exchange interaction, phase transitions, magnetic excitations, etc. A central question of ferromagnetism in MG is the possible description by the theory of itinerant magnetism and hence the relation of magnetism to the electronic band structure. The more basic properties in the field of mechanical properties of metallic glasses are related to insight into the structure and the behavior at an atomic level. Nowadays the investigation and understanding of strength, ductility, toughness, deformation, and fracture of metallic glasses are of great interest. The developments in basic research of both the magnetic and mechanical properties are covered by two separate chapters in [1.1].

1.3 Theoretical Approaches

Any theoretical treatment of a specific property of a metallic glass has to cope with the difficult problem of electronic structure and "lattice dynamics" of disordered solids. The state of the art in this domain of condensed matter physics has recently been reviewed in a very comprehensive way by *Ziman* [1.14]. The basic Hamiltonian of a metallic system is of the form

$$\mathscr{H} = T_{\text{ion}} + T_{\text{el}} + V_{\text{ion-ion}} + V_{\text{el-el}} + V_{\text{el-ion}}\,, \tag{1.1}$$

where T denotes the kinetic energies of ions and conduction electrons and the V terms stand for the interactions among the ionic cores, the conduction electrons, and between the two subsystems. For simple metals the latter interaction would be given in terms of pseudopotentials, whereas for TM and RE ions one may use a muffin-tin atomic potential of the type used in band structure calculations. Starting from (1.1) one can aim at dealing with two kinds of problems:

i) In the spirit of the adiabatic approximation the positions of the ions are considered to be given, fixed parameters. One then calculates equilibrium structure and transport properties of the interacting *electrons* moving in the static external potential due to the ionic cores. (For some problems it may be important to generalize this concept in order to take into account possible energy transfer between electrons and "lattice vibrations" of the ions; see the discussion of resistivity below.) In order to get rid of the dependence of the results of such a calculation on the specific ionic configuration chosen, a suitable configurational average has to be performed at the end. This introduces second-, third-, and higher order ionic correlation functions characterizing the random ionic structure. Typical examples of such ensemble averaged electronic quantities are electronic self-energies for pseudopotential systems [1.15] or Ziman's formula [1.16] for the resistivity of simple liquid metals.

ii) In order to calculate, on the other hand, effective interactions and dynamical properties of the *ion system*, it is useful to eliminate the electronic degrees of freedom. The total free energy of the metal is evaluated as a function of the ionic coordinates (treating the ions classically), which were the input for i). This allows one, for example, to express the static energy as a sum of a homogeneous electron gas term, single-ion self-energies and interionic potentials between pairs, triples, etc., of ions. Unfortunately, this procedure can only be realized in a quantitative way for pseudopotentials, i.e., for simple metals, for which it is described at various places in the literature [1.17]. Normally the analysis goes as far as including ionic pair potentials (linear electronic screening). Already heterovalent alloys of simple metals present serious problems due to the strong inhomogeneities of the electron gas. Nevertheless, the pair potential approach to thermodynamics and ionic dynamics of simple metals is very useful.

Obviously the final goal would be to achieve a self-consistent combination of i) and ii), in the sense that a given ionic structure determines the electronic structure, this latter giving rise to effective interionic forces which again determine the ionic structure one started with. Such a concept is realized to some extent in Chap. 6 for simple MGs. Many of the known MGs contain, however, transition and rare earth metals.

After this rather general description of a first principle approach to the treatment of a metallic glass we would like to mention some specific aspects of concrete calculations, which are found in this book.

1.3.1 Electronic Properties

First we should emphasize again that it is only on the basis of weak pseudopotentials that quantitatively reliable calculations can be performed on the same systematic footing (e.g., second-order perturbation expansion with respect to $V_{\text{el-ion}}$) for the three interesting quantities, electronic density of states, electrical resistivity, and effective ionic pair potentials.

The latter are obtained by using the best available dielectric function of the homogeneous interacting electron gas ("linear screening"). The pair potentials then show their characteristic long-range oscillations, replacing the monotonic attractive tail of, say, a Lennard-Jones potential. There amplitudes may be somewhat overestimated in the literature because finite mean free paths effects on the screening electrons due to $V_{\text{el-ion}}$ are usually neglected. These oscillations seem to play an important part in the discussion of stability and forming ability of simple MGs; see below.

The corresponding Born approximation for the electrical resistivity ϱ, appropriate for metallic glasses, has been derived by *Baym* [1.18]. When T is greater than the Debye temperature, his result goes over into the well-known Ziman formula widely used for liquid metals.

Here we want to stress one important fact: in the framework of Ziman's approach the T dependence of ϱ is governed by that of the structure factor $S(q)$. Negative temperature coefficients of ϱ (NTC) are obtained when $2k_F$ lies in the vicinity of k_p, the position of the main peak of $S(q)$, which corresponds to about two or somewhat less electrons per ion. As a matter of fact a great majority of MGs show NTCs, at least at not too low temperatures. This strongly supports the idea that glassy alloys typically have compositions which yield about this number of conducting electrons per ion. This will be discussed again in connection with stability and formation of simple MGs. Negative TCs of resistivity seem to occur quite generally in disordered metallic systems with a relatively high resistivity, i.e., with a short electronic mean free path (Mooij correlation). Various attempts have been made to explain this fact from a tight-binding point of view, rather than a perturbation calculation for quasifree electrons. In such a framework NTCs seem to arise as some kind of precursors of localization due to disorder. At very low T some MGs show Kondo-like minima in $\varrho(T)$ (see Chaps. 7 and 8) which are still not well understood.

Calculations of the density $N(E)$ of electronic states on the same level of perturbation theory (usually the electron Green's function is calculated self-consistently to second order) show that $N(E)$ should in general deviate rather weakly from the free electron parabola. However, some small dip is expected for energies E corresponding to a wave vector $k \equiv \sqrt{2mE}$ of the order of $k_p/2$. Since k_p roughly replaces the first (spherical averaged) shell of reciprocal lattice vectors of a close-packed crystal structure this dip is reminiscent of the gap expected in a crystal at that energy, at least in some lattice directions. Unfortunately there seem to be no detailed experimental data confirming or disproving such a local minimum in $N(E)$ and showing its exact location.

Turning to alloys containing TMs or REs the theoretical problems are much more difficult. As to the electrical resistivity, Baym's result has been generalized by replacing the pseudopotential by a suitable ionic scattering matrix (for amorphous metals see Chap. 7, for liquid metals see, e.g., [1.19]). Numerical calculations on this basis for liquid TMs and REs have led to good agreement with experiment. As an important point in this t matrix approach the basic idea of the T dependence of ϱ through $S(q)$ remains valid. So NTCs for such alloys again point to there being about 1.5 to 2 effective conduction electrons per ion.

The *d* band density of states for TM alloys are usually obtained by some tight-binding calculations taking the distribution of neighbor distances (which determine the values of the hopping terms) from some structural model, or by coherent potential-type approximations in a muffin-tin approach (see [1.1]). Such calculations can be used to investigate typical differences between $N(E)$ for crystalline and glassy structures, which directly influence observable quantities like magnetic susceptibility χ, superconducting transition temperature T_c, etc. For the purpose of a unified approach to equilibrium electronic structure (characterized by density of states data) and electronic transport (estimate of number of conduction electrons, etc.) it is highly desirable to develop and improve methods taking into account *s*- and *d*-like states and their hybridization on the same footing.

Other aspects of the electronic properties of metallic glasses, like magnetism and superconductivity, are discussed in the specific chapters of this book and in [1.1]. A very important and challenging aspect is also the extreme corrosion resistance of certain metallic glasses. However, compared to the knowledge of surface physics of crystalline metals, little is known about the surface of noncrystalline metals.

1.3.2 Ionic Properties

Here the first aim is, of course, to understand the disordered static structure of a glassy alloy on the basis of the interionic interactions. Since the most important part of metallic pair potentials is a relatively strong repulsive core most of the structure models are based on hard spheres. Neglecting all details it has in fact turned out that metallic glasses (like liquid metals for that matter) are quite well described by dense random packings of hard spheres with large numbers of neighbors, in contrast to amorphous semiconductors where, due to covalent bonds, the idea of random networks is more appropriate. Thus, for many calculations, the use of Percus-Yevick structure factors for hard sphere mixtures with additive diameters is a good approximation. However, there exist important deviations from random packing, to be discussed in the various chapters devoted to structure and structure determination. They concern the number of nearest neighbors of different species, the nonadditivity of nearest neighbor distances, short-range order, etc.

Rigid ionic model structures correspond in some sense to the structure of a (classical) metallic glass at $T=0$. At finite T the ions will vibrate around their equilibrium positions. Moreover diffusion processes and structural rearrangements can take place. A simple theoretical approach to the vibrational properties is to consider the harmonic approximation by introducing spring constants between the ions. Although the equations of motion are linear, the problem of evaluating the dynamical structure factor $S(q, \omega)$ or vibrational densities of states is still formidable due to the disorder in the spring constants and the ionic positions. Thus one has to resort to simple approximations (Debye models, continued fraction representations, etc.) or to perform dynamic computer simulations. Detailed knowledge of $S(q, \omega)$ is not only of interest by itself, but it is also required to evaluate the electrical resistivity according to Baym's formula. Thus $S(q, \omega)$ provides a link between ionic dynamics and low-temperature electrical transport. Concerning the main features of the density correlation function $S(q, \omega)$, current calculations predict, besides an elastic peak for all q (corresponding to elastic scattering of neutrons from a rigid ionic configuration) a longitudinal collective mode. Its dispersion would look like the phonon-roton spectrum of superfluid helium showing a minimum for $q \approx k_p$. For such wave vectors the excitations would, however, already be strongly damped. The transverse mode, not directly seen by neutrons, is predicted to have a rather monotonic frequency versus wave vector behavior.

In the limit of long wavelengths even a disordered system will behave like an elastic continuum in which longitudinal and transverse sound waves propagate, whose velocities are determined by the elastic constants. Concerning the latter the shear moduli of many MGs are smaller than those of corresponding crystalline alloys by up to 30%. It is not quite clear whether such a drastic decrease can be understood by simply taking the appropriate average over the interionic spring constants (corresponding to the sums over neighbor shells in a crystal) or whether one ought to take into account local rearrangements of the random packing which can be induced by the elastic strain of sound waves. Such Debye-type relaxation processes have been mentioned to show up also directly in the frequency and temperature dependence of sound absorption.

The low T anomalies of metallic glasses (mentioned in Sect. 1.2.2), which are in many respects very similar to those found in insulating glasses, have been explained by the same model, i.e., by postulating the existence of structural units which can tunnel between two energetically almost degenerate configurations. The details of these processes are as obscure as for insulating glasses, but their existence in MGs seems to rule out any origin related specifically to the bonding properties of nonconducting glasses. They rather seem to be a very general consequence of the high degree of (quasi-) degeneracy in configuration space of a random structure, as opposed to a crystal lattice of specified symmetry, which – given the density – is fully determined, up to rigid translations in space. The major difference between dielectric and conducting glasses is that the motion of these units, represented by "two-level systems", is coupled to the conduction electrons in the latter case. This additional interaction, besides the two-level system phonon interaction, has been used to clarify the typical differences between the two glasses with respect to the low-temperature anomalies (see Chap. 8), but a detailed understanding of the consequences of this interaction for the low T transport properties of metallic glasses is still lacking.

It was proposed some time ago for dielectric glasses (see, e.g., the review in [1.20]) – and the point may equally be relevant to metallic glasses – that there exists a broad distribution of such "bistable" units with very different potential energy curves leading to different tunnelling energies. Thus some fraction of such systems would give rise to the anomalies at low *T*, where quantum mechanical tunnelling is the dominant process, whereas other systems (being possibly of a different geometrical nature) would get accross the barrier between the energy minima by (classical) thermally activated and stress-induced jumps. Therefore the specific frequency and temperature dependence of sound velocities and ultrasonic absorptions in glasses, which deviates from a slow variation produced by anharmonic interactions, may have a similar origin at high and at low temperatures.

One of the most challenging theoretical problems, however, seems to be the understanding of the ability of a certain metallic alloy to form a glass upon quenching, and of its thermal stability against recrystallization. It would evidently be very desirable to predict these properties in a quantitative manner on the basis of simple criteria and calculations. The problem has a thermodynamic (equilibrium) and a kinetic (dynamic) aspect. Thermodynamic calculations should provide information about the free energy of the glassy phase of a metal, which can be compared with the energy of possible stable crystalline phases. Since the amorphous phase is at best metastable, the "kinetic" aspect is, however, at least as important: one should determine the energetically favorable paths and the barrier heights in phase space between the glassy and competing stable phases.

The kinetic aspects of glass forming have been investigated in connection with the rapid cooling process through which the glass is obtained. One of the important ingredients of such an approach is the temperature dependence of the viscosity in the *T* range between supercooled liquid and metastable glassy solid [1.21]. This quantity is a measure for the ability of the system to "escape" from the liquidlike disordered state during quenching. Obviously, the higher the quench rate the easier it is to obtain a truly amorphous metal by this procedure. It is an interesting open question whether it will be possible to quench pure elemental metals into glasses, provided that still higher cooling rates can be achieved.

In order to judge beforehand whether an alloy of a given composition is likely to form a glass, various criteria have been introduced. First, the size ratio of the various constituent ions is important. This criterion was initially developed for metal-metalloid glasses on the basis of getting as dense a packing as possible for a mixture of hard spheres of different radii.

However, the basic importance of ionic sizes – defined in a proper way for the corresponding metallic environment of an ion – has again shown up in first principle pseudopotential calculations (see Chap. 6). Second, chemical bonding has been invoked which would prefer specific numbers of nearest neighbors in order to be saturated. Since metallic glasses show genuine metallic behavior – similar to that of liquid metals – in their important physical properties, the existence and relevance of covalent bonds in MGs seems rather doubtful, even when metalloids like Ge or Si are involved (which are metallic in the liquid state).

The first criterion which was based on considerations taking into account the specifically metallic behavior of MGs was put forward by *Nagel* and *Tauc* [1.22]. They had observed that – as discussed above – many metallic glasses have negative $d\varrho/dT$, which in terms of Ziman's approach to the calculation of ϱ meant $2k_F \approx k_p$. The authors suggested that the same conditions for the basic electronic ($2k_F$) and structural (k_p) wave numbers would also ensure stability of the amorphous alloy, since then the Fermi energy would lie in the local minimum of the density of electronic states mentioned before. It is an open question, both theoretically and experimentally, whether such a minimum can really be seen and whether E_F lies exactly there when $2k_F \approx k_p$. Since the argument leading to such a minimum is based on nearly free electron perturbation theory one would anyhow expect it only in the *s*-type partial density of states of the conduction electrons which in general cannot be measured separately. Recent specific heat data for Pd–Si seem rather to point to a high density of states near E_F [1.23].

In any case, the deviations of the (*s*-like) conduction band density from the free electron value are expected to be relatively small; therefore it would be desirable to discuss the influence of varying the concentration of conduction electrons (and hence $2k_F$) on structure and stability of a MG on a feature which does not only appear as a small perturbation on top of a large free electron quantity. Such an entity is the effective interionic pair potential, the oscillations of which have a wave vector of about $2k_F$. It is then easy to see that, if $2k_F \approx k_p$, the positions of the maxima of the pair correlation function (i.e., the preferred first, second, etc., neighbor distances) coincide with the positions of the consecutive minima in the potential. This "matching" of pair potential and pair correlation functions is certainly favorable for the stability of the disordered structure. Therefore the criterion $2k_F \approx k_p$ may be of some relevance for determining the favorable alloy composition for glass forming, especially where $2k_F$ changes rapidly with concentrations, as in strongly heterovalent alloys. For glasses involving TM elements one should certainly also discuss the influence of *d* states on the binding energy.

1.4 Special Features

1.4.1 Applications

The main interest in MG was initiated by possible technical applications based on outstanding mechanical, magnetic, electrical, and chemical properties. Many of the published papers come from industrial laboratories.

The mechanical properties suggest applications as strengthening fibers in composite materials for constructional, aeronautical, or sporting uses, and reinforcement of concrete or similar materials. Strong ribbons can be used for simple filament winding to reinforce pressure vessels or to construct large flywheels for energy storage. High hardness and no grain boundaries present the possibility of excellent cutting edges for cutting tools, in particular razor blades. There are possible applications for several kinds of springs made from MGs.

The magnetic properties suggest applications as core materials in inductive components for electronic circuits; in power transformers to replace conventional grain-oriented Fe–Si alloys; and in motors, as flexible magnetic shielding material, as recorder heads, as sensors, actuators, mechanical filters, and delay lines.

The electrical properties may lead to applications such as resistance thermometers and resistance heaters at low temperatures and precision resistors with zero temperature coefficient. The superconducting ribbons are insensitive to radiation damage and might therefore favor the application for nuclear fusion technology.

The good resistance to corrosion suggests chemical, surgical, and biomedical uses. For corrosion protection in general forms other than ribbons would be necessary.

There are other possible applications such as brazing foils, emission cathodes, electrical fuses, and hydrogen storage. A new field might be opened by forming complex shapes and compact samples with powders of MG by explosion techniques in a few millionths of a second.

1.4.2 Cross-Disciplinary Relations

Among the scientists interested in MG are metal physicists, metallurgists, solid-state physicists, chemists, and engineers. All these have different background, scientific language, and experience. Moreover, they all would like to keep as much as possible from the original fields they were working in. They bring expertise together from the fields of crystalline metals, dielectric glasses, polymers, and all kinds of liquids. Nevertheless, there is great hope that the cross-disciplinary relations can help to develop the field of metallic glasses. Much can be learned by investigating the relations – similarities and differences – of conducting and insulating glasses, polymers, glassy and liquid alloys, and glassy and crystalline alloys.

There is great hope that a fruitful collaboration will clarify our understanding of MGs in the near future.

References

1.1 H.-J.Güntherodt, H.Beck (eds.): *Metallic Glasses II*, Topics in Applied Physics (Springer, Berlin, Heidelberg, New York in preparation)

1.2 E.M.Breinan, B.H.Kear, C.M.Banas: Phys. Today **29**, 45 (1976)

1.3 R.Clampitt, M.G.Scott, K.L.Aithen, L.Gowland: In *Rapidly Quenched Metals III*, ed. by B. Cantor (The Metals Society, London 1978)

1.4 S.Davis, M.Fischer, B.C.Giessen, D.Polk: In *Rapidly Quenched Metals III*, ed. by B. Cantor (The Metals Society, London 1978)
1.5 G.S.Cargill III: In *Solid State Physics*, **30**, ed. by H.Ehrenreich, F.Seitz, D.Turnbull (Academic Press, New York 1975) p. 227
1.6 J.Dixmier, J.F.Sadoc: In *Metallic Glasses*, ed. by J.J.Gilman, H.J.Leamy (American Society for Metals, Metals Park, OH 1976) p. 97
1.7 G.S.Cargill III: In *Liquid and Amorphous Metals*, ed. by E.Lüscher, H.Coufal, NATO Conf. Series E, Vol. 36 (Sijthoff and Noordhoff, Alphen 1980)
1.8 C.N.J.Wagner: J. Non Cryst. Solids **31**, 1 (1978);
C.N.J.Wagner, H.Ruppersberg: "Applications of Nuclear Techniques to the Studies of Amorphous Metals", At. Energy Rev., Topical Issue ed. by U.Gonser (IAEA, 1980)
1.9 F.Spaepen: J. Non Cryst. Solids **31**, 207 (1978)
1.10 M.Jonson, S.Girvin: Phys. Rev. Lett. **43**, 1447 (1979)
1.11 T.R.McGuire, R.J.Gambino, R.C.O'Handley: In *Hall Effect and its Applications*, ed. by C.L.Chien, C.R.Westgate (Plenum, New York 1980) p. 137
1.12 C.D.Graham, Jr., T.Egami: In *International Conference on Magnetism*, ed. by W.Zinn, G.M.Kalvius, E.Feldtkeller (North-Holland, Amsterdam 1980) p. 1325
1.13 R.Alben, J.I.Budnick, G.S.Cargill III: In *Metallic Glasses*, ed. by J.J.Gilman, H.J.Leamy (American Society for Metals, Metals Park, OH 1976) p. 304
1.14 J.M.Ziman: *Models of Disorder* (Cambridge University Press, Cambridge 1979)
1.15 L.E.Ballentine: Adv. Chem. Phys. **31**, 263 (1975)
1.16 J.M.Ziman: Philos. Mag. **6**, 1013 (1961)
1.17 N.W.Ashcroft, D.Stroud: *Solid State Physics*, **33**, ed. by H.Ehrenreich, F.Seitz, D.Turnbull (Academic Press, New York 1978) p. 2
1.18 G.Baym: Phys. Rev. **135**A, 1691 (1964)
1.19 O.Dreirach, R.Evans, H.-J.Güntherodt, H.U.Künzi: J. Phys. F**2**, 709 (1972)
1.20 S.Hunklinger, W.Arnold: In *Physical Acoustics*, Vol. 12, ed. by R.N.Thurston, W.P.Mason (Academic Press, New York 1976) p. 155
1.21 H.A.Davies: In *Rapidly Quenched Metals* III, ed. by E.Cantor (The Metals Society, London 1978) p. 1
1.22 S.R.Nagel, J.Tauc: Phys. Rev. Lett. **35**, 380 (1975)
1.23 U.Mizutani, K.T.Hartwig, T.B.Massalski, R.W.Hopper: Phys. Rev. Lett. **41**, 661 (1978)

Bibliography

1. R.W.Cahn: Contemp. Phys. **21**, 43 (1980)
2. R.Cahn: Metallurgist Engl. Transl. 309, (June, 1978)
3. B.Cantor (ed.): Proceedings of the 3rd International Conference on Rapidly Quenched Metals (The Metals Society, London 1978)
4. G.S.Cargill III: In *Solid State Physics* **30**, ed. by H.Ehrenreich, F.Seitz, D.Turnbull (Academic Press, New York 1975) p. 227
5. P.Chaudhari, B.C.Giessen, D.Turnbull: Scientific American **242**, 84 (1980)
6. P.Chaudhari, D.Turnbull: Science **199**, 11 (1978)
7. H.S.Chen: Mater. Sci. Eng. **25**, 59–69 (1976)
8. H.S.Chen: Rep. Prog. Phys. **43**, 353 (1980)
9. H.S.Chen: In *The Structure of Non-Crystalline Materials*, ed. by P.H.Gaskell (London) Taylor and Francis 1977) p. 79
10. R.M.J.Cotterill: American Scientist **64**, 430 (1976)
11. P.Duwez: Annu. Rev. Mater. Sci. **6**, 83 (1976)
12. B.C.Giessen, C.N.J.Wagner: In *Liquid Metals, Chemistry and Physics*, ed. by S.Z.Beer (Dekker, New York 1972)
13. J.J.Gilman: In *Crystal Growth and Materials*, ed. by E.Kaldis, H.J.Scheel (North-Holland, Amsterdam 1977) p. 728

14. J.J.Gilman: Science **208**, 856 (1980)
14a. J.J.Gilman: Phys. Today **28**, 46 (1975)
14b. J.J.Gilman, H.J.Leamy (eds.): *Metallic Glasses* (American Society for Metals, Metals Park, OH 1978)
15. N.J.Grant, B.C.Giessen (eds.): *Rapidly Quenched Metals*, Part I (MIT Press, Cambridge, MA 1976);
Part II: Mater. Sci. Eng. **23** (1976)
16. H.-J.Güntherodt: Festkörperprobleme (Adv. Solid State Phys.) XVII, 25 (1977)
17. H.-J.Güntherodt, H.Beck, P.Oelhafen, K.P.Ackermann, M.Liard, M.Müller, H.U.Künzi, H.Rudin, K.Agyeman: In *Electrons in Disordered Metals and at Metallic Surfaces*, ed. by P.Phariseau, B.L.Gyorffy, L.Scheire (Plenum, New York 1979)
18. H.Jones: Rep. Prog. Phys. **36**, 1425 (1973)
19. R.A.Levy, R.Hasegawa (eds.): Proceedings of the Second International Symposium on Amorphous Magnetism, RPI, Troy, NY (Plenum, New York 1977)
20. E.Lüscher, H.Coufal (ed.): *Liquid and Amorphous Metals*, NATO Conf. Series E, Vol. 36 (Sijthoff and Noordhoff, Alphen 1980)
21. R.Mehrabian, B.H.Kear, M.Cohen (eds.): *Proceedings of the Conference on Rapid Solidification Processing* (Claitor's, Baton Rouge, LA 1978)
21a. Sci. Rep. Res. Inst. Tohoku Univ., Ser. A **25** (6), 200–244 (1975); **26** (1), 1–99 (1976); **26** (4–5), 185–294 (1977); **27** (2), 97–258 (1979); Suppl. (1978)
22. D.Turnbull: J. Phys. Paris C-**4**, 1 (1974)
23. H.Warlimont: Phys. Technol. **11**, 28 (1980)
24. H.Warlimont: Z. Metallkd. **69**, 212 (1978)
25. D.Weaire: Contemp. Phys. **17**, 173 (1976)

2. Metallic Glasses–Historical Background

P. Duwez

With 1 Figure

By the time this book is published twenty-one years will have passed since a very thin foil of the first metallic glass was obtained. It was indeed in September 1959 that, as part of a research program whose purpose was far remote from metallic glasses, an alloy containing 75 at.% Au and 25 at.% Si rapidly quenched from the liquid state appeared to be amorphous. The purpose of the research program initiated in June 1959 was to obtain a solid solution in binary alloys of Cu and Ag. The fact that the CuAg phase diagram is a eutectic type is in contradiction to the generally accepted Hume-Rothery rules. Since the separation of the homogeneous liquid into Cu-rich and Ag-rich phases during cooling is a rate process, I thought that by cooling the melt very rapidly, the Cu and Ag atoms would not have time to cluster and would be forced into a nonequilibrium solid solution. A very simple piece of equipment was rapidly put together which incorporated all the elements which lead to what is referred to now as the "gun technique". The first firing was a success combined with a failure. The very small pieces of quenched foils collected on the copper substrate were just enough material for a Debye-Scherrer pattern which was clearly that of a single-phase, face-centered solid solution. That was the success. The failure was the destruction of the rapidly put together apparatus partially made of glass tubings. The shock pressure was too high and about half of the apparatus disintegrated, sending hot broken pieces into the laboratory. After that, some time was spent on designing the next apparatus so that it could be used more than once.

The next alloy to be quenched could have been another system in which complete solubility would be predicted but does not occur under normal conditions. I was more interested however in finding out what would happen in a system in which the two components cannot form a solid solution under any circumstances because for example they have different crystal structures and very different valences. For some more or less logical reasons, alloys of Cu, Ag, or Au with the tetravalent semimetals Si and Ge were chosen, and among these Ag–Ge alloys with Ge concentrations up to about 40 at.% were quenched first. An increase in the solubility of Ge in Ag from 9.6 to 13.5 at.% was obtained, as expected from the previous results on CuAg solid solutions. An unexpected and surprising result however occurred when a quenched alloy containing 23 at.% Ge was found to be a hexagonal close-packed phase which does not exist in equilibrium. This was the first nonequilibrium crystalline phase obtained by liquid quenching, and since then more than 100 such phases have been

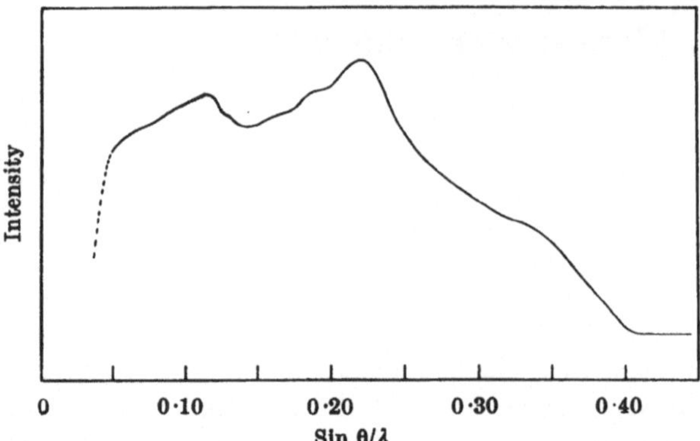

Fig. 2.1. Microphotometer trace of Debye-Scherrer diffraction pattern of liquid quenched Au–Si alloy [2.1]

reported. The next move was a lucky one leading to the first metallic glass. The alloy system Au–Si is very similar to Ag–Ge and it was expected that a hexagonal close-packed phase would also be obtained in Au–Si alloys. Instead the x-ray diffraction pattern of the quenched Au-20 at.% Si alloy indicated the absence of crystallinity. But was the alloy really amorphous or were we overenthusiastic about some rather questionable results? Personally I was not convinced that the x-ray diffraction pattern, reproduced here in Fig. 2.1, was unquestionably typical of an amorphous solid, and I intended to delay the publication until additional evidence became available. The two graduate student Ph.D. candidates who had been working with me, being young and ambitious, did not agree with me and were impatient to publish the results. I finally agreed if I were listed as the third author in the order Klement, Willens, and Duwez thinking I would be somewhat shielded from expected and probably justified adverse reactions from the experts in the field of amorphous solids. The title however was carefully worded "Non-crystalline Structure in Solidified Gold-Silicon Alloys" thus avoiding the word "amorphous". The manuscript was sent to Nature and published without critical comments from the reviewers [2.1]. Anybody familiar with x-ray diffraction by solids would not accept this pattern as a proof of the amorphous nature of an alloy; it might at least be a mixture of an amorphous phase and one or several microcrystalline phases. In fact it was stated in the paper that the amorphous foil was quite unstable and that crystallization could have occurred during the 3 h exposure in the Debye-Scherrer camera. It certainly did occur since the foil was completely transformed to a nonequilibrium crystalline state after 24 h at room temperature.

Soon after the paper on Au–Si was published *Cohen* and *Turnbull* [2.2] pointed out that the amorphous alloy composition was close to that of a very low melting eutectic point in the equilibrium diagram and this is a favorable condition for glass formation in metallic as well as ionic systems. This simple condition provided a very powerful guide in the search for alloy systems susceptible to be quenched into a glass. In consulting a compilation of phase diagrams it was found that quite a few binary alloys between a metal and a semimetal (similar to Au–Si) had a eutectic point at a rather low temperature compared to that of the metal. In order to avoid the thermal instability found in Au–Si amorphous alloys, Au should be replaced by a metal having a higher melting point. With these two conditions in mind the choice was limited to a few alloys and the most promising system appeared to be Pd–Si with Si concentration around 20 at.%. In contrast with Au–Si, the Pd–Si diffraction pattern left no doubt that this alloy was really amorphous [2.3]. Shortly after amorphous PdSi alloys were reported, two very important papers by *Chen* and *Turnbull* were published describing results of specific heat measurements which established for the first time the existence of a glass transition temperature in both Au–Ge–Si and Pd–Si liquid-quenched amorphous alloys [2.4, 5]. The existence of a glass transition temperature provided a convincing proof that the liquid-quenched amorphous alloys were indeed glassy and the name of metallic glasses was completely justified. It is regrettable that the term glass has also been used in connection with other amorphous alloys obtained by vapor deposition, electrolysis, sputtering, etc., because by definition a glass is a solid which does not crystallize during cooling from the liquid state. Only the amorphous liquid-quenched alloys satisfy this definition. The structure of a given amorphous alloy obtained by melt quenching and by any one of the other methods might be the same and, if so, the term glass would just refer to the method of preparation. So far there is no definite proof that the structure of an amorphous alloy does not depend on the method of preparation. A very important not clearly answered problem is whether or not a glass transition temperature in a given amorphous alloy exists only if the alloy has been prepared by liquid quenching. It is hoped that some critical experiments will settle this question in the near future.

In 1965 a search was initiated in my laboratory for a ferromagnetic metallic glass. It was theoretically predicted that ferromagnetism could exist in an amorphous solid and the properties of ferromagnetic amorphous vapor-deposited thin films of AuCo had been reported in the literature [2.6]. Since PdSi alloys were extensively studied at that time, the simplest way to obtain a ferromagnetic metallic glass was to replace some of the Pd atoms in PdSi by a ferromagnetic atom. This was found to be possible, and the alloys remained amorphous up to 5 at.% Fe, 12 at.% Co, and 15 at.% Ni. These alloys were relatively weak ferromagnets because of the low concentration of magnetic atoms [2.7]. The next logical step was to find a metallic glass in which, for example, Fe would be the main constituent, and this led to an interesting sequence of events. With the low-temperature eutectic as a guide, the first choice was the binary Fe–P system. The quenched foils of a $Fe_{75}P_{25}$ alloy were not completely amorphous and the diffraction patterns indicated clearly the presence of some microcrystals in an amorphous matrix. When the quenching of additional specimens was resumed they were found to be completely amorphous. It was soon discovered that the graduate student (S.C.H.Lin) had forgotten to insert a small alumina crucible in the gun and the Fe–P alloys had been molten in graphite. Obviously the liquid Fe–P alloys dissolved the equilibrium concentration of carbon at the melt temperature and the alloys were actually ternary $Fe_{75}P_{15}C_{10}$ alloys. And this is how the first strongly ferromagnetic metallic glass was discovered [2.8].

The original Fe–P–C alloy is the prototype for all the ferromagnetic metallic glasses which have been extensively studied during the last ten years. It is indeed a fact that a general formula for the chemical composition of all these alloys can be written $T_{(75-85)}M_{(25-15)}$ in which T refers to the sum of all the transition metals and M is the sum of all the metalloids. Some alloys may contain as many as five elements, such as $Ni_{49}Fe_{29}P_{14}B_6Si_2$ and others only two, like $Fe_{80}B_{20}$. Up to 1967, all the metallic glasses reported in the literature contained a certain percentage of metalloid atoms and it was thought that this was a necessary condition for obtaining a glass. In 1967 the metalloid alloying elements became just a favorable but not necessary condition when alloys of Nb and Ta with Ni were reported to be amorphous by *Ruhl* et al. [2.9]. It should be pointed out that in the metal-metal metallic glasses, one of the atoms is always appreciably smaller than the other; and this condition is also present in the metal-metalloid case.

By 1975 practically all the interesting properties of crystalline alloys had been obtained in metallic glasses, with the exception of superconductivity, and I started a search for superconducting glass at Caltech with two graduate students, Johnson and Poon. The first obvious approach was to formulate an alloy composition in which there would be a high concentration of a superconducting element, for example Nb, combined with a metalloid leading to a low melting eutectic. No really favorable alloy compositions were found and the search was extended to alloys containing two metals, but having a low eutectic point. The La–Au phase diagram appeared to satisfy the low eutectic condition. A $La_{80}Au_{20}$ alloy was amorphous after liquid quenching and superconducting with $T_c = 3.5$ K [2.10]. As of now the number of superconducting metallic glasses is about 12 and their transition temperatures range from 1.5 to about 10 K [2.11]. It is not sure that these rather low transition temperatures are due to the absence of a lattice and so far there is no clear theoretical reason why metallic glasses could not have transition temperature in the same range as crystalline alloys. The incentive for carrying out further studies of superconducting glasses is that their properties are not adversely affected by high radiation fields, which is not the case of most superconducting crystalline alloys.

During the last twenty years, the interest in metallic glasses has increased steadily. If the progress made during that time could be measured in some arbitrary units, it is probable that the curve would be close to an exponential. Obviously an exponential increase cannot continue forever and it is probable that the development of metallic glasses will follow the familiar nucleation and growth pattern which has been observed for many fields of science and technology. I do not think that we have reached the inflexion point of the *S* curve yet and it may be another twenty years before a plateau is reached. There is still widespread interest in both theoretical and experimental studies of metallic glasses, and it will not be long before the subject becomes an important chapter in standard textbooks of solid-state physics and physical metallurgy. About ten years ago the potential importance of metallic glasses as a class of new materials with unusual physical properties was recognized by industry. It will take a long time before production of metallic glasses can be measured in tons but it is encouraging to see that an increasing number of industrial research centers are involved in studies of metallic glasses.

References

2.1 W. Klement, Jr., R. H. Willens, P. Duwez: Nature **187**, 869 (1960)
2.2 M. H. Cohen, D. Turnbull: Nature **189**, 131 (1961)
2.3 P. Duwez, R. H. Willens, R. C. Crewdson: J. Appl. Phys. **36**, 2267 (1965)
2.4 H. S. Chen, D. Turnbull: J. Chem. Phys. **48**, 2560 (1968)
2.5 H. S. Chen, D. Turnbull: Acta Metall. **17**, 1021 (1969)
2.6 S. Mader, A. S. Nowick: Appl. Phys. Lett. **7**, 57 (1965)
2.7 C. C. Tsuei, P. Duwez: J. Appl. Phys. **37**, 435 (1966)
2.8 P. Duwez, S. C. H. Lin: J. Appl. Phys. **38**, 4096 (1967)
2.9 R. C. Ruhl, B. C. Giessen, M. Cohen, N. J. Grant: Acta Metall. **15**, 1693 (1967)
2.10 W. L. Johnson, S. J. Poon, P. Duwez: Phys. Rev. B **11**, 150 (1975)
2.11 P. Duwez, W. L. Johnson: J. Less Common Met. **62**, 215 (1978)

3. Structural Study by Energy Dispersive X-Ray Diffraction

T. Egami

With 12 Figures

The energy dispersive x-ray diffraction (hereafter denoted as EDXD) technique is an alternative method of structural study to that of conventional angular scanning x-ray diffraction. It produces the atomic pair distribution function, or the radial distribution function (RDF), just as the conventional method, but in a much shorter time. It is also capable of determining the structure factor in a wider wave vector space. Unfortunately the procedures of data analysis of the EDXD method are more complex than those of the conventional method which is already by no means trivial, so that the EDXD method is only beginning to be used more or less on a trial basis, and the results obtained are quite limited. Nevertheless, it is likely that the method will be more frequently used in the future because of its significant advantages in certain fields of application.

As the name indicates, the EDXD method is a technique to obtain structural information by *x-ray spectroscopy*. A continuous spectrum of x-ray (white x-ray) is used as a radiation source, and the energy spectrum of the x-ray beam is studied. Such a spectroscopic measurement can be made either by a single-crystal monochromator and an ionization counter when the incident beam is as strong as the synchrotron radiation, or in most cases by a standard x-ray spectroscopic system involving a semiconductor x-ray (γ-ray) detector and a multichannel pulse height analyzer. The concept of an energy dispersive diffraction measurement has been known for a long time. However, it became practical only after the development of a semiconductor detector with a sufficient energy resolution (e.g., [3.1]). The first EDXD measurements were reported in 1968, by *Giessen* and *Gordon* [3.2] and by *Buras* et al. [3.3], using lithium drifted silicon detectors. A number of investigators have tried to use the method since then, primarily for the purpose of fast structural determination of crystalline solids, taking advantage of its high data collection rate. They had rather limited success due to the relatively poor resolution in the $\boldsymbol{q}$ space (e.g., [3.4]). The problem of the resolution is far less severe for examining liquids or amorphous solids. *Prober* and *Schultz* [3.5] demonstrated in 1975 that the structure of liquid Hg can be successfully determined by the EDXD method. However, when the method is applied to elements with atomic number much less than that for Hg, a rather involved data analysis becomes necessary in order to correctly treat the Compton scattering intensity. Such a method of data reduction was developed only recently [3.6, 7] and will be discussed here in some detail. The power of the EDXD method was most clearly demonstrated

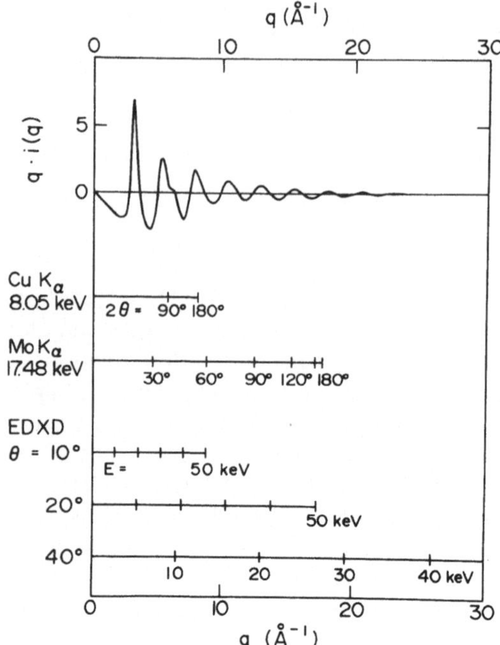

Fig. 3.1. Typical interference function (top) and range of q space covered by the conventional method with Cu or Mo radiation, and by the EDXD method at various diffraction angles

by the study of structural relaxation in metallic glasses [3.6]. The results will be discussed in relation to a recent theoretical study of structural defects in metallic glasses.

3.1 Principles

As described in the previous section, the diffraction vector $\boldsymbol{q}$ is related to the energy of the incident x-ray E[keV] by

$$\begin{aligned}|\boldsymbol{q}| &= (4\pi/hc)\cdot\sin\theta\cdot E\\ &= 1.014\cdot\sin\theta\cdot E,\end{aligned} \tag{3.1}$$

where θ is the diffraction angle. In the conventional angular scanning method, a monochromatic x-ray source is used, and θ is varied to scan the $\boldsymbol{q}$ space. The structural information is contained in the variations in the diffracted beam intensity with θ. Since $\sin\theta \leqq 1$, the energy of the source radiation E determines the maximum value of $\boldsymbol{q}$ accessible by this method, as shown in Fig. 3.1. When Cu K_α radiation is used as a source, only the first two peaks in the structure factor $i(\boldsymbol{q})$ can be observed. With Mo K_α or Ag K_α radiation, the maximum value of $\boldsymbol{q}$ can be extended up to 17 or 19 Å^{-1}. But no characteristic radiations are readily available to probe the $\boldsymbol{q}$ space beyond that limit.

The EDXD method utilizes the converse of this method; the diffraction angle θ is fixed during the measurement and E is used as a variable, using white x-ray radiation as a source. Since the white x-ray radiation up to 50 keV is readily available, a very wide $\boldsymbol{q}$ space can be covered by the EDXD method as shown in Fig. 2.1. The structural information is contained in the spectral variations of the diffracted beam. The principal advantages of the method are the higher rate of data collection due to the total intensity of the white radiation being an order of magnitude greater than that of the characteristic radiation, and the increased signal-to-noise ratio because of the parallel counting of photons with different energies.

3.2 Equipment

3.2.1 System Design

Even though the diffraction angle θ is fixed during one diffraction measurement in the EDXD method, the EDXD system should have a capacity to vary θ. This is because several diffraction measurements have to be carried out at different values of θ in order to cover a large area in the $\boldsymbol{q}$ space, since the energy range which yields reliable information is limited as will be discussed later. The EDXD system can most simply be constructed by a conversion of a conventional diffractometer, by replacing a proportional counter by a semiconductor detector. For studying liquids, a vertical $\theta-\theta$ goniometer is the best [3.8], while for most of the solid samples, a horizontal $\theta-2\theta$ goniometer is more conventient. As shown in Fig. 3.2, the x-ray tube should be mounted with the tube axis tilted away from the diffraction plane by 45° in order to avoid the effect of polarization. It should be kept in mind that the EDXD method utilizes relatively higher energy ranges of x-ray, up to 40 or 45 keV. The absorption by the sample is small so that the penetration depth is large. The air attenuation is small. Therefore it is better to use a system with a large arm length to avoid a loss of angular resolution, particularly in studying polymers or organic liquids and oxides with light elements.

3.2.2 X-Ray Source

The source x-ray for the EDXD measurement can be obtained either from a regular x-ray tube with W or Mo target, a rotating anode x-ray generator, or from synchrotron radiation. In the conventional method, the diffraction data are obtained by *intensity modulation* of the incident x-ray beam by the sample, so that any intensity fluctuation of the source beam appears as a noise in the data. The EDXD method is based upon *spectral modulation* of the incident x-ray, so that intensity fluctuations do not influence the data, as long as the source spectrum averaged over the measurement period remains the same from

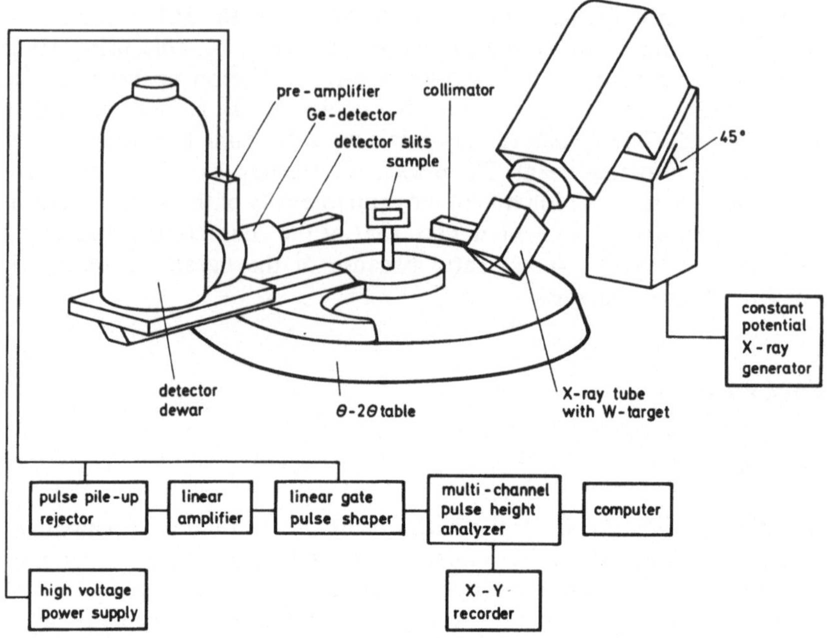

Fig. 3.2. Schematic diagram of the EDXD equipment

one run to another. Therefore, one does not require an expensive source intensity stabilization system; an ordinary primary voltage stabilizer is usually sufficient. One can even do without a constant potential unit, if it were purely to improve the signal-to-noise ratio. The use of a constant potential unit, however, is beneficial because it increases the intensity of the higher energy portion of the spectrum. It is usually helpful to use a primary beam filter to reduce the fluorescent radiations from the sample, by reducing the spectral intensity of the primary beam near the absorption edges of the sample.

The use of a rotating anode x-ray source would greatly speed up the measurement, particularly at high angles. Synchrotron radiation is an excellent light source from many aspects [3.9]. Its potential application will be discussed in the later sections.

3.2.3 Optical System

The conventional optical devices, collimators, and slits with Ta blades, are largely acceptable for use in the EDXD system. However, at a higher energy range (above 35 keV), a partial transmission of the x-ray through the edges of the Ta blades becomes observable. Therefore the primary beam spectrum needs to be corrected for such a modulation by the optical system, every time the optical resolution is changed. As has been mentioned earlier, the penetration

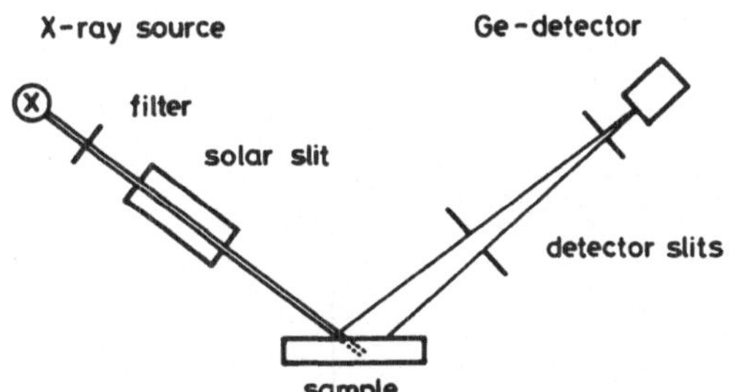

Fig. 3.3. Schematic diagram of the optical system

depth in the sample is appreciable in many cases. The use of a conventional optical system with equal resolutions for incident and diffracted beams leads to a complex absorption correction [3.5, 10] which strongly depends on the details of the optical system and alignment. Such a correction is very difficult to apply accurately, and tends to introduce errors. The system we use has a wide slit right after the sample (Fig. 3.3), so that the absorption factor is correctly described by $1/\mu(E)$, where $\mu(E)$ is the absorption coefficient, for the reflection geometry. This results in some loss of angular resolution if the penetration depth is not smaller than the incident beam width.

3.2.4 Detector System

A standard x-ray spectroscopic system, such as those used in energy dispersive chemical analysis or γ-ray radiation analysis, consists of a semiconductor detector, an amplifier, and a multichannel pulse height analyzer (MCA). As a photon detector, a lithium drifted Si or Ge detector, or an intrinsic Ge detector can be used. The Si(Li) detector has a detection efficiency which rapidly decreases above 20 keV, while the Ge detectors have a dip in the efficiency above the absorption edge of Ge (11.1 keV). The energy range in which the structural determination can be made is determined by the source radiation and the sample, i.e., the maximum energy of white radiation and the location in the energy spectrum of the characteristic and the fluorescent radiations. The type of detector chosen should depend upon this energy range. In studying the transition metal-metalloid metallic glasses, we used the combination of an x-ray tube with a W target and an intrinsic Ge detector (Princeton Gamma-Tech. IG–25U). The energy range of 20–40 keV was used for analysis. An MCA with 1024 channels (Tracor-Northern TN-1705) was used for photon counting which was interfaced with a minicomputer (PDP-11/10) with floppy disks and real-time software. The semiconductor detectors presently available have an energy resolution down to $\sim$150 eV at 7 keV and $\sim$230 eV at 40 keV. The resolution determination can be made relatively easily by studying the line-widths of several fluorescent radiations. Such a relatively poor resolution presents problems in accurate lattice constant measurements, but is sufficient for the structural study of most amorphous solids and liquids. A typical

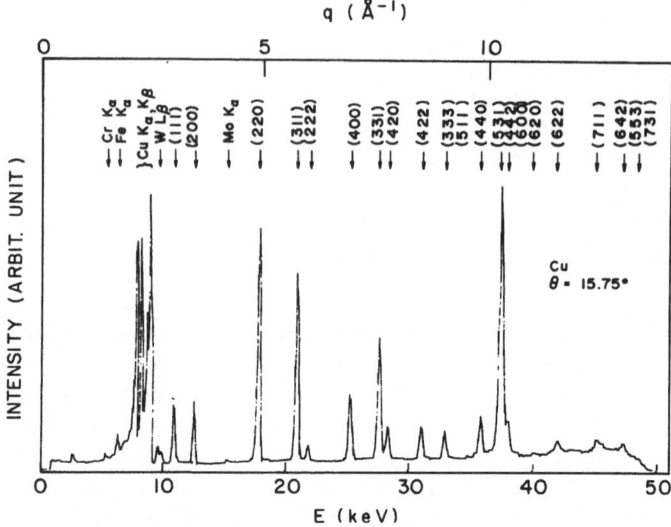

Fig. 3.4. Diffraction pattern of polycrystalline Cu

diffraction pattern of a crystalline solid is shown in Fig. 3.4. The pulse shaping time of a linear amplifier recommended by the producer is generally 6–10 μs, which means that the actual pulse width is up to 40 μs. Therefore a pulse pile-up rejector has to be used when the count rate approaches $10^3\ \mathrm{s}^{-1}$. The time resolution of the pulse pileup rejector is about 300 ns at present, so that some pileups are unavoidable. The maximum count-rate, using a very short pulse shaping time, is less than $2\times10^5\ \mathrm{s}^{-1}$ now. For the semiconductor detector to be used effectively with synchrotron radiation, the maximum count rate has to be improved appreciably.

3.3 Data Reduction

3.3.1 Theory

The spectral intensity measured by a photon detector is given by, at energy E,

$$I_t(E)=C(E)\,[I_{\mathrm{inel}}(E)+I_{\mathrm{el}}(E)+I_{\mathrm{m}}(E)+I_{\mathrm{a}}(E)]+I_{\mathrm{s}}(E)\,, \tag{3.2}$$

where $C(E)$ is the combined detection efficiency, $I_{\mathrm{inel}}(E)$ is the inelastic scattering intensity, $I_{\mathrm{el}}(E)$ is the elastic scattering intensity, $I_{\mathrm{m}}(E)$ is the multiple scattering intensity, $I_{\mathrm{a}}(E)$ is the air scattering intensity, and $I_{\mathrm{s}}(E)$ is the spurious photon counts noise. $C(E)$ is given by

$$C(E)=Q(E)\cdot\exp[-\mu_{\mathrm{A}}(E)\cdot R_{\mathrm{s}}]\cdot B(E)\,, \tag{3.3}$$

where $Q(E)$ is the detector efficiency, $\mu_A(E)$ is the absorption coefficient of air, R_s is the distance from the sample to the detector, and $B(E)$ is a geometrical factor including the energy-dependent modulation by the detector slits. Also,

$$\begin{aligned} I_{\mathrm{inel}}(E) &= \int A(E,E')\langle F_c(q')\rangle P(E,E',\theta) I_p(E') S(E,E') dE' \\ I_{\mathrm{el}}(E) &= A(E,E)\,[i(q)\langle f(q)\rangle\langle f^*(q)\rangle + \langle f(q) f^*(q)\rangle] \cdot \\ &\quad P(E,E,\theta)\cdot I_p(E), \end{aligned} \tag{3.4}$$

where $A(E,E')$ is the absorption factor; $P(E,E',\theta)$ is the polarization factor; $I_p(E)$ is the primary beam intensity, including the air attenuation, filter transmissibility, and collimator modulation; $S(E,E')$ is the Compton shift factor, Compton profile; $F_c(q)$ is the total Compton scattering intensity; $f(q)$ is the scattering factor; $i(q)$ is the structure factor (interference function); $\langle \ldots \rangle$ is the compositional average; and

$$q' = (4\pi/hc)\sin\theta \cdot E'. \tag{3.5}$$

The Compton profile $S(E,E')$ has a width due to the momentum distribution of electrons, but in practice below 40 keV it can be replaced by

$$S(E,E') \rightarrow \delta(E' - E - \Delta E), \tag{3.6}$$

$$E + \Delta E = E/(1 - 0.00392 \cdot E \cdot \sin^2\theta), \tag{3.7}$$

E being measured in keV. For an infinitely thick sample and in reflection geometry,

$$A(E,E') = 1/[\mu(E) + \mu(E')], \tag{3.8}$$

where $\mu(E)$ is the absorption coefficient of the sample. The polarization factor is given by

$$\begin{aligned} P(E,E',\theta) &= (E/E' + E'/E - \sin^2 2\theta)/2 \\ &\quad - \sin^2 2\theta \cdot \pi(E')/2 \end{aligned} \tag{3.9}$$

which represents the Klein-Nishina formula [3.11] except for the prefactor $(E/E')^2$. This factor, which is just the Breit-Dirac recoil factor, is cancelled by the contraction of the spectrum due to the Compton shift [3.6]. The photons with incident energy from E' to $E' + dE'$ converge after Compton scattering into a smaller energy window from E to $E + dE$, so that the intensity is increased by

$$dE'/dE = (E'/E)^2 \tag{3.10}$$

which exactly cancels the Breit-Dirac recoil factor. $P(E,E',\theta)$ is reduced to the classic Thomson formula when $E' = E$. The difference is a relativistic effect, but is of the order of $(\Delta E/E)^2$, so that it is negligible in most cases. The multiple scattering intensity is given by an equation similar to the Warren formula [3.12], but it is a little complicated because of the Compton shift and will not be given here [3.13]. The spurious photon counts consist of a photon pile-up noise, escape peaks, a thermal white noise, and a bremsstrahlung from the sample. The latter two are mostly negligible. The structural information is contained in $i(q)$ which yields the radial distribution function (RDF) by

$$G(r) = 4\pi r[\varrho(r) - \varrho_0] = \frac{2}{\pi}\int_0^\infty i(q)\sin qr \cdot q dq\,. \tag{3.11}$$

Thus a fairly complex procedure is required to obtain $i(q)$ out of the original EDXD data, although some factors can be neglected in the actual structure study as described below.

3.3.2 Determination of the Parameters

Out of the many quantities required for data reduction using (3.2–10), some can be theoretically calculated, while others have to be experimentally determined. The values of $f(q)$ and $F_c(q)$ are obtained from the standard sources [3.14–16]. The anomalous dispersion f' and f'' had been readily available only for several energies corresponding to the frequently used characteristic radiations [3.17], but recently they were calculated as continuous functions of E between 15 and 45 keV for most of the elements [3.18]. There is a difficulty in determining the absorption coefficient $\mu(E)$, since neither experimental nor theoretical values are sufficiently accurate. This difficulty can be averted by a procedure described later.

Other quantities have to be determined more or less experimentally. The detector efficiency $Q(E)$ of an intrinsic Ge detector can be determined from the intensities of the diffraction peaks and escape peaks using a single crystal as a sample. A typical value of $Q(E)$ is ~0.92 at 20 keV, ~0.99 at 40 keV. Such a slowly changing $Q(E)$, however, can usually be incorporated into $I_p(E)$. The decrease in the value of $Q(E)$ for Si(Li) detectors above 20 keV is more difficult to determine with accuracy. The values of $\pi(E)$ can be determined by simply rotating the diffraction plane by 90° with respect to the x-ray tube axis, but care has to be taken to include the effect of the Compton shift self-consistently [3.6]. It is best to use the 45° mounting to avoid this difficulty. Attempts have been made to determine $I_p(E)$, using a pinhole [3.2], or a reduced radiation intensity [3.10]. Neither of these methods is likely to succeed; a small angle scattering and attenuation by the pinhole and a change in the generator circuit may produce a distortion in the spectrum, and more importantly, the values of $\mu(E)$ and other corrections are not known accurately enough to allow reliable data analysis. It appears that the best method is to use the diffraction data of the sample to be studied [3.5], and to determine $C(E)$; $I_p(E)/\mu(E)$ self-consistently

[3.6]. From (3.2–9), if $\pi(E)=0$, one obtains

$$\begin{aligned} & C(E)\cdot[I_{\text{inel}}(E)+I_{\text{el}}(E)] \\ & =P(\theta)\cdot([i(q)\langle f(q)\rangle\langle f^*(q)\rangle+\langle f(q)f^*(q)\rangle]\tilde{I}_{\text{p}}(E) \\ & \quad +\langle F_{\text{c}}(q')\rangle\cdot[C(E)/C(E')]\cdot\{2\mu(E')/[\mu(E)+\mu(E')]\}\tilde{I}_{\text{p}}(E')), \end{aligned} \tag{3.12}$$

where $P(E,E',\theta)$ was denoted $P(\theta)$ neglecting the relativistic effect and

$$\tilde{I}_{\text{p}}(E)=C(E)\cdot I_{\text{p}}(E)/2\mu(E), \tag{3.13}$$

$$E'=E+\Delta E. \tag{3.14}$$

If one takes data at reasonably high angles, say around $\theta=30°$, $i(q)$ is much smaller than unity. Furthermore, if one chooses two angles so that the oscillations in $i(q)$ are out of phase to each other, it becomes possible to determine $\tilde{I}_{\text{p}}(E)$ by assuming $i(q)=0$, and averaging the two values of $\tilde{I}_{\text{p}}(E)$ obtained from the two data using (3.12). If one uses data from very high angles, say 50°, the oscillation in $i(q)$ is indeed absent from the data. However, then (3.6) is no longer valid, and the analysis becomes very complicated. Note that in this method small errors in the values of $C(E)$ or $\mu(E)$ do not affect the results, since they are compensated by the corresponding changes in $\tilde{I}_{\text{p}}(E)$. In fact one needs to determine only $C(E)/C(E')$ and $2\mu(E')/[\mu(E)+\mu(E')]$, both of which are close to unity. The former can be neglected (replaced by unity) in case an intrinsic Ge detector is used, and the latter can be calculated theoretically.

The effect of multiple scattering is more serious for the EDXD method than for the conventional method, because an x-ray of a higher energy range is used, and $\mu(E)$ is smaller. For Fe base metallic glasses $I_{\text{m}}(E)$ is as large as 10% of the total scattering at $E=40$ keV and $\theta=40°$. For an accurate determination of $I_{\text{m}}(E)$, it is necessary to know $i(q)$, since a sharp first peak in $i(q)$ does produce a small change in the values of $I_{\text{m}}(E)$, compared to those obtained by assuming $i(q)$ to be zero. However, such an effect can also be absorbed in $\tilde{I}_{\text{p}}(E)$ in the self-consistent analysis described above, and hardly affects the outcome. $I_{\text{a}}(E)$ is generally small, typically 1% of the total intensity at 20 keV at θ above 20°, again due to the use of high-energy x-rays. It becomes more significant at low angles, being proportional to $1/\sin^2\theta$. The effect of pulse pileups is appreciable when there is strong fluorescent radiation from the sample. It can amount to several percent at high angles since the fluorescent radiation increases slightly with increasing θ while the scattering intensity decreases sharply. It is advisable to reduce the fluorescent radiation as much as possible by an incident beam filter, or sometimes by a detector filter. In the latter case, the values of $C(E)$ have to be changed accordingly. The escape peaks amount to about 3% at 20 keV and 1% at 30 keV. These contributions [$I_{\text{m}}(E)$, $I_{\text{a}}(E)$, and $I_{\text{s}}(E)$] are of second order as long as the self-consistent determination of $\tilde{I}_{\text{p}}(E)$ is employed. Neglecting them altogether [3.6] could introduce errors of several percent in

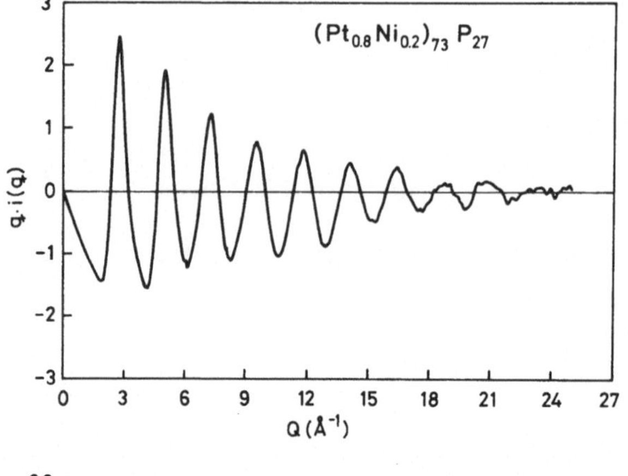

Fig. 3.5. Structure factor of $(Pt_{0.8}Ni_{0.2})_{73}P_{27}$ [3.13]

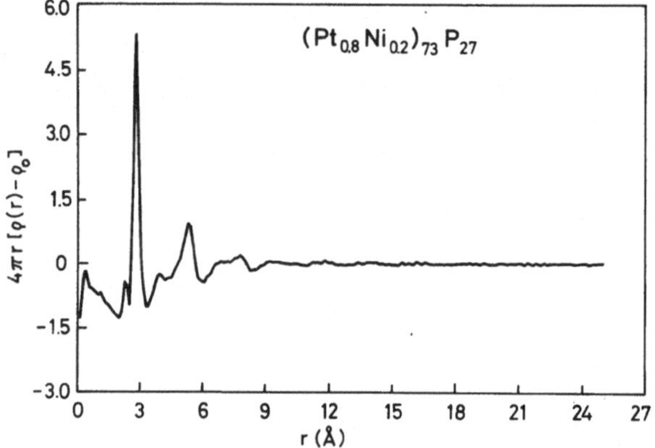

Fig. 3.6. RDF $G(r)$ of $(Pt_{0.8}Ni_{0.2})_{73}P_{27}$ [3.13]. A small peak inside the first peak corresponds to the $P-Pt$ peak. This prepeak cannot be resolved by the conventional method

the low q region, for the transition metal base metallic glasses. Although such an error could be of a similar order of magnitude as the other experimental errors such as the deviation in the absorption factor from $1/\mu(E)$ due to the geometry of the sample, it is desirable that these corrections be made, particularly in dealing with lighter elements.

3.3.3 Structural Determination

The structure factor $i(q)$ can now be obtained by (3.2–9), using the parameters determined as described above. Usually it is not possible to use all the energy ranges from zero to the applied excitation voltage. First, one should avoid the energy range where there are strong fluorescence peaks. Furthermore, if a W tube is used, L characteristic lines appear below 10 keV, so that the analysis can be made using only the spectrum above 10 keV. Also the high energy end of the spectrum should not be used, since $\tilde{I}_p(E)$ drops rather sharply and makes accurate structural determination difficult. For instance, the distribution in the

Compton shift may have significant effects there. For the structural determination of the Fe–Ni base metallic glasses, using white radiation from a W tube with the applied voltage of 48 keV, the energy range between 20 and 40 keV was chosen. The lower limit of the range was set rather high to avoid the two photon peaks of fluorescence radiations. It is necessary to determine $\tilde{I}_p(E)$ up to about 43 keV, to use this energy range for structure determination, because of the Compton shift.

If we use this energy range, the minimum number of runs required for a quick survey is three, say at 30°, 15°, and 7.5°. However, then there would hardly be any overlaps in the q space between the runs. In order to produce accurate structural data, it is always preferred to keep substantial overlaps in the $\boldsymbol{q}$ space between the runs to check the self-consistency. Six to eight runs, therefore, are required for a complete structural study. The final result is produced by a synthesis of data at various diffraction angles. It is unavoidable that some small arbitrariness enter in this process of synthesis. A check of the self-consistency is the only way to prevent errors due to such arbitrariness.

The structure factor $q \cdot i(q)$ and the RDF $G(r)$ of a metallic glass $(Pt_{0.8}Ni_{0.2})_{73}P_{27}$ are shown in Figs. 3.5 and 3.6. The corrections for I_m, I_a, and I_s are all incorporated in this analysis, resulting in excellent self-consistency in the data. The structure factor $q \cdot i(q)$ is resolved up to 22 Å^{-1}, showing nine peaks. The RDF $G(r)$ shows no structure beyond 9 Å, indicating a large disorder in this alloy. At the same time, the absence of spurious peaks beyond 9 Å other than rapidly oscillating noise proves the accuracy of the data.

3.4 Results of Structure Study

3.4.1 Structural Relaxation Observed by EDXD

The structure factor $i(q)$ and the RDF of $Fe_{40}Ni_{40}P_{14}B_6$ obtained by the EDXD are shown in Figs. 3.7 and 3.8. The RDF clearly describes the well-known features of the dense random packing structure, including the split second peak (see Chap. 5). The structure study of metallic glasses, however, should not end here. Metallic glasses exhibit a wide variety of physical properties, depending upon the composition and heat treatment, while their structures, presented as RDFs, are often quite similar. The structure study should be able to relate these relatively small differences in the RDF to the variations in the physical properties, in order to gain understanding of the role of structure on the properties. The EDXD method is a powerful technique for such a purpose, since the high statistical accuracy of the method makes it relatively easy to determine the small differences.

The lower halves of Figs. 3.7 and 3.8 show the changes in the structure factor and the RDF, respectively, which were caused by heat treating the sample at a temperature below the glass transition temperature. Since the

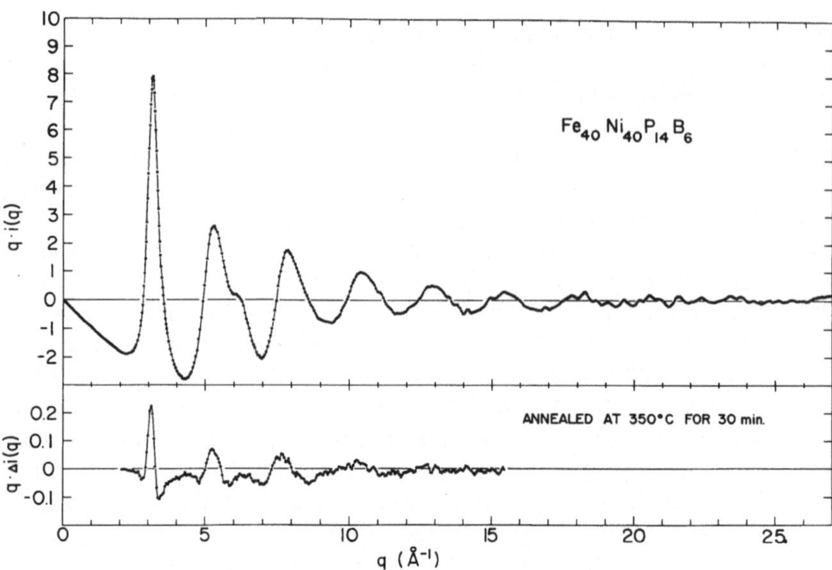

Fig. 3.7. Structure factor of $Fe_{40}Ni_{40}P_{14}B_6$ (above) and the difference in $q \cdot i(q)$ caused by annealing at 350 °C for 30 min (below) [3.6]

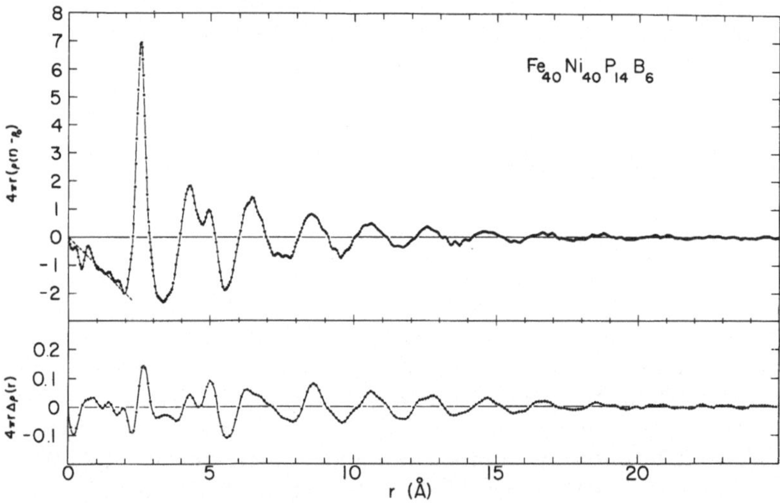

Fig. 3.8. RDF $G(r)$ of $Fe_{40}Ni_{40}P_{14}B_6$ (above) and the difference caused by annealing at 350 °C for 30 min (below) [3.6]

metallic glasses are obtained by a rapid quenching of the melt, the as-quenched structure is not fully stabilized, and an annealing which does not induce crystallization causes a relaxation toward a more stable (yet metastable) state. This structural relaxation process results in changes in many physical properties of metallic glasses [3.6, 7, 19]. The effect is sometimes drastic. For instance the diffusivity can be decreased by several orders of magnitude [3.20]. The results presented in Figs. 3.7 and 3.8 revealed the following characteristics of the changes in the short-range order which occur during structural relaxation:

1) The positions of the peaks and valleys in the RDF do not shift significantly. There is no overall contraction or expansion. The atomic positions simply become better defined.
2) The relative changes in the first and second peak heights are 2–3%, but the changes in the shoulder of the second peak and the third to fifth peaks are of the order of 10%. Therefore, the structural relaxation is caused by multiatomic motions, rather than the single atom diffusion.
3) The change in the second peak of $i(q)$ is different for the case of structural relaxation from crystallization, at least in the Fe–Ni–P–B alloy. During relaxation, the second peak height is increased while the shoulder height is decreased. The change due to crystallization is the opposite; the peak height is reduced while the shoulder grows into a Bragg diffraction peak. Therefore it can be concluded that *structural relaxation is not an incipient crystallization process.* The structure approaches a glassy metastable state during the relaxation, rather than a crystalline state.

These structural changes can be described in terms of the structural defects which are annealed out during structural relaxation. The structural defects in amorphous solids cannot be so readily defined as in crystalline solids, because an ideal or perfect structure cannot be uniquely defined. However, a recent analysis of the internal stresses on the atomistic level and the atomic site symmetry indicated that a physically meaningful definition of the structural defects can be derived [3.21]. The atomic level stresses in a central force solid are given by [3.22]

$$\tau_i^{\alpha\beta} = \frac{1}{2\Omega} \sum_j \frac{\left.\dfrac{d\phi(r_{ij})}{dr_{ij}}\right|}{|\boldsymbol{r}_{ij}|} r_{ij}^{\alpha} r_{ij}^{\beta}, \tag{3.15}$$

where α and β denote the Cartesian components, i and j are the atomic sites, Ω is the atomic volume, $\phi(r)$ is the central force interatomic potential, and $\boldsymbol{r}_{ij}$ is the distance between the ith and jth atoms. Note that this is not the *force* acting on the ith atom which is always zero in equilibrium. An example of the distribution of the atomic site pressure, which is one-third of the trace of this stress tensor, calculated for a model dense random packing structure [3.23] is shown in Fig. 3.9. The pressure shows a significant variation from site to site. One may also expand the total energy with respect to a displacement of the ith atom, $\Delta\boldsymbol{r}_i$,

$$E(i) = E_0 + \sum_{n=1}^{\infty} \frac{d^nE}{d\Delta r_i^n} \frac{|\Delta\boldsymbol{r}_i|^n}{n_i}. \tag{3.16}$$

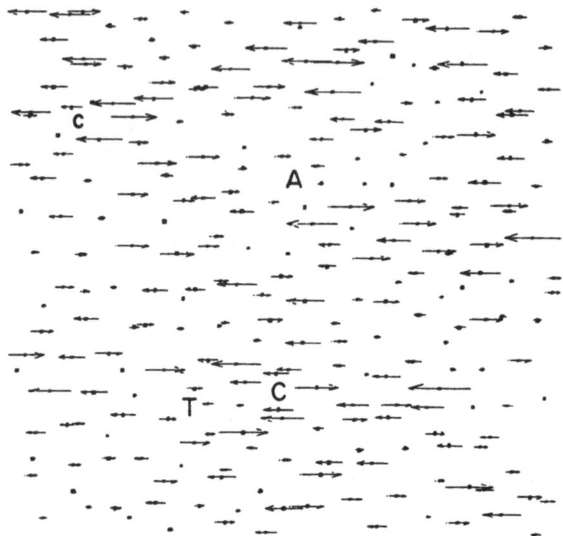

Fig. 3.9. Atomic site pressure in a DRP model. The length of the arrow is proportional to the magnitude of the pressure, with the arrow pointing to the left indicating compression. A: relaxed region, C: compressed region, T: tensile region [3.21]

The coefficients are dependent on the direction of $\Delta \boldsymbol{r}_i$, so that

$$\frac{d^n E}{d\Delta r_i^n} = \sum_{l=0}^{\infty} \sum_{m=-l}^{l} \varepsilon_n^{l,m}(i)\, Y_l^m(\theta_i, \phi_i), \tag{3.17}$$

where θ_i and ϕ_i are the azimuthal angles determining the direction of $\Delta \boldsymbol{r}_i$, and Y_l^m the spherical harmonics. The coefficients $\varepsilon_n^{l,m}(i)$ are defined as the site symmetry coefficients. Significant correlations between the stresses and the symmetry coefficients were found.

The analysis of a model structure revealed that the stresses and the symmetry coefficients are also spatially correlated, and there are clusters of atoms (10–20 atoms), which have either small stress and high symmetry, or large stress and low symmetry. The structural defects can be defined as the latter regions. It was found that the defects are made not only of low-density dilated regions (n-type defects which resemble free volumes) but also of high-density compressed regions (p-type defects which may also be called anti-free volumes), and in either case accompanied by strong shear stress fields. The partial RDFs from the low-density regions, relaxed regions, and high-density regions all look similar, with a clearly split second peak, except for some phase shift of the oscillations (Fig. 3.10; [3.24]). The split of the second peak of the total RDF is less pronounced, because of the overlap of these shifted peaks. The difference between the RDF of the defect regions and that of the relaxed regions (Fig. 3.11) strongly resembles the EDXD result in Fig. 3.8.

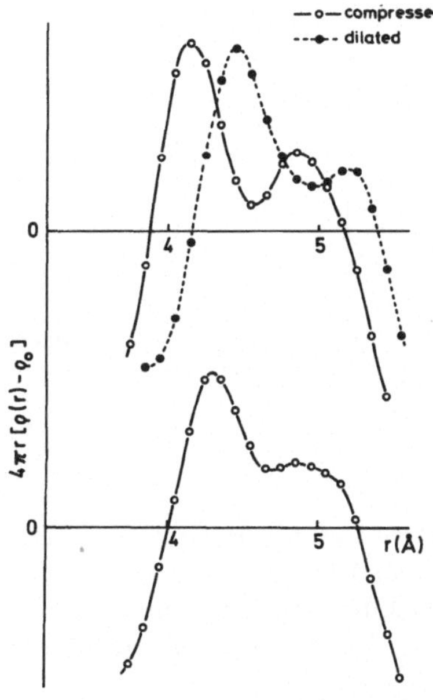

Fig. 3.10. Partial RDFs from the low-density (dilated) regions and from high-density (compressed) regions (above) and the total RDF (below). Only the vicinity of the second peak is shown [3.24]

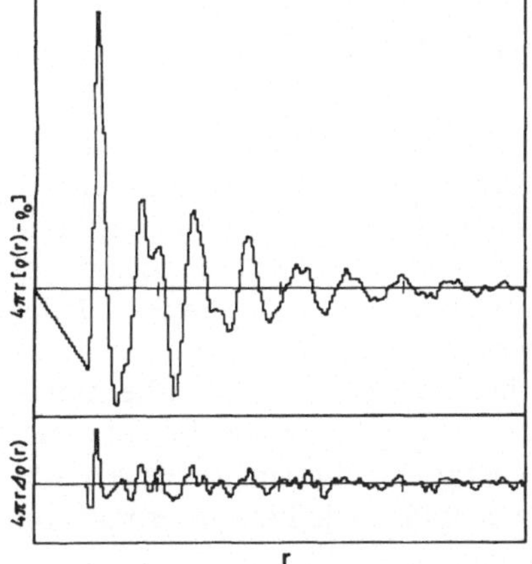

Fig. 3.11. Partial RDF from the relaxed regions (above) and the difference in the RDF between from the relaxed regions and from the defect regions (below) [3.24]

Similar changes in the structure factor and the RDF were found in metallic glasses of different compositions [3.25] and in computer simulations of the structural relaxation [3.26, 27]. It appears that these features are very general characteristics of the dense random packing structure.

3.4.2 Kinetics of Structural Relaxation

Structural relaxation is a kinetic process which becomes faster at higher temperatures. By taking advantage of the high data collection rate of the

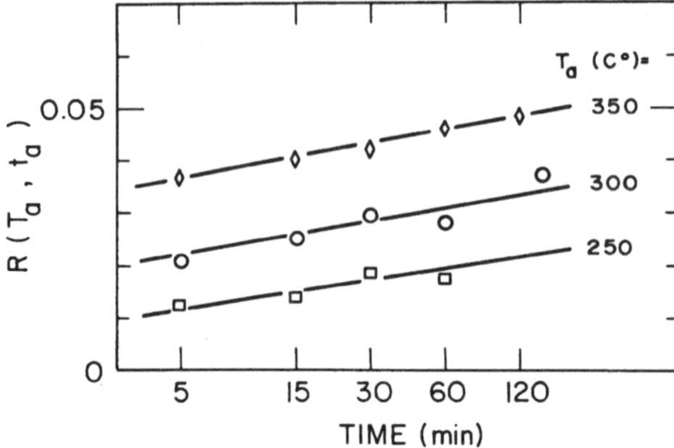

Fig. 3.12. Kinetics of the relaxation parameter R defined by (3.18) [3.6]

EDXD method, the kinetics of the structural relaxation was directly studied [3.6, 28]. It is, however, quite time consuming and tedious to determine the whole $\Delta i(q)$ with sufficient accuracy at various annealing times and temperatures. Instead, a relaxation parameter R was defined, making use of the characteristic behavior of the second peak in $i(q)$, as

$$R(T,t)=\frac{1}{q_2-q_1}\int_{q_1}^{q_2} q|\Delta i(q)|dq(q_1=5\,\text{Å}^{-1}, q_2=6.15\,\text{Å}^{-1}), \tag{3.18}$$

where T is the annealing temperature and t is the annealing time, and was determined directly from the EDXD measurements as shown in Fig. 3.12. It was found that $R(T,t)$ follows the logarithmic ($\ln t$) kinetics. The logarithmic kinetics results when the activation energy has a wide and constant spectrum of distribution. It is, however, quite artificial to assume the presence of independent relaxation centers and neglect interactions among them, while the structural defects are spatially extended and their density is high. An alternative picture is to consider primarily the interaction among the defects. For the sake of simplicity we consider a single activation energy E_a. This can be viewed as a weighted average of a spectrum of activation energies. The first-order reaction equation for the relaxation is

$$dx/dt=b(x-x_\infty)e^{-E_a/kT} \tag{3.19}$$
$$R=x-x_0\,.$$

If E_a is independent of t, this would simply yield an exponential kinetic equation. However, E_a is not likely to be a constant, but it should increase with time as the structure becomes more relaxed. Such a correlation can be represented, by retaining the lowest order, as [3.6, 7]

$$E_a=\alpha x\,. \tag{3.20}$$

Equations (3.19) and (3.20) then yield, if $\alpha(x_\infty - x_0)/kT \gg \alpha(x-x_0)/kT \gg 1$,

$$R(T,t) = (kT/\alpha)\ln(t/\tau_0 + 1)$$
$$\tau_0 = (1/b)\exp(\alpha x_0/kT). \tag{3.21}$$

This result shows a good fit to the data, as shown in Fig. 3.12. Most of the kinetic processes observed in metallic glasses show a logarithmic time dependence [3.29–31], partly because the structural relaxation controls most of these kinetic processes. In particular, the viscosity around the glass transition temperature T_g was found to follow the kinetics directly related to the structural relaxation [3.34]. The macroscopic stress relaxation time at around T_g is of the order of a second, but the viscosity η was found to change more slowly and linearly with time,

$$\eta = \frac{\gamma}{T}(t+t_0). \tag{3.22}$$

For $Fe_{40}Ni_{40}P_{14}B_6$, γ was found to be 6×10^{12} poise K/s in the temperature range of 641–666 K. Note that η is proportional to $(\partial R/\partial t)^{-1}$. Furthermore, the diffusive jump frequency calculated from η using the Einstein-Stokes equation is by about 10^3 higher than estimated from $\partial R/\partial t$. Thus it may be concluded that the structure relaxation occurs through the annihilation or the recombination of the structural defects of opposing characters, such as the n-type and p-type defects, and the mean free path or the dispersion length [3.21] of the defect is about 30 atomic distances, which is in agreement with the estimates by *Chen* et al. [3.20], but is greater than that by *Spaepen* [3.35]. The viscous flow in solids takes place via the directional diffusive flow of these defects. Therefore, the viscosity is sensitively dependent upon the thermal history of the sample [3.20, 36] and the annealing sometimes results in embrittlement [3.30].

3.4.3 Structural Defects and Physical Properties

The structural defects appear to influence many of the physical properties of metallic glasses, in addition to the viscous properties mentioned above. The diffusive motion of the defects is controled by the maximum potential barrier heights within the mean free paths of the defects, while the local motion of the defects occurs over lower potential barriers. Therefore the defects are permitted to move locally, say within few atomic distances, at temperatures too low for long-range diffusion to occur. Such a local motion of the defects may be responsible for the internal friction [3.37, 38] and the magnetic aftereffect (e.g., [3.39]). Furthermore, the low-temperature anomalies of the glasses, explained in terms of the two-level tunnelling states [3.40, 41], are likely to be caused by such a local motion of the defects. As the origin of the two-level systems, various single atom defects have been considered (e.g., [3.42, 43]). However, single atom defects tend to have high potential barriers for motion. The potential barriers for the structural defects are the Peierls potential [3.44] of a wider sense; therefore they are estimated to be roughly proportional to $\exp(-2\delta/a)$ where a is the atomic distance and δ is the size of the defect (core) [3.45]. The potential barrier height for the two-level systems active below $1K$ is of the order of $10^{-2}-10^{-4}$ eV. The defect size δ is then estimated to be about 3–5 atomic distances, within the range of the observed size of the defect as described earlier.

It is also instructive to discuss the tendency of glass formation in terms of the properties of the structural defects. In short, the ease of glass formation may be interpreted to imply small formation energies and large migration energies of defects. The physical and chemical effects on glass formation can be more readily understood from such a point of view. The physical effect is principally the effect of atomic size. An appropriate difference in the atomic sizes would help relax the atomistic stresses in the binary alloys and reduce the formation energies of the defects. This would mean a smaller value of b in (3.19), hence a smaller value of x_0 for a given τ_0, i.e., in an identical quenching condition. The chemical effect is the effect of stabilizing the defect by alloying, or by introducing a larger number of different elements. This is sometimes referred to as the "confusion principle". On first sight, the chemical effect appears to increase the defect formation energy, since the defect tends to disturb the chemical, or compositional, short-range order (CSRO). This, however, will be a minor effect, since the CSRO within the defect can readily be optimized by a small short-range atomic motion. Instead, the chemical effect would most significantly be observed as the increase in the defect migration energy. If the defect were to move, it has to disrupt the CSRO, while the CSRO would have no time to adjust itself during the motion of the defect. Therefore, this effect increases the value of α, and again results in a smaller value of x_0. The alloying, or the chemical effect, of course, would not stabilize the glassy phase unless the heat of compound formation is positive. It would then result in phase separation and generally in destabilization of the glassy phase. An evidence of the alloying effect increasing the defect migration energy can be found in the embrittlement behavior. In the Fe-metalloids ternary alloys, the embrittlement is most pronounced when the contents of two metalloid elements are roughly equal [3.46]; therefore the effect of CSRO is maximum. The glassy alloys are also easiest to produce with such compositions.

Thus in either case, the value of x_0 is reduced. In other words, the density of the defects in the as-quenched state, which is proportional to $x_\infty - x_0$, is increased. The relation between structural relaxation and crystallization has not been investigated; however, since the crystallization temperature T_x and the glass transition temperature T_g are close to each other in the metallic glasses, we may reasonably assume that crystallization takes place at approximately the same level of structural relaxation. The reduced value of x_0, then, directly implies the increased stability of the glassy phase.

3.5 Further Applications

The EDXD method could be used, other than for the study of the structural relaxation discussed above, in studying various details of the structure of metallic glasses, such as the effect of composition, temperature, or pressure. It is considerably more difficult to study small differences in the structure between two samples with different nominal compositions, because the effect of the geometry of the sample sometimes interferes severely. The EDXD method is superior to the conventional method in that regard, since the penetration depth is larger. The effect of composition on the structure [3.7], particularly with respect to the glass-forming ability, is an interesting field of study still hardly explored in depth. The effect of temperature is also of considerable interest, particularly in relation to the transport properties (Chap. 6). The EDXD method is particularly suited to the structural study under pressure [3.9], since the diffraction angle is fixed and the measurement requires only small diffraction windows.

The use of synchrotron radiation as a source in the EDXD method [3.9] is likely to produce substantial progress in the structural study. The extremely high intensity of the radiation will shorten the measurement time by several orders of magnitude, and will make the determination of the compositional short-range order through the use of the anomalous dispersion [3.47–49] much easier. The high radiation intensity, however, added to the nature of the synchrotron radiation, in that it consists of pulsed bursts of photons rather than a constant stream of photons, may cause the difficulty that a semiconductor detector becomes saturated. In such a case, one can use a crystal monochromator of medium resolution to carry out the EDXD measurement in a much shorter time than with the conventional x-ray source. Also for the study of the CSRO, a high-resolution crystal monochromator will be required to study the structure using the energy range in the vicinity of the absorption edge of the sample.

References

3.1 H.P.Klug, L.E.Alexander: *X-ray Diffraction Procedures for Polycrystalline and Amorphous Materials*, 2nd ed. (Wiley, New York 1974)
3.2 B.C.Giessen, G.E.Gordon: Science **159**, 973–975 (1968)
3.3 B.Buras, J.Chwaszczewska, S.Szarras, Z.Szmid: Rep. 894–11–PS, Inst. Nucl. Res. Warsaw (1968)
3.4 M.Mantler, W.Parrish: "Energy Dispersive x-ray Diffractometry", in *Advances in x-ray Analysis*, Vol. 20, (Plenum, New York 1976) pp. 171–186
3.5 J.M.Prober, J.M.Schultz: J. Appl. Crystallogr. **8**, 405–414 (1975)
3.6 T.Egami: J. Mater. Sci. **13**, 2587–2599 (1978)
3.7 T.Egami: J. Appl. Phys. **50**, 1564–1569 (1979)
3.8 C.N.J.Wagner: "Diffraction Analysis of Liquid and Amorphous Alloys", in *Advances in x-ray Analysis*, Vol. 12 (Plenum, New York 1969) pp. 50–71

3.9 B.Buras, J.Staun Olsen, L.Gerward, G.Will, E.Hinze: J. Appl. Crystallogr. **10**, 431–438 (1977)
3.10 Y.Murata, K.Nishikawa: Bull. Chem. Soc. Jpn. **51**, 411–418 (1978)
3.11 O.Klein, Y.Nishina: Z. Phys. **52**, 853–868 (1929)
3.12 B.E.Warren: *X-ray Diffraction* (Addison-Wesley, Reading 1968)
3.13 S.-W. Aur, T.Egami: Unpublished
3.14 J.A.Ibers, W.C.Hamilton (eds.): *International Tables for x-ray Crystallography*, Vol. 4 (Kynoch Press, Birmingham 1974)
3.15 D.T.Cromer, J.B.Mann: J. Chem. Phys. **47**, 1892–1893 (1967)
3.16 D.T.Cromer: J. Chem. Phys. **50**, 4857–4859 (1969)
3.17 D.T.Cromer, D.Liberman: Los Alamos Scientific Laboratory Rpt. LA–4403 (1970)
3.18 Y.Waseda, T.Egami: Unpublished
3.19 T.Egami: Mater. Res. Bull. **13**, 557–560 (1978)
3.20 H.S.Chen, L.C.Kimering, J.M.Poate, W.L.Walter: Appl. Phys. Lett. **32**, 461–463 (1978)
3.21 T.Egami, K.Maeda, V.Vitek: Philos. Mag. A**41**, 883–901 (1980)
3.22 M.Born, K.Huang: *Dynamical Theory of Crystal Lattices* (Clarendon, Oxford 1954)
3.23 K.Maeda, S.Takeuchi: J. Phys. F **8**, L283–288 (1979)
3.24 D.Srolovitz, K.Maeda, V.Vitek, T.Egami: Philos. Mag. (to be published)
3.25 Y.Waseda, T.Egami: J. Mater. Sci. **14**, 1249–1253 (1979)
3.26 P.H.Gaskell: J. Non-Cryst. Solids **32**, 207–224 (1979)
3.27 K.Suzuki, T.Fukunaga: Sci. Rep. RITU A**27**, 110–117 (1979)
3.28 T.Egami, T.Ichikawa: Mater. Sci. Eng. **32**, 293–295 (1978)
3.29 C.D.Graham, Jr., T.Egami, R.S.Williams, Y.Takei: AIP Conf. Proc. **29**, 216–217 (1976)
3.30 R.S.Williams, T.Egami: "Temper Embrittlement of Amorphous Alloys", in *Rapidly Quenched Metals III*, Vol. 1, ed. by B.Cantor (The Metals Society, London 1978) pp. 214–219
3.31 H.H.Lieberman, C.D.Graham, Jr., P.J.Flanders: IEEE Trans. MAG-**13**, 1541–1543 (1977)
3.32 H.S.Chen: J. Appl. Phys. **49**, 4595–4597 (1978)
3.33 Y.-N.Chen, T.Egami: J. Appl. Phys. **50**, 7615–7617 (1979)
3.34 P.M.Anderson III, A.E.Lord, Jr.: Mater. Sci. Eng. **44**, 279–284 (1980)
3.35 F.Spaepen: Acta Metall. **25**, 407–415 (1977)
3.36 R.S.Williams, T.Egami: IEEE Trans. MAG-**12**, 927–929 (1976)
3.37 H.S.Chen, J.J.Leamy, M.Barmatz: J. Non-Cryst. Solids, **5**, 444–448 (1971)
3.38 M.Barmatz, H.S.Chen: Phys. Rev. B**9**, 4073–4083 (1974)
3.39 M.Moser, H.Kronmüller: J. Mag. Mag. Mater. **19**, 275–277 (1980)
3.40 P.W.Anderson, B.I.Halperin, C.M.Varma: Philos. Mag. **25**, 1–9 (1972)
3.41 W.A.Phillips: J. Low Temp. Phys. **7**, 351–360 (1972)
3.42 W.A.Phillips: J. Non-Cryst. Solids **31**, 267–283 (1978)
3.43 D.A.Smith: Phys. Rev. Lett. **42**, 729–732 (1979)
3.44 R.Peierls: Proc. Phys. Soc. (London) A**52**, 34–37 (1940)
3.45 A.Seeger, P.Schiller: "Kinks in Dislocation Lines and Their Effects on the Internal Friction in Crystals", in *Physical Acoustics*, Vol. IIIA, ed. by W.P.Mason (Academic Press, New York 1966) pp. 361–495
3.46 M.Naka, T.Masumoto: J. Non. Cryst. Solids (to be published)
3.47 J.G.Ramesh, S.Ramaseshan: J. Phys. C**4**, 3029–3033 (1971)
3.48 Y.Waseda, S.Tamaki: Philos. Mag. **32**, 951–960 (1975)
3.49 T.Egami, R.S.Williams, Y.Waseda: "Structure of Amorphous Alloys Studied by Energy Dispersive x-ray Diffraction Method", in *Rapidly Quenched Metals III*, Vol. 2, ed. by B.Cantor (The Metals Society, London 1978) pp. 318–324

4. EXAFS Studies of Metallic Glasses*

J. Wong

With 19 Figures

EXAFS (extended x-ray absorption fine structure) is the oscillatory modulation of the absorption coefficient on the high energy side of an x-ray absorption edge of a constituent atom in a system. This mode of spectroscopy has been realized in recent years to be a very powerful tool for probing the local atomic environment of all states of matter, particularly since the advent of very intense synchrotron radiation in the x-ray region. In this chapter the physical mechanism associated with the EXAFS phenomenon is described in light of the single scattering approximation. The use of synchrotron radiation as a light source, data analysis, and unique features of EXAFS as a structural tool are discussed. EXAFS studies on some metallic glasses are reviewed. Information derived in the dynamic disorder and local structure of both the metalloid and metal constituents in these glasses are described in some detail.

4.1 Background

EXAFS is an acronym for extended x-ray absorption fine structure coined by *Lytle* in [4.1]. Experimentally it is associated with the oscillatory modulation of the absorption coefficient on the high-energy side of an x-ray absorption edge of a given constituent atom in a material. In Fig. 4.1 an example is given for the EXAFS of Fe metal above its K absorption edge at 7112 eV [4.2]. When an x-ray beam passes through a medium, its intensity is attenuated exponentially according to the classical absorption equation

$$I = I_0 \exp(-\mu x), \tag{4.1}$$

where I and I_0 are the transmitted and incident intensities, respectively, μ is the linear absorption coefficient, and x is sample thickness. In general, μ is a function of the photon energy. When the x-ray energy $h\nu$ becomes equal to or greater than the binding energy E_b of a core electron, the latter is emitted by a photoelectric process from the atom with kinetic energy E, conserving energy in the process:

$$E = h\nu - E_b. \tag{4.2}$$

* Manuscript completed in December, 1979.

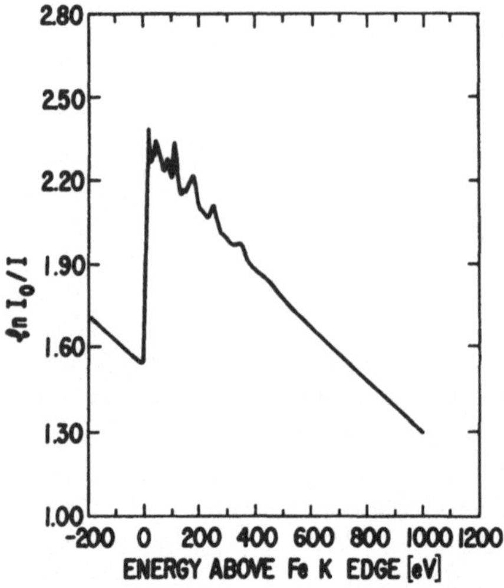

Fig. 4.1. Experimental EXAFS spectrum of Fe metal taken with synchrotron radiation above the K absorption edge of Fe at 7112 eV

In the case of Fe, when $h\nu = 7112$ eV the binding energy of the innermost K electron in Fe, a sharp rise in μ occurs giving rise to a characteristic K absorption edge shown in Fig. 4.1. On the high-energy side of the absorption edge, μx exhibits fluctuations with increasing photon energy extending to a few hundred eV beyond the edge. These oscillations are now theoretically understood to be a final state electron effect arising from the interference between the outgoing photoejected electron and that fraction of itself that is backscattered from the neighboring atoms. The interference is directly reflective of the net phase shift of the backscattered electron in the vicinity of the central excited atom, which is largely proportional to the product of the electron momentum k and the distance traversed by the electron. The type of central absorbing atom as well as backscattering neighboring atoms (i.e., their positions in the Periodic Table) also play a significant role in the interference event. As a result, EXAFS has now been realized to be a powerful structural tool for probing the atomic environment of matter [4.3–5], particularly since the advent of intense, continuous synchrotron radiation in the x-ray region [4.6]. Excellent general reviews on EXAFS [4.7–10] and its applications to disordered solids [4.11, 12] catalysts [4.13], and biological molecules [4.14, 15] have recently been given by a number of authors.

Historically, the fine structure above x-ray absorption edges had been reported in as early as 1920 by *Fricke* [4.16] and *Hertz* [4.17] working with the K edges of Mg-, Fe-, and Cr-containing compounds, and with L edges of Cs to Nd, respectively. Progress was slow prior to 1970 (see [4.10] for a more detailed account of development prior to 1930), primarily because the physical processes associated with EXAFS were not well understood (hence no adequate theory to account for the observed spectra), and because the experiments were tedious to perform prior to the availability of synchrotron radiation (this point is discussed later).

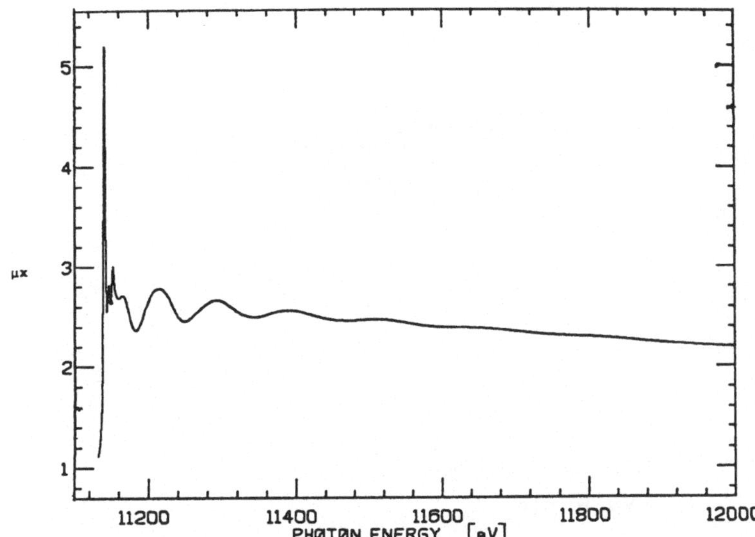

Fig. 4.2. Experimental EXAFS spectrum of Ge in $GeCl_4$ molecule taken with synchrotron radiation above the K absorption edge of Ge at 11103.3 eV (after [4.23]). The fine structure arises from backscattering of the Ge K photoelectron by the Cl atoms bonded to Ge in the molecule

Various early theories [4.18–20] had been proposed to explain the EXAFS and these can be broadly classified into two categories: long-range order (LRO) and short-range order (SRO). The LRO theories [4.18, 19] require the existence of lattice periodicity characteristic of crystalline solids and assume transition to quasistationary states to explain the fine structure. However, as pointed out by *Azaroff* [4.20a] and *Stern* [4.3], the LRO theories do not adequately predict the shape of the experimental absorption curve, since the dominant matrix element effects are neglected. The early Kronig theory [4.18] also failed to explain EXAFS in gases and amorphous materials.

On the other hand, there is ample experimental evidence supporting the SRO approach. EXAFS has been observed in simple gaseous molecular systems such as $GeCl_4$ [4.21] and, more recently, Br_2 [4.22]. The Ge EXAFS above its K edge in $GeCl_4$ taken with synchrotron radiation [4.23] is shown in Fig. 4.2. The fine structure arises from backscattering of the Ge K photoelectron by the four Cl atoms bonded to Ge in the tetrahedral molecule. In another early study, *Van Nordstrand* [4.24] observed a close similarity in the EXAFS spectra of a series of chromium, manganese, and cobalt crystalline compounds to those of their aqueous solutions, and concluded that the region of influence in an EXAFS event extended only 4–5 Å from the center of the atom being excited. In solids, perhaps the first convincing experiments to demonstrate the SRO effects associated with EXAFS were performed by *Nelson* et al. [4.25]. They measured the EXAFS above Ge K edge in glassy GeO_2 out to 350 eV and

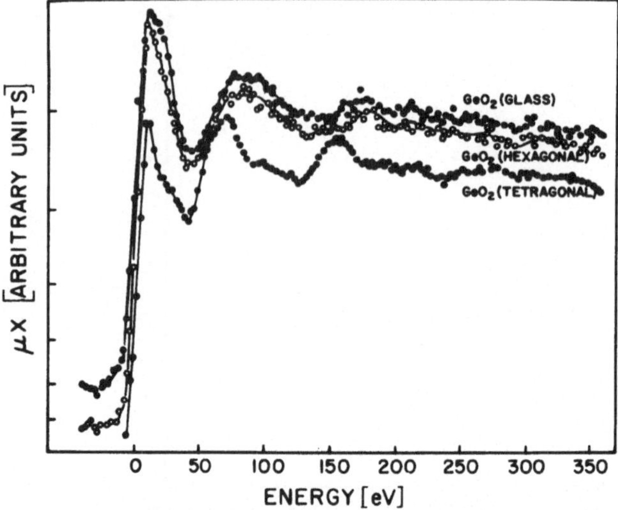

Fig. 4.3. Experimental K edge EXAFS spectrum of Ge in crystalline and glassy GeO_2 taken by *Nelson* et al. [4.24]. The energy is labeled with reference to the K edge of Ge taken as zero

compared it to those of the hexagonal and tetragonal crystalline polymorphs shown in Fig. 4.3. The extended absorption fine structures of glassy and hexagonal GeO_2 (in which Ge is 4-coordinated by oxygen) were very similar, but differed notably from that of tetragonal GeO_2 (in which Ge is 6-coordinated by oxygens). These observations were later reconfirmed by *Lytle* [4.1] who extended the measurement to 1100 eV beyond the K absorption edge of Ge. Since no long-range order, (i.e., lattice periodicity) exists in the amorphous phase, it is necessary to conclude that the fine structure is most strongly influenced by the arrangement of neighboring atoms about the germanium. The next-nearest neighbors appear to have little or no effect on the EXAFS spectra of disordered systems. This was demonstrated by the similarity of the Ge fine structure in a series of CaO-Al_2O_3-SiO_2-GeO_2 glasses [4.26].

Furthermore, K absorption fine structure measurements for some simple oxides of Ti, V, Cr, Mn, Fe, and Ge led to the conclusion [4.27] that the factors most significant in controlling the fine structure are bond character, primary coordination number, and valence; long-range order appears to be of little or no consequence.

In this chapter, the physical mechanism associated with EXAFS in terms of a single scattering approximation is first discussed. This is followed by EXAFS experimentation, data acquisition using the synchrotron radiation source at Stanford, data reduction, and analysis to extract structural information about the central absorbing atom. Finally applications of EXAFS to structural problems in a variety of metallic glasses studied to date are given.

4.2 EXAFS Fundamentals

As noted above the observation of x-ray absorption fine structure has been known for over half a century. Recent revival of interest in EXAFS began with the work of *Sayers* et al. [4.28] in 1970. They showed that using a single scattering approximation the observed fine structure oscillations may be understood in terms of the interference between the outgoing photoelectron wave in the vicinity of the central atom and that portion of it backscattered from neighboring atoms. Furthermore, it should be possible to invert the problem and obtain distances r_j from a Fourier analysis of EXAFS data. In particular, they performed a Fourier transform of EXAFS data in k space for crystalline and amorphous Ge and showed that peaks in the transforms correspond to various atomic shells [4.29]. It is now recognized that analysis of the EXAFS can yield not only the distance but also the number and type of nearest neighbor about the central atom.

Another milestone in the development of EXAFS is the availability of synchrotron radiation in the x-ray region at Stanford in 1974 [4.6]. The 10^4–10^5 increase of intensity in tunable x-rays over a broad spectral region enables EXAFS spectra with excellent signal-to-noise ratio to be obtained in a matter of minutes. In the past 5 years or so there has been growing appreciation of EXAFS as a new structural tool for studying a wide variety of materials for which conventional techniques like x-ray diffraction are less useful or impossible.

4.2.1 The Physical Mechanism

The attenuation of x-rays traversing through a medium occurs by three principal modes: scattering, pair production, and photoelectric absorption. In the EXAFS regime photoelectric absorption dominates the attenuation process, resulting in the total absorption of a photon which, in turn, gives its full energy to electrons according to (4.2). To understand the mechanism that gives rise to the EXAFS spectrum we consider the K edge fine structure. In the dipole approximation [4.30] the probability of x-ray absorption is given by

$$p=2\pi^2e^2(\omega c^2m)^{-1}|M_{fs}|^2\varrho(E_f)\,, \tag{4.3}$$

where $M_{fs}=\langle f|\boldsymbol{\varepsilon},\boldsymbol{p}|s\rangle$, $|s\rangle$ is the K shell s state, $|f\rangle$ is the final unoccupied state of $\boldsymbol{p}$ symmetry, $\varrho(E_f)$ is the density of states per unit energy at the energy E_f of the final state, $2\pi\omega$ is the frequency of the x-ray, $\boldsymbol{p}$ is the momentum operator, and $\boldsymbol{\varepsilon}$ is the electric field vector of the x-ray. For x-ray energies well above the edge $\varrho(E_f)$ gives a monotonic contribution and can be approximated by that of a free electron of energy $E=\hbar^2k^2(2m)^{-1}+E_0$. Here E_0 is the energy of $k=0$ free electrons and is the effective mean potential experienced by an excited electron.

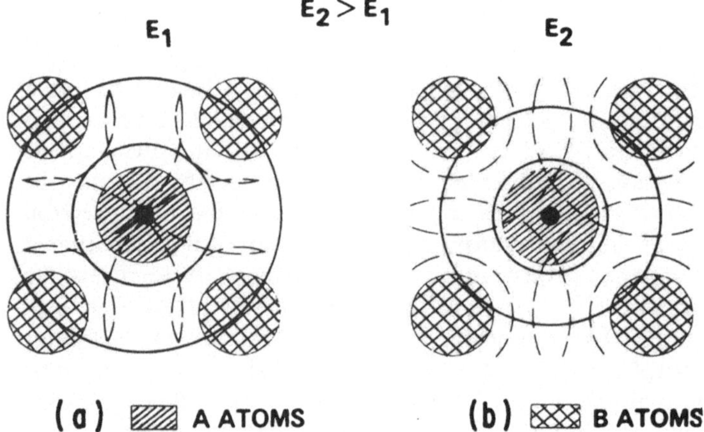

Fig. 4.4a, b. Schematic representation of an EXAFS event. The excited electronic state is centered about the A atom. The solid circles represent the crests of the outgoing part of the electronic state. The surrounding B atoms diffract the outgoing part as shown by the dashed circles. Constructive interference is represented in (**a**) and destructive interference in (**b**)

It is often called the threshold energy. With this assumption for $\varrho(E_f)$, the only remaining factor that can contribute to the EXAFS signal is M_{fs}. Now, the initial state $|s\rangle$ is fixed and does not vary with ω. The final state $|f\rangle$, however, varies with ω and produces the fine structure.

Further, the wave function $|f\rangle$ is a sum of two contributions. If the atom is isolated, the excited photoelectron would be in a solely outgoing state from the center atom as shown schematically in Fig. 4.4 by the outgoing solid rings. In this case, M_{fs} exhibits no fine structure and the x-ray absorption coefficient would vary monotonically with ω. This is the case for a monatomic gas like Kr, where the spectrum [4.23] beyond the K edge at 14.32 keV follows a decay predicted by the photoelectric effect and which reveals no fine structure (Fig. 4.5).

If now the excited atom is surrounded by other atoms, as in a molecule or condensed phase, whether liquid, glassy, or crystalline the outgoing electron is scattered by surrounding atoms producing incoming waves, depicted by the dashed lines in Fig. 4.4. These incoming or backscattered waves can constructively or destructively interfere with the outgoing wave near the origin where $|s\rangle$ exists. In Fig. 4.4a, the amplitudes of the outgoing and backscattered waves add at the central A atom site, leading to a maximum in the x-ray absorption probability. In Fig. 4.4b, the x-ray energy has been increased to E_2, leading to a shorter photoelectron wavelength for which the outgoing and backscattered waves interfere destructively at the absorbing A atom site with a resulting minimum in the absorption. This interference gives rise to an oscillatory variation in M_{fs} as ω is varied, changing the electron wavelength and thus the phase between the outgoing and backscattered waves.

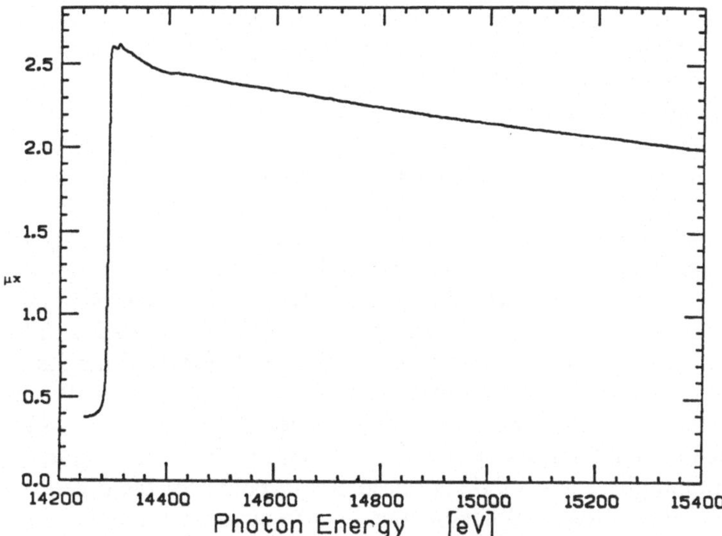

Fig. 4.5. *K* edge EXAFS spectrum of Kr gas (after [4.23])

Constructive interference increases M_{fs} while destructive interference decreases M_{fs} from the isolated atom value.

The total absorption $\mu(k)$ above the absorption edge is then given by

$$\mu(k)=\mu_0(k)[1+\chi(k)]\,, \tag{4.4}$$

where $\mu_0(k)$ is the smooth varying portion of $\mu(k)$ and physically corresponds to the absorption coefficient of the isolated atom. (Low frequency and very small amplitude oscillations have been observed in the normal EXAFS region. The atomic origins of these EXAFS-like spectra have been discussed by *Holland* et al. [4.31].) The fine structure $\chi(k)=[\mu(k)-\mu_0(k)]/\mu_0(k)$ is, therefore, due to interference between backscattered and outgoing photoelectron waves in the photoabsorption matrix element.

4.2.2 The Single Scattering Approximation

Based on the physical ideas discussed in the previous section, *Sayers* et al. [4.28] derived the first successful working theory of EXAFS. This was subsequently modified by *Stern* [4.3] to a more general form and further refined by others [4.32, 33]. For an unoriented specimen, the fine structure above the *K* or L_1 edge, can be described by

$$\chi(k)=-\frac{1}{k}\sum\frac{N_j}{r_j}\exp(-2r_j/\lambda)\exp(-2\sigma_j^2k^2)t_j(2k)\sin[2kr_j+\delta_j(k)]\,, \tag{4.5}$$

where $k=[2m(E-E_0)/\hbar^2]^{1/2}$ is the wave vector of the ejected photoelectron of energy E and E_0 is the inner potential or threshold energy caused by the atomic potentials and represents the threshold above which the kinetic energy must be added to determine the total energy E. The summation is over shells of atoms which are at distance r_j from the absorbing atom and contain N_j atoms (the coordination number). λ is the mean free path of the photoelectron. The second exponential containing σ_j^2 is a Debye-Waller type term where σ_j^2 is not the usual mean-square vibrational amplitude of an atom, but is the mean-square *relative* positional fluctuation of the central and backscattering atoms. The fluctuations may be static (structural disorder) or dynamic (thermal) in origin. In this form the resultant EXAFS is a sum of sine waves with periods $2kr_j$ from each jth shell with an amplitude which represents the number of neighbors modified by an envelope due to the scattering amplitude, the Debye-Waller damping, and the mean-free-path damping. Besides the usual $2kr_j$ which accounts for the phase difference of a free electron making the return trip to the neighbor, additional phase shifts $\delta_j(k)$ are needed to account for the potentials due to both the central atom and backscatterers. The factor r_j^2 arises from the product of the amplitudes of the outgoing and backscattered waves, both of which decay as r_j^{-1} because of their spherical nature. For a single crystal specimen, the factor $3\cos^2\theta_j$ has to be included in the summation, where θ_j is the angle the jth neighbor makes with the polarization vector of the x-ray. This factor averages to 1 for polycrystalline or amorphous materials. Conceptually, EXAFS may be considered a mode of electron diffraction where now the source of electrons is generated from within the particular atomic species participating in the absorption event.

The derivation of (4.5) is based on the following assumptions: i) The atomic radius is small enough for the curvature of the incident wave on the neighboring atoms to be neglected so that the incident wave may be approximated by a plane wave. This is achieved mathematically by replacing the Hankel function by its asymptotic form [4.33, 34] which in turn yields the factor k^{-1} in (4.5). ii) Only single scattering by the neighboring atoms in included. These assumptions have been examined in some detail for the case of fcc Cu by *Lee* and *Pendry* [4.33]. They first treated the electron scattering using a spherical wave expansion to take account of the finite size of the atoms. The effects are quite large but appear to make quantitative but not qualitative changes on the single-scattering description. As the size of the scattering atom increases significant deviations in both phase and amplitude are noted between the spherical wave calculation and the asymptotic plane wave approximation [4.35]. *Lee* and *Pendry* [4.33] further showed that multiple scattering becomes important when an inner shell atom shadows an outer shell atom. This is the situation for scattering from the fourth shell in the monatomic fcc and the fifth shell in the bcc structures in which these shells are shadowed by the first shell atoms and are thus strongly affected by forward scattering due to the first shell. Since the path length for multiple scattering must be larger than that for the dominant first shell interaction, in closely packed disordered systems like metallic glasses (see Sect. 4.3.1), the EXAFS signal arises predominantly from backscattering of nearest neighbors so that multiple scattering is zero.

4.2.3 Synchrotron Radiation as Light Source for EXAFS Experiments

We have seen that EXAFS deals with the fine attenuation, 5–10% in relative magnitude, on the high-energy side of a steep absorption edge. To ensure that we are measuring an EXAFS signal, good signal counting statistics must be obtained to yield high signal-to-noise ratio (>300:1). To achieve this with a conventional x-ray tube and a flat dispersing crystal, an experimental scan such as one of those shown in Fig. 4.3 for GeO_2 polymorphs typically takes a couple of weeks. The procedure is tedious and time consuming. Source instability over such an extended period of operation adds to the problem. With the advent of accessible synchrotron radiation sources in the x-ray region, particularly that at the Stanford Synchrotron Radiation Laboratory (SSRL) [4.6], a second important milestone has been reached in the advance of EXAFS spectroscopy.

Synchrotron radiation is emitted as the major loss mechanism from charged particles such as electrons and positrons, in circular motion at relativistic energies. The properties of synchrotron light emitted from electrons with velocities near that of the light are drastically different from the classical dipole radiation [4.36] and constitute the importance of synchrotron radiation as a new and powerful light source [4.9, 31–40]. The properties have been measured by *Elder* et al. [4.41] and studied theoretically by *Schwinger* [4.36], and may be summarized [4.6b] as follows: i) continuous spectral distribution from the infrared to the x-ray region which is ideal as a light source for uv and x-ray spectroscopies; ii) higher intensity permitting the use of monochromators with narrow band pass; iii) plane polarized, with the electric vector in the orbital plane of the circulating particles; iv) extremely high collimation which is important to lithography of submicron structure; and v) sharply pulsed time structure.

SSRL is a national facility for uv and x-ray research using synchrotron radiation from the storage ring SPEAR at the Stanford Linear Accelerator Center. SPEAR operates in the electron-positron colliding beam mode with stored beam energy E varying from 1.5 to 4.0 GeV and with up to 35 mA in each beam. SPEAR has a radius of curvature of 12.7 m. The spectral distribution of synchrotron radiation from SPEAR [4.6] is shown in Fig. 4.6 with the electron-beam energy E as the parameter from 1.5 to 4.5 GeV. As can be seen, it is an intense, continuous distribution extending from the infrared and up into the x-ray region. With this spectral distribution and the transmission characteristic of the Be window assembly, useful x-ray flux in the range of 3.5 keV to approximately 20 keV for a 3.0 GeV electron beam is available in SPEAR. This permits the EXAFS above the K absorption edges of Ca to Ru and above the L edges of higher Z elements to be measured.

Compared with the bremsstrahlung output of a 50 kW standard x-ray tube, synchrotron radiation is higher in intensity by a factor of 10^4 to 10^5. This

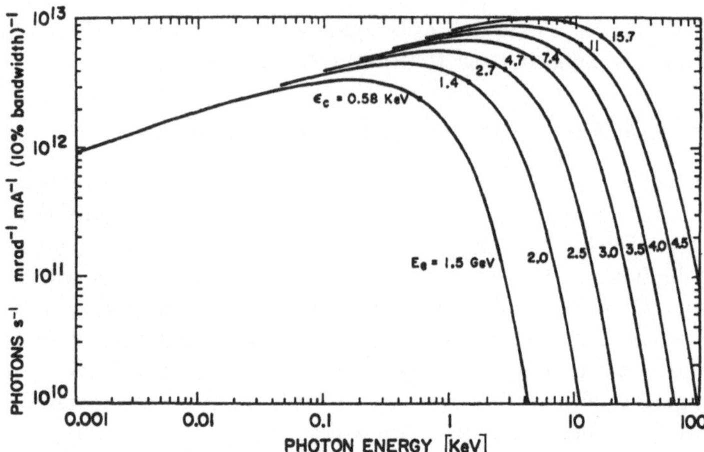

Fig. 4.6. Spectral distribution of synchrotron radiation from SPEAR having a radius of curvature of 12.7 meters (after [4.66])

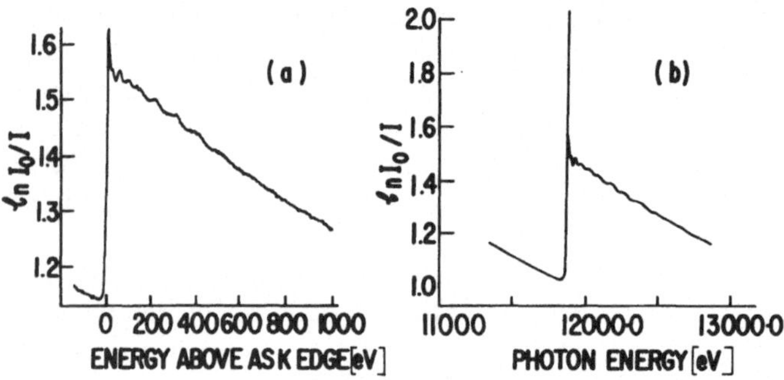

Fig. 4.7a, b. *K* edge EXAFS spectra of As in As_2Te_3 glass. Taken (**a**) with a conventional x-ray tube and (**b**) synchrotron or radiation. Spectrum (**a**) is a summation of seven scans; each took 3 days of continuous scanning. Spectrum (**b**) was taken in 1.3 hr. The sharp white line in the synchrotron radiation spectrum is due to higher resolution (1 eV) compared with ~8 eV for the data taken with the x-ray tube (after [4.42])

reduces the measurement time for a typical EXAFS experiment from a couple of weeks or more to an hour or less. In Fig. 4.7, we compare the *K* edge EXAFS spectrum of As in glassy As_2Te_3 taken by *Pettifer* [4.42] with a conventional x-ray tube and synchrotron radiation. The spectrum shown in Fig. 4.7a is the summation of 7 individual scans each of which took 3 days of continuous scanning. The synchrotron spectrum shown in Fig. 4.7b was taken in 1.3 h. The product of increase of resolution *x* measurement time *x* signal-to-noise ratio of the two spectra shows an improvement of $\sim 3 \times 10^4$ of the synchrotron data over those obtained using a conventional source [4.42].

In Fig. 4.8, the EXAFS apparatus used at SSRL is shown schematically [4.43]. The x-ray beam from the SPEAR vacuum chamber passes successively

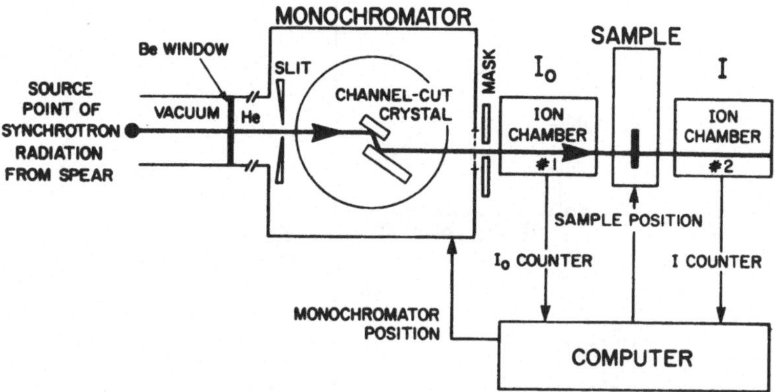

Fig. 4.8. Schematic diagram of EXAFS experimental apparatus at Stanford Synchrotron Radiation Laboratory (after [4.43])

through a He chamber, a slit, a channel-cut crystal monochromator, a mask, ion chamber No. 1 which measures I_0, a sample, and ion chamber No. 2, which measures I. The experiment is controlled by a PDP-11/05 minicomputer with interfaces to control the angle of the crystal monochromator, digitize the ion chamber currents, store, and plot the ratio I_0/I. Further experimental details associated with EXAFS measurements are described elsewhere [4.4, 23, 43, 44]. Since its opening in 1974, the request for beam time for x-ray experiments at SSRL has been doubled every year, as judged from the number of work proposals submitted to Stanford for beam time. This has led to a recently revived interest [4.45] in in-house laboratory EXAFS apparatus constructed using more powerful rotating anode sources for x-ray generation in conjunction with curved-crystal optics [4.46, 47]. As a consequence of the demand of the scientific community, various synchrotron radiation facilities over the world are now being updated and built as dedicated sources for synchrotron radiation research [4.48].

4.2.4 Data Analysis

Experimentally the EXAFS spectrum shown in Fig. 4.1 appears as low-intensity oscillations (relative to the jump at the absorption edge) superimposing on the smooth atomic absorption background which decays with increasing energy above the absorption edge. The fine structure according to (4.4) is therefore given by $\chi(k)=[\mu(k)-\mu_{0(k)}]/\mu_0(k)$, where $\mu(k)$ is the total absorption measured above the edge and $\mu_0(k)$ is the smooth atomic contribution. The first step in data reduction is to subtract this contribution. This can be achieved by using a smooth line routine developed by *Lytle* [4.49]. The routine utilizes the

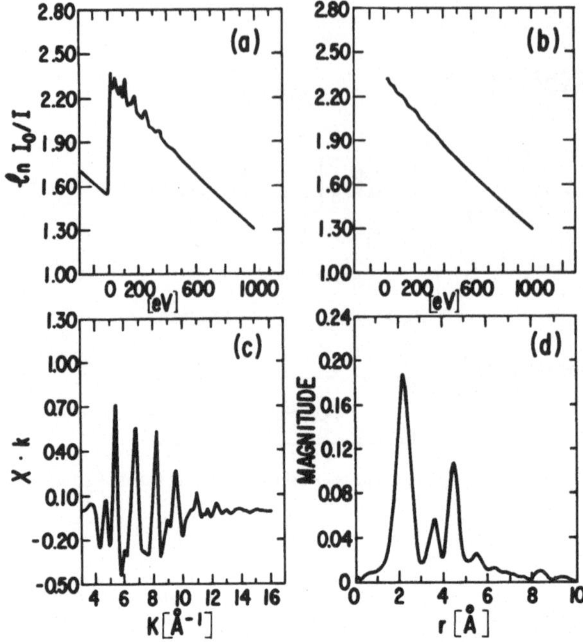

Fig. 4.9a–d. Graphical representation of a typical EXAFS data analysis. (**a**) Raw experimental K edge EXAFS spectrum for Fe; (**b**) polynominal smooth line fit of background absorption above absorption edge; (**c**) normalized $\chi \cdot k$ vs k plotted with a Hanning window applied to the first and last 10% of the k space data; and (**d**) Fourier transform of (**c**) according to (4.7)

so-called variable width sliding box-car window alogarithm. It begins by taking the last m number of raw data points at the high-energy end of the data set and least-squares fitting with a linear polynomial. Here, m is an odd integer and may take a value in the range 51–201 for a typical data set of 500 points. The mean value at the $1/2(m+1)^{\text{th}}$ point, i.e., the midpoint of the set of m points, is then determined. The routine then picks the next set of m points by just moving 1 point towards the low-energy end and the new set of m points is then least-square fitted. The process is repeated point by point all the way to the absorption edge. The result is shown graphically in Fig. 4.9 for the case of Fe. Figure 4.9a is the raw experimental K edge EXAFS spectrum of a 5 μm thick Fe foil taken at room temperature. The spectrum was recorded at SSRL with SPEAR running at 2.6 GeV electron energy and beam current ~30 mA. Figure 4.9b is a plot of the smooth line resulted from the sliding box-car window fitting.

The next step is to subtract the smooth line in Fig. 4.9b from the raw data 30 eV above the edge. Below this energy, many-body interactions [4.50], band effects [4.51] and coulombic distortion of excited state wave function [4.52] dominate, so that the oscillation are non-EXAFS in origin. The difference is then normalized to the step jump to yield χ according to (4.4). Also the energy space is converted to k space using $k = \sqrt{0.263E}$, where E is the energy eV about the first inflection point at the absorption edge. This procedure yields the normalized χ which is then weighted by k to yield the familiar $\chi \cdot k$ vs k plot given in the literature. The k weighting or more generally the k^n weighting will be discussed below in conjunction with Fourier analysis.

Returning to (4.5), the expression for $\chi(k)$ can be Fourier transformed to yield a radial structure function $\phi(r)$ which contains structural information about the absorbing atom. The Fourier inversion represents a significant step in the development of EXAFS technique and converts it from a qualitative effect to a quantitative effect [4.4, 29]

$$\begin{aligned}\phi(r) &= (2\pi)^{-1/2}\int \chi(k)\exp(2ikr)dk \\ &= \sum_j N_j \int dr'/r'^2 T(r-r')\exp[-2(r-r'_j)^2/\sigma_j^2]\,. \end{aligned} \tag{4.6}$$

As seen from (4.6), the Fourier transform of the EXAFS, $\phi(r)$, consists of a sum of radial peaks located at r_j and determines the spatial variation of the scattering matrix. Since in actual practice an EXAFS spectrum is taken over a finite energy range (hence k space), the Fourier transform that is actually taken is

$$\phi(r) = (2\pi)^{-1/2}\int_{k_{min}}^{k_{max}} W(k)k^n\chi(k)\exp(2ikr)dk\,, \tag{4.7}$$

where k_{max} and k_{min} are the maximum and minimum k values of the usable experimental data. k^n is a weighting function used to compensate for amplitude reduction as a function of k [4.4, 53] especially for low Z scatterers. Values of $n=1$, 2, and 3 have been suggested by *Teo* and *Lee* [4.53] for back-scatterers with $Z>57$, $36<Z<57$ and $Z<36$, respectively. Also as noted by *Stern* et al. [4.5] the χk^1 transform is rather sensitive to k_{min} in the region between the origin and r_1, the first peak in $\phi(r)$. The χk^3 transform may be approximated to a pseudocharge density which is rather insensitive to k_{min} and E_0. The k^3 transform weights less at low k and more at high k, where the EXAFS effect is better approximated by the single scattering expression given in (4.5), but experimentally of poorer quality because of poorer signal-to-noise ratio.

The factor $W(k)$ on the right-hand side of (4.7) is a window function which when multiplied to the integrand converts our finite data set to an infinite set which is necessary for Fourier transform. This is done by choosing functions which smoothly set the raw data points to zero at k_{min} and k_{max}. An example of $W(k)$ is a Hanning function [4.54] defined in terms of k as follows:

$$W(k) = 1/2\{1-\cos 2\pi[(k-k_{min})/(k_{max}-k_{min})]\}\,. \tag{4.8}$$

It is easily seen that $W(k)=0$ at $k=k_{min}$ and $k=k_{max}$. This window function is applied to the first and last 10% of the normalized Fe data discussed above, and plotted as $\chi\cdot k$ vs k in Fig. 4.9c. The Fourier transform so obtained is shown in Fig. 4.9d. Here we note that the transform is made with respect to $\exp(2ikr)$ *without including* the phase shift $\delta_j(k)$. This has the effect of shifting all the peaks in $\phi(r)$ closer to the origin to $r_j - \delta'$ where δ' is some average of the first derivative of $\delta_j(k)$ with respect to k. In Fig. 4.9d, the first peak is the nearest-neighbor position in bcc Fe shifted to 2.24 Å. The crystallographic value from diffraction is 2.482 Å, so that δ' for $j=1$ is 0.242 Å.

In general the effect of $\delta_j(k)$ on the transform may empirically be corrected for by measuring the EXAFS spectrum of a standard or model compound of known structure. Indeed, as in the case of complex biomolecules, a number of such model compounds are used on a trial and error basis to deduce a model of the unknown structure. Alternatively, one can use theoretical values of $\delta_j(k)$ such as those of *Teo* and *Lee* [4.53] in the Fourier transform and obtain r_j directly.

Besides interatomic distances r_j, EXAFS contains other structural information such as the coordination number N_j and type of jth atoms in the shell at r_j and their relative mean square disorder σ_j^2 about the average distance r_j. These structural parameters may be obtained by measuring the EXAFS of model compounds under identical conditions and using "transferability" of phase shifts [4.57]. The structure of the unknown is then modelled by curve fitting procedures so as to arrive at a calculated EXAFS that best fits the experimental values. A useful way to perform modelling is in combination with back Fourier transforming, especially in systems where the coordination shells are well separated in r space or whose $\phi(r)$ is dominated by the nearest-neighbor shell as in the case of amorphous materials. The shell-by-shell back transforming enables the envelope $t_j(2k)\exp(-2\sigma_j^2k^2)\exp(-2r_j/\lambda)$ to be determined for each jth shell. Various curve fitting routines for extracting structural parameters from EXAFS data have been prescribed and the reader is referred to the literature for specific details [4.5, 11, 34, 58, 59].

4.2.5 Unique Features of EXAFS

In this section we illustrate the unique features of EXAFS as a structural probe by way of an example. In Fig. 4.10a the EXAFS above the K edges of both Fe and Ni in an fcc Ne–Ni alloy containing 45 at.% of Ni are shown. These were obtained in one experiment by first tuning the synchrotron radiation near the K absorption edge of Fe at 7.11 keV, scanning the Fe EXAFS over the 1000 eV range, and continuing scanning another 1000 eV beyond the Ni K edge to obtain its EXAFS. Note that Fe and Ni are separated by 2 units in atomic number, yet their K absorptions are far apart in energy so that the EXAFS of Fe is not overlapped by the onset of the K absorption of Ni. This in turn means that structural information extracted from analyzing each EXAFS spectrum is atom specific in the sense that the central atom is defined; hence, the origin of each of the $\phi(r)$ is known. This clearly demonstrates atomic selectivity of the EXAFS technique for studying multiatomic systems.

Using the data reduction procedure described in the previous section, the normalized EXAFS for Fe is obtained and plotted as χk vs k in Fig. 4.10b. This

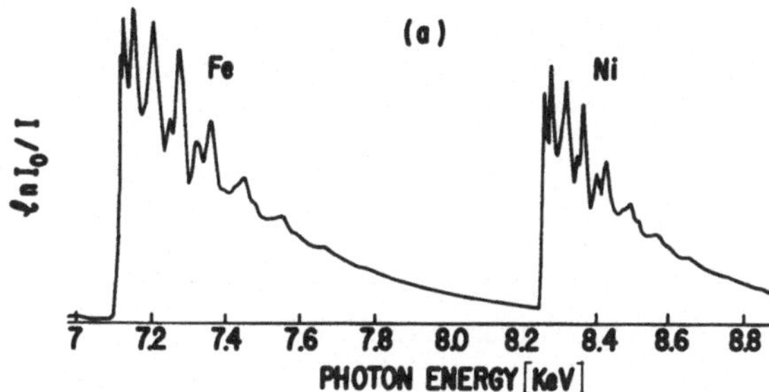

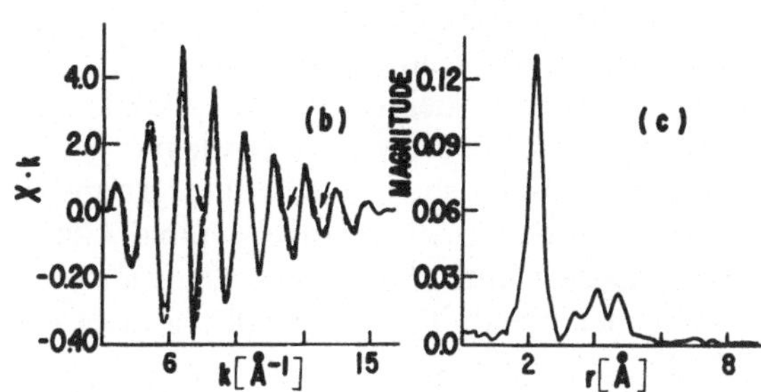

Fig. 4.10a–c. fcc Fe–Ni alloy containing 45 at.% Ni. (**a**) Raw EXAFS spectra above K edges of Fe and Ni; (**b**) normalized $\chi \cdot k$ vs k for Fe EXAFS (solid line) and inverse transform of first Fourier peak in the region 1–3 Å shown in (**c**); (**c**) Fourier transform of (**b**). The satellite features (arrows) in the full EXAFS in (**b**) are not reproduced in the inverse transform since they arise from second and higher shell contributions which are filtered off in the inverse transform

is then Fourier transformed with respect to $\exp(2ikr)$ to obtain $\phi(r)$ which is shown in Fig. 4.10c. The radial structure function is again dominated by a strong nearest-neighbor peak, but higher neighbor shells are also visible to 5 Å. The oscillations in the low r side of the first peak are due to termination errors of the transform. Compared with the transform shown in Fig. 4.9d for pure Fe which is bcc and has 8 nearest neighbors, the transform pattern shown in Fig. 4.10c for Fe in the Fe–Ni alloy is quite different. The latter in fact is characteristic of an fcc structure like Ni which has 12 nearest neighbors. This can also be seen directly in the new spectra in Fig. 4.10a, in that the Fe EXAFS patterns is isomorphic with that of Ni and is a direct consequence of the alloying effect which results in structuring the Fe atoms in an fcc lattice.

We now back transform the first shell from the region 1–3 Å in r space to k space to obtain the EXAFS contribution due to the nearest neighbors. The inverse transform $\chi_i k$ for the first shell is shown in Fig. 4.10b by the dotted line. It is seen that at least 90% of the contribution comes from the nearest-neighbor shell of atoms. The shoulders marked by arrows in the full spectrum was not reproduced in the back transform as they are due to second and higher shells which are filtered off in the back transform. This clearly demonstrates the predominate contribution of the nearest neighbor to the EXAFS intensity even in crystalline materials. This sample further exemplifies the power of EXAFS as a short range order probe for atomic environment in matter.

The combined features of atomic specificity and high sensitivity to local structure make EXAFS a unique tool for probing the short-range order in amorphous materials, metal environment in biomolecules, structure around ions in solution, catalysts, and surface structure, for which conventional techniques such as x-ray diffraction are less useful or impossible.

4.3 Metallic Glasses

Amorphous metallic alloys formed by splat quenching [4.60] the corresponding liquid at rates $\sim 10^6$ deg·s^{-1} constitute a new class of glassy solids. They exhibit high electrical conductivity, ferromagnetism, and superconductivity [4.61] that are not found in conventional inorganic and organic polymeric glasses. Of interest too are their high ductility in bending, corrosion resistance, and low coercivity. These properties, not found in their crystalline counterparts, have aroused considerable technological [4.62] as well as academic [4.63] interest in recent years. It has now been established that certain groups of related alloys exist for which this metastable phase is relatively stable. Glass-forming metallic alloys may be classified into two types: I) transition metal+metalloid and II) metal+another metal. Type I metallic glasses are exemplified by the well-known $Pd_{80}Si_{20}$, $Au_{80}Ge_{20}$, $Ni_{80}P_{20}$ glasses and more recently the $(Fe, Ni)_{80}(P, B)_{20}$ ferromagnetic glasses produced by Allied Chemical. The composition range of glass formation in this type of alloy falls in a narrow range of 15 to 25 at. % of metalloid where usually a deep eutectic exists in the phase diagram. Type II metallic glasses may be generically represented by TM′–TM″, RE–TM″, and M′–M″ where TM′ is an early transition metal such as Ti, Zr, Nb; TM″ is a late transition metal such as Fe, Co, Ni, Cu; RE is a rare-earth metal such as Gd, Tb, Dy; and M′–M″ are simple metal binaries such as $Be_{70}Al_{30}$ [4.64] and $Mg_{70}Zn_{30}$ [4.65]. Contrasted to Type I alloys, the glass-forming composition of TM′–TM″ alloys occurs in the middle of the binary and is usually wider than for Type I alloys. In the case of splat-quenched $Nb_{100-x}Ni_x$, x ranges from 40 to 70 at. % [4.66] and can be expanded to 30–85 at. % of Ni by vapor quenching [4.67]. More detailed listings of both types of glass-forming metallic systems prepared to date have been given by *Takayamana* [4.68] and *Donald* and *Davies* [4.69].

Metallic glasses may also be prepared by electrolytic and vapor deposition. These techniques and physicochemical properties of metallic glasses are discussed in other chapters of this book. In this chapter we focus our attention on the structural aspects of this class of glassy solids.

4.3.1 Structure of Metallic Glasses – A Brief Survey

To date most of the characterizations of atomic arrangements in metallic glasses have been made using conventional techniques of x-ray diffraction [4.70, 71]. These measurements provide a statistically averaged description of the arrangements of atoms which are used for testing three-dimensional

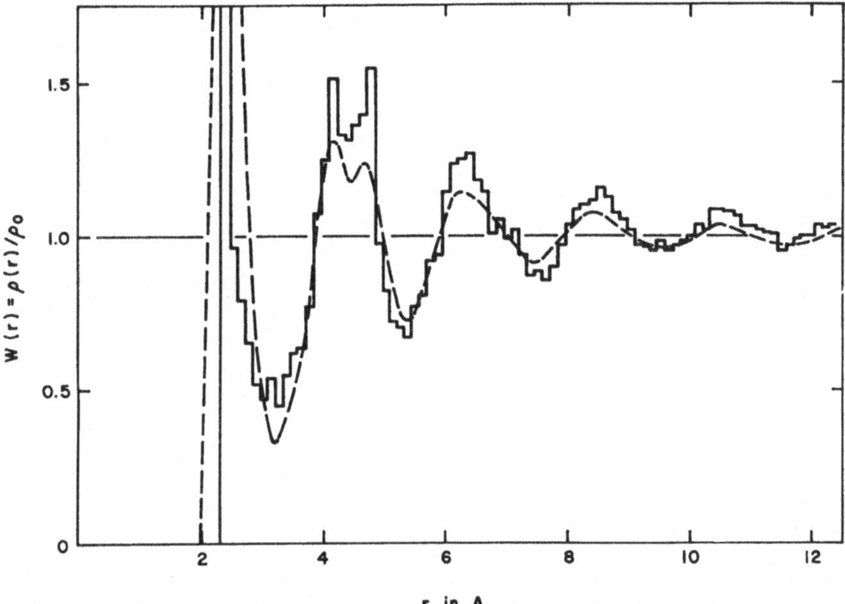

Fig. 4.11. Comparison of Finney's atomic distribution function $W(r)$ for DRP with a sphere diameter of 2.42 Å and Cargill's experimental $W(r)$ for amorphous $Ni_{76}P_{24}$ (after [4.79])

structural models. The x-ray interference functions $i(k)$ of metal-metalloid glasses exhibit a shoulder on the high k side of the second peak. Upon Fourier transforming the intensity data, the atomic distribution function $W(r)$ as well as the radial distribution function (RDF) contain a split double peak beyond the first maximum (see Fig. 4.11). The number of neighbors in the first coordination sphere varies from 10 to 13, depending on the system. The ratio of second to first near neighbor average distances is ~ 1.7. *Cohen* and *Turnbull* [4.72] on the basis of their free volume theory [4.73] suggested that the random packing of hard spheres proposed by *Bernal* [4.75] for the structure of monatomic liquids might be a good representation of the atomic arrangement in monatomic glasses. A high-resolution radial distribution function for this Bernal type of dense random packing (DRP) has been determined by *Finney* [4.75], while *Cargill* [4.76] has considered this DRP as a structural model for these amorphous alloys and suggested that to a first approximation the structure could be represented by a random mixture of metal and metalloid in a Bernal DRP of hard spheres of one size. Good agreement is obtained between the atomic distribution function obtained by *Finney* for DRP with a sphere diameter of 2.42 Å and that obtained experimentally with x-ray by *Cargill* for amorphous $Ni_{76}P_{24}$. In particular, both have the split second peak evident in Fig. 4.11.

The doublet observed in the second peak in the RDF is a characteristic *structural feature* of glassy metals. A geometrical interpretation of its origin has

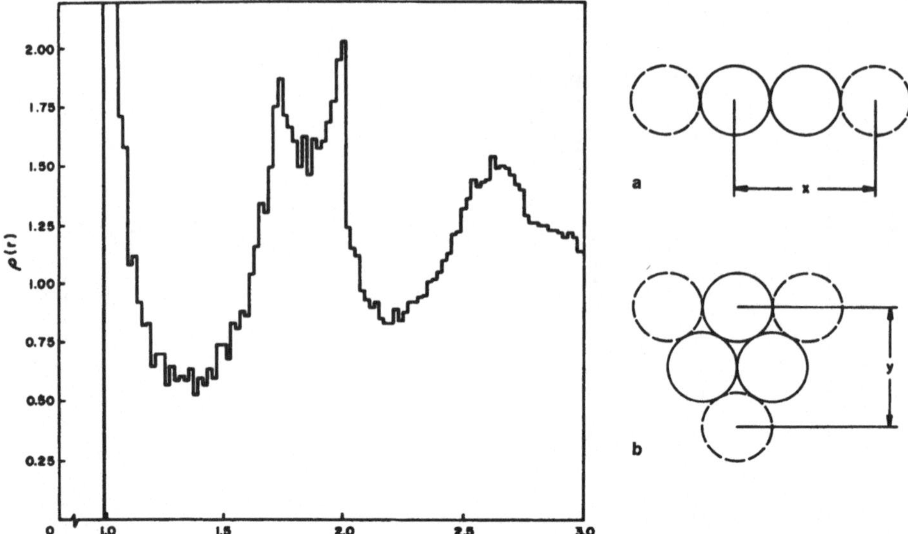

Fig. 4.12a, b. The radial distribution function $\varrho(r)$ calculated by *Finney* [4.75] for dense random packing to three sphere diameters. (**a**) and (**b**) are geometrical interpretations of the characteristic distances giving rise to the two relative maxima of the second peak in the RDF

been proposed by *Finney* [4.75]. In Fig. 4.12 a plot of Finney's $\varrho(r)$ to three sphere diameters for DRP is shown. The sphere diameter is taken as 1.00. The first peak which is very sharp is due to sphere in contact. The peak at 1.99 diameter is attributed to the separation x between three collinear spheres shown in Fig. 4.12a. Slight departure from exact collinearity shifts the maximum to below 2.00 diameters. This is substantiated by the sharp drop in $\varrho(r)$ slightly above 2.00. The peak at 1.73 is attributed to the characteristic distance y resulting from the arrangement of four atoms into coplanar equilateral triangles sharing an edge shown in Fig. 4.12b. Thus the split second peak simply arises from the short-range structure characteristic of DRP.

Contrary to what is generally observed, in $(Pt_{100-x}Ni_x)_{75}P_{25}$ amorphous alloys with $20 \leqq x \leqq 60$, the RDF does not exhibit a double peak beyond the first maximum. The ratio of second to first near-neighbor atomic distances is 1.86 ± 0.02 Å, a value which is larger than that (~ 1.7) observed for most metallic glasses. However, as pointed out by *Polk* [4.77] this is not inconsistent with the RDF of DRP since there is more than one sphere size present in the Pt–Ni–P alloys. In fact, for these alloys, the average of the two relative maximum of the second peak is ~ 1.85 Å so that the second peak in Pt–Ni–P is merely broadened not shifted.

Bernal's DRP structure [4.74] may be described as an assembly of polyhedra holes with atoms at the vertices and with edge lengths that are at most $\sim 15\%$ larger than the hard sphere diameter. Five types of holes were identified by *Bernal*. These are: i) tetrahedron, ii) octahedron, iii) trigonal prism with three half-octahedral caps, iv) archimedian antiprism with two half-octahedral caps, and v) tetragonal dodecaheron. (*Whittaker* [4.78] recently re-examined Bernal's model and identified four more shapes not mentioned by *Bernal*.) These holes are surrounded by 4, 6, 9, 10, and 8 atoms, respectively. Their frequency of occurrence is 86, 6, 3.8, 0.5 and 3.7%, respectively. Based on these findings, *Polk* [4.79] proposed a structural model for transition metal-metalloid glass having 80 at.% metal. In this model, the transition metal would form a Bernal DRPHS and the metalloid would fill in the large 8- and 9-fold coordinated holes inherent in the structure. This simple model postulates i) that each metalloid is surrounded by 8–9 metal atoms, ii) that there is no metalloid-metalloid nearest neighbor, and iii) that the metal-metalloid distance will be smaller than the sum of the atomic radii of the atoms. These points will be discussed later in light of recent EXAFS results on metallic glasses. The DRPHS model for metallic glass has further been elaborated by *Cargill* [4.70] and more recently by *Finney* [4.80] who stressed the role of soft potentials and density in more realistic modelling of metallic glass structures.

4.3.2 EXAFS Results

As noted in the discussion given in Sect. 4.2, most of the papers report theoretical treatment of EXAFS or experiments devoted to testing various theoretical aspects of the technique. Structural determinations of amorphous solids have been surveyed by *Hayes* [4.11] and *Pettifer* [4.12]. In the area of metallic glasses, there are only a few EXAFS studies reported in the literature to date. Systems include ferromagnetic a-(Fe–Ni–)$_{80}$(P, B)$_{20}$ alloys [4.81, 82], superconducting a-Nb_3Ge [4.83], magnetic a-RFe_2 compounds [4.84] where R = Tb and Dy, and a well-chosen system a-$Pd_{80}Ge_{20}$ [4.85] for probing the metalloid environment in metal-metalloid glasses. (a denotes amorphous in the above systems).

In EXAFS experiments the structural parameters of interest are the distance of separation r_j of the jth shell atoms from a chosen central atom participating in the photoelectron absorption event, the number N_j of these jth atoms, and their relative disorder (static or thermal) σ_j. The type of jth atoms can be identified in favorable cases from the backscattering amplitude $t_j(2k)$ in conjunction with a shell-by-shell back Fourier transform and curve fitting discussed in Sects. 4.2.4 and 4.2.5. Since EXAFS is dominated basically by local environment about a central atom, structural information derived for amorphous solids like metallic glasses will be most detailed on the nearest neighbors for which $j=1$.

In x-ray diffraction, the Debye-Waller factor $\bar{\mu}_j^2$ is the mean-squared amplitude of vibration of an atom perpendicular to the diffracting plane and moderates the intensity of the Bragg diffraction lines according to $I=I_0 \exp(-\bar{\mu}_j^2 k^2)$. For EXAFS, the scattering given in (4.5) may be rewritten

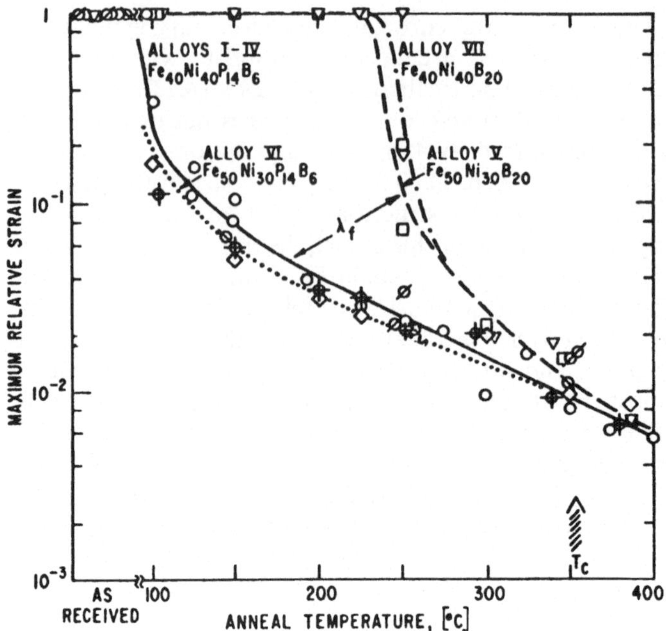

Fig. 4.13. Relative strain for fracture after 2 h isochronal anneal steps for alloys with and without phosphorus (after [4.90])

as $\chi=\chi_0 \exp(-2\sigma_j^2 k^2)$, where σ_j^2 is the relative displacement between the absorbing atom and its jth coordination shell. Since EXAFS involves a two-center average, $\sigma_j^2=2\bar{\mu}_j^2$ and hence $\chi=\chi_0 \exp(-4\bar{\mu}_j^2 k^2)$. In other words, an EXAFS even is $\exp(-3\bar{\mu}_j^2 k^2)$ times more sensitive than the corresponding x-ray scattering event in measuring thermal disorder which is dynamic in character.

a) Dynamic Disorder

A variety of ferromagnetic metallic glasses of general composition (Fe, Co, Ni)$_{\sim 80}$ (P, C, B, Si, Al)$_{\sim 20}$ are extremely strong in the as-cast state, and deform plastically in bending without fracturing [4.86–88]. However, when heated for relatively short times at temperatures several hundred degrees below their glass transition temperatures T_g, these alloys become brittle [4.89–91]. Auger analysis revealed regions of high phosphorus concentration on the fractured surface, indicative of phosphorus segregation at these low annealing temperatures. By removing phosphorus from the alloys, the stability towards embrittlement is greatly enhanced as seen in Fig. 4.13. Taking advantage of the atom selectivity of the EXAFS technique, changes in the local structure and dynamics about both the ferromagnetic metal atoms in a series of $Fe_{40}Ni_{40}P_{20-x}B_x$ glasses were studied as a function of phosphorus content and

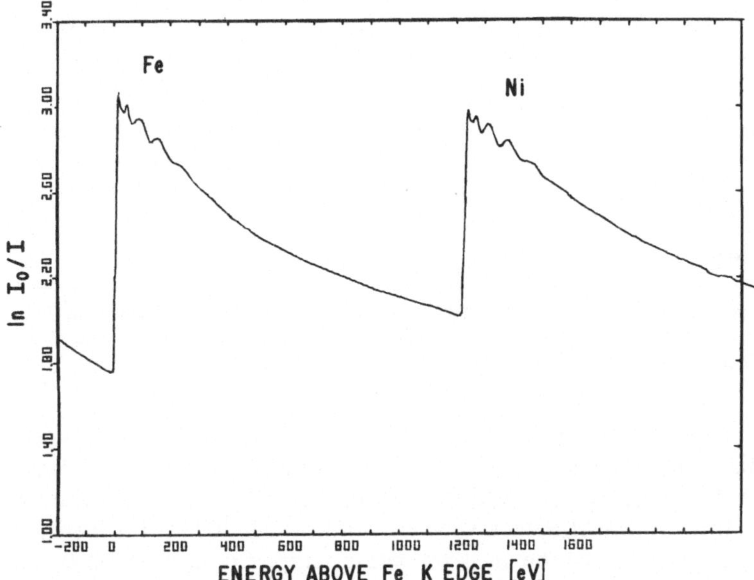

Fig. 4.14. *K* edge EXAFS spectra of Fe and Ni in $Fe_{40}Ni_{40}B_{20}$ glass at 77 K. The energy is labeled with reference to the *K* edge of Fe taken as zero

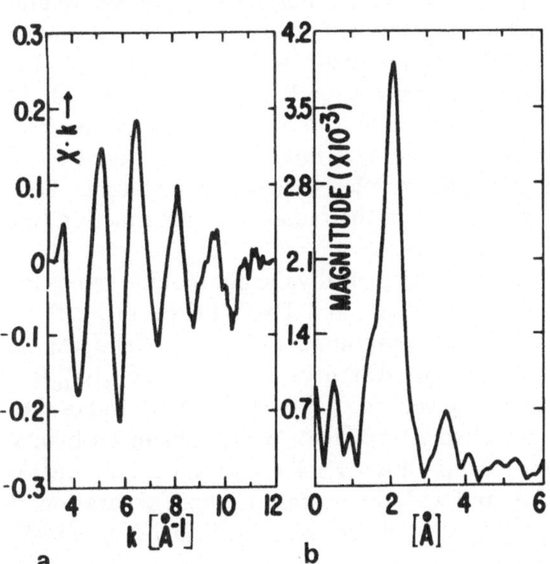

Fig. 4.15. (**a**) Normalized $\chi \cdot k$ vs k and (**b**) corresponding Fourier transform for Ni in glassy $Fe_{40}Ni_{40}B_{20}$ at 77 K (after [4.82])

annealing temperature. Spectra were recorded at 77 and 300 K and a typical scan is shown in Fig. 4.14.

In Fig. 4.15, the normalized χ_k vs k and its corresponding Fourier transform are given for the case of Ni in glassy $Fe_{40}Ni_{40}B_{20}$ above its K absorption edge at 8332.8 eV. The Fourier transform which yields a radial structure function is

Table 4.1. Metal-metal and metal-metalloid distances [in Å] in glassy $Fe_{40}Ni_{40}B_{20}$ and $Fe_{40}Ni_{40}P_{20}$

	$Fe_{40}Ni_{40}B_{20}$				$Fe_{40}Ni_{40}P_{20}$		
	r'	Δr	r		r'	Δr	r
Fe–B	1.75	—	—	Fe–P	1.70	0.56	2.26
Fe–Fe	2.02	—	—	Fe–Fe	2.02	0.61	2.63
Ni–B	1.75	0.30	2.00	Ni–P	1.70	0.60	2.30
Ni–Ni	2.02	0.49	2.51	Ni–Ni	2.02	0.45	2.47

r' = position of peak or shoulder in radial structure function; Δr = phase shift values from Ni_3B, Fe_3P, and Ni_2P. Actual distance $r = r' + \Delta r$. Errors in the metal-metal and metal-metalloid distances are estimated to be 0.05 and 0.10 Å, respectively.

characterized by a dominant peak 2.02 ± 0.02 Å. The position of this radial peak in all the radial structure functions about both the Fe and Ni atoms in the series of amorphous $Fe_{40}Ni_{40}B_{20-x}P_x$, where $x = 0$, 10, 14, and 20, remains invariant with temperature of EXAFS measurement at 77 and 300 K. Furthermore annealing for 2 hours below the glass transition temperature at 197, 250, and 325 °C did not alter the position of the first radial peaks of Fe and Ni in the $Fe_{40}Ni_{40}B_{20}$ alloy.

Similar EXAFS data were obtained for crystalline Fe_3P, Fe_2P, Ni_3B, and Ni_2P compounds. The radial structure functions of these compounds all show a doublet feature at ~1.70 and ~2.3 Å. Comparison with the crystallographic data shows that the low-distance peak corresponds to the metal-metalloid separation in these compounds. In the amorphous alloys, there is also evidence, though much weaker because of disorder and interference of nonphysical ripples at low r arising from termination effects, that the metal-metalloid distance appears as a shoulder at ~1.65 Å in the radial structure function (see Fig. 4.15).

For distance determination we obtain empirically the phase shifts values for Ni–B, Ni–Ni in Ni_3B; Ni–P, Ni–Ni in Ni_2P; and Fe–P, Fe–Fe in Fe_3P by comparing similar EXAFS data on these compounds with the crystallographic data [4.92]. The average nearest-neighbor distances in ternary $Fe_{40}Ni_{40}B_{20}$ and $Fe_{40}Ni_{40}P_{20}$ glassy alloys are given in Table 4.1. These values are consistent with results on the partial structure factor measurement on binary Fe–P [4.93] and Ni–P [4.94] amorphous alloys: Fe–P = 2.38 Å, Fe–Fe = 2.61 Å, Ni–P = 2.35 Å, and Ni–Ni = 2.55 Å, and with the nearest neighbor separation of 2.54 Å in amorphous $Fe_{40}Ni_{40}P_{14}B_6$ alloy determined by *Egami* et al. [4.95] using the energy dispersive technique.

From (4.6) we see that the magnitude of the jth peak is given by $N_j(r_j^2\sigma_j)^{-1}\exp(-2r_j/\lambda)T_j(2k)$. For $j=1$, r_1 remains invariant at 2.02 ± 0.02 Å as we discussed above. If we assume no change in the nearest-neighbor coordination number N_1, and invariance of λ and T_1 in the temperature range 77 to 300 K, then $M_1(\sigma_1)^{-1}$, so that the change in peak magnitude of the first peak is directly related to a change in disorder $\Delta\sigma^2$ about the nearest-neighbor coordination shell:

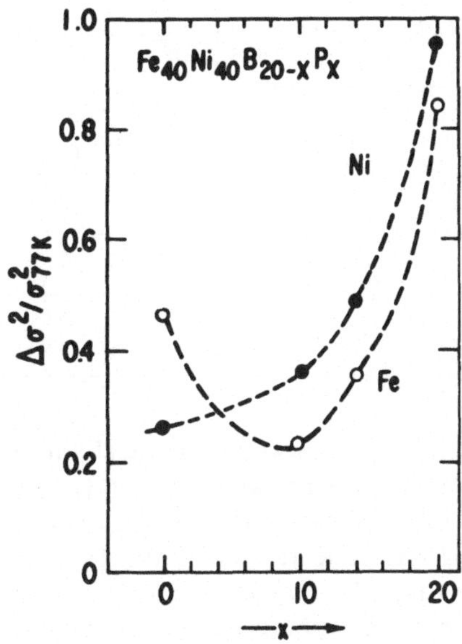

Fig. 4.16. Fractional change in disorder about the Fe and Ni atoms in glassy $Fe_{40}Ni_{40}B_{20-x}P_x$ as a function of phosphorus content

$\Delta\sigma^2 = \sigma^2_{300\,K} - \sigma^2_{77\,K}$. For the first shell, one can then define the fractional change in thermal disorder with respect to a low temperature state, say, that at 77 K, by

$$\Delta\sigma^2/\sigma^2_{77\,K} = (\sigma^2_{300\,K}/\sigma^2_{77\,K}) - 1 = (M^2_{77\,K}/M^2_{300\,K}) - 1 \,. \tag{4.9}$$

This quantity for the Fe and Ni atoms in $Fe_{40}N_{40}B_{20-x}P_x$ alloys is plotted vs x in Fig. 4.16. With increased phosphorus content in the alloy, $\Delta\sigma^2/\sigma^2_{77\,K}$ for Ni increases monotonically from a value to 0.26 at $x=0$ to 0.95 at $x=20$. For the Fe atoms, however, $\Delta\sigma^2/\sigma^2_{77\,K}$ decreases initially with phosphorus addition, and increases above $x=10$, attaining a value of 0.84 at $x=20$. This suggests that when $x \geqq 10$, addition of phosphorus enhances dynamic decoupling of both metal atoms with their nearest-neighbor coordination shell. The dynamic disorder data are consistent with Auger spectroscopic observation of phosphorus segregation [4.91] associated with mechanical enbrittlement in phosphorus-containing alloys.

The reverse order of $\Delta\sigma^2/\sigma^2_{77\,K}$ in $Fe_{40}Ni_{40}B_{20}$ (i.e., Ni Fe) and in $Fe_{40}Ni_{40}P_{20}$ (i.e., Fe < Ni) is interesting. This may be indicative of a preferential coordination of the metalloid with the metal atoms in these two ternary alloys. In other words, in the boron-containing $Fe_{40}Ni_{40}B_{20}$ glass, boron coordinates preferentially with Ni, forming a more tightly bound shell around the Ni atoms so that there is dynamic coupling of the Ni atoms with their neighboring environment, hence, a smaller fractional change in thermal disorder. Similarly in $Fe_{40}Ni_{40}P_{20}$, phosphorus coordinates preferentially with Fe giving rise to a lower $\Delta\sigma^2/\sigma^2_{77\,K}$ value of 0.84. A stronger preferential coordination of phosphorus with Fe may be responsible for the initial decrease in $\Delta\sigma^2/\sigma^2_{77\,K}$ for Fe from $x=0$ to $x=10$ shown in Fig. 4.16.

Table 4.2. Magnitude of first Fourier peak of Fe and Ni in amorphous $Fe_{40}Ni_{40}B_{20}$ alloy as a function annealing temperature. Annealed samples were measured at 77 and 300 K

Annealing Temp. [°C]	Fe($r_1=2.02\pm0.02$ Å) Magnitude ($\times10^{-3}$)		Ni($r_1=2.02\pm0.02$ Å) Magnitude ($\times10^{-3}$)	
	77 K	300 K	77 K	300 K
As cast	2.67	2.14	3.78	3.52
197	2.84	2.26	4.03	3.23
249	3.14	2.30	4.24	3.44
323	3.50	2.34	5.10	3.94

To gain structural insight into the effect of annealing on mechanical embrittlement, we have measured the room temperature and liquid nitrogen temperature EXAFS of Fe and Ni in samples of glassy $Fe_{40}Ni_{40}B_{20}$ alloy annealed for 2 h each at 197, 240, and 323 °C. The results of the Fourier transform of the normalized EXAFS in the range $k=3$ to 12 Å^{-1} are given in Table 4.2. Again, the position of the dominant peak for both Fe and Ni remains invariant at 2.02 ± 0.02 Å with temperature (i.e., 77 or 300 K) as well as thermal history (i.e., annealing temperature). At either 77 or 300 K, the magnitude of the peak, however, increases with increasing annealing temperature, implying structural ordering in the nearest-neighbor coordination shell about the metal atoms without changing the average distance of separation.

Using (4.9) and the magnitude data given in Table 4.2, we can calculate the fractional change in disorder $\Delta\sigma^2/\sigma^2_{77\mathrm{K}}$ at each of the Fe and Ni sites as a function of annealing temperature. The results are plotted in Fig. 4.17. For the Ni atoms, the variation of $\Delta\sigma^2/\sigma^2_{77\mathrm{K}}$ is gradual with annealing temperature. For the Fe atoms, however, there is a break near the temperature for the onset of embrittlement for this alloy at $\sim$200 °C. The larger and abrupt change in the Fe atoms is reflective of the "uneasiness" or metastability of Fe in an fcc-like environment and may largely be responsible for the embrittlement of Fe-containing amorphous alloys below the glass transition.

b) Metalloid Environment

The Polk model for metal-metalloid glass structure is heavily focused on the metalloid environment and specifies its local structure in some detail. A number of diffraction studies [4.94, 96–98] have been attempted to determine the metalloid environment in these glasses with the special effort of detecting metalloid-metalloid avoidance. The results, though consistent with Polk's postulate, usually are not definitively conclusive. This is due to a variety of experimental reasons such as lack of a complete set of partial distribution functions (e.g. ϱ_{MM}, ϱ_{MX}, and ϱ_{XX} in a binary M–X glass) or low signal-to-noise ratio due to low metalloid concentration used. Perhaps the most conclusive experiment is that of *Hayes* et al. [4.85] on a well chosen $Pd_{80}Ge_{20}$ system.

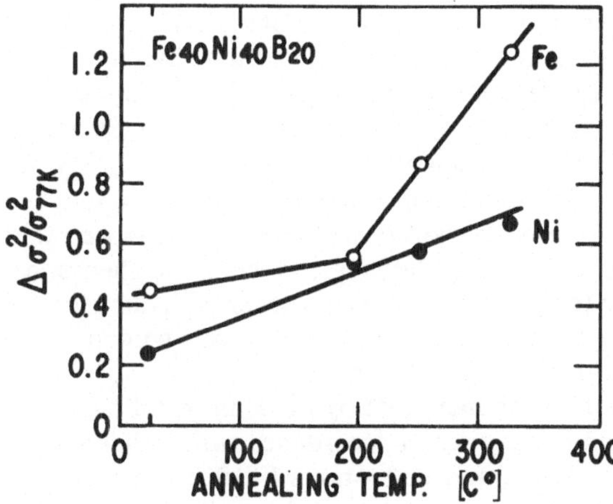

Fig. 4.17. Fractional change in disorder about the Fe and Ni atoms in glassy $Fe_{40}Ni_{40}B_{20}$ as a function of annealing temperature (after [4.82])

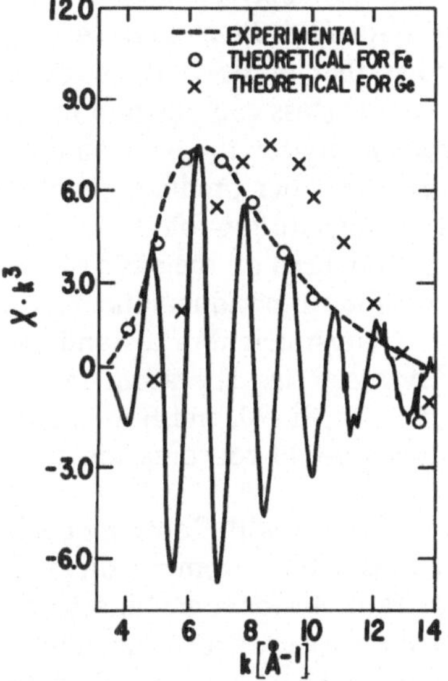

Fig. 4.18. Inverse k^3 transform of the first radial structure peak in the range 1–3 Å for Ge *K* edge EXAFS in $Fe_{80}Ge_{10}B_{10}$ glass taken at room temperature. The experimental envelope is compared with those calculated by *Teo* and *Lee* [4.53] for Fe (circles) and Ge (crosses). The good agreement with the calculated Fe backscattering amplitude indicates that Ge in the glass is surrounded mainly by Fe atoms and there are no nearest Ge–Ge neighbors

Using the atom selectivity power of EXAFS, they measured the fine structure above the Ge *K* edge in amorphous $Pd_{80}Ge_{20}$ and deduced in a single experiment conclusively the environment of Ge in the glass. Their findings can be summarized as follows. The Ge in Pd–Ge glass is coordinated by 8.6 ± 0.5 Pd atoms as nearest neighbors at a distance of 2.49 ± 0.1 A (rms halfwidth). By simulation, they found that there is no Ge–Ge nearest neighbor pair. Furthermore, the Ge environment in arc quenched $Pd_{78}Ge_{22}$ glass was invariant to that in sputtered amorphous $Pd_{80}Ge_{20}$.

A similar EXAFS experiment has recently been carried out in a ferromagnetic $Fe_{80}B_{10}Ge_{10}$ glass [4.99]. The Ge which substitutes for half of the

B in the binary glass was used as an EXAFS probe of the metalloid environment. Fourier transform of the normalized EXAFS above the Ge *K* edge yields predominantly a single peak centered at ~2.40 Å (phase-shift corrected). This radial peak is back Fourier transformed to *k* space and is plotted in Fig. 4.18. Using the values calculated by *Teo* and *Lee* [4.53], we plotted the theoretical backscatter amplitude function for both Fe and Ge normalized to the maximum of the back-transformed envelope. It is seen that the observed first shell envelope coincides quite nicely with the theoretical backscattered amplitude of Fe (open circles) and not at all with that for Ge (crosses), which exhibits a maximum at higher *k*. Again, it can be concluded that there is no Ge–Ge pair in the $Fe_{80}B_{10}Ge_{10}$ glass.

The 8- to 9-fold coordination of the metalloid by metal in metal-metalloid glasses appears to be a common occurrence in accordance with Polk's model. However, there are exceptions. *Suzuki* and co-workers [4.100] using a high-resolution pulsed neutron diffraction technique showed that the coordination of Si by Pd atoms in Pd–Si glasses varies from 6 to 7 with decreasing Si content and extrapolates to 9 at pure hypothetical amorphous Pd. *Boudreaux* [4.101] also observed in his computer simulation studies that in Fe–B glass the B coordination by Fe atoms is no higher than 7 over the glass composition range of 24 to 15 at.% B. There is, however, a discrepancy between the neutron data and computer calculation for Pd–Si glass, in that the latter predicts a higher than 8-fold coordination of Si by Pd atoms. The change from 6-fold to 8- to 9-fold coordination around the metalloid may be rationalized on the basis of size ratio of the metalloid and metal [4.102]. The larger the atom radius ratio, r_X/r_M, the more metal atoms can be accommodated as nearest neighbor around the metalloid atom before excess coulombic repulsion sets in. The atom radius ratios are 1.01, 1.00, 0.96, and 0.77 for the P–Fe, Ge–Pd, Si–Pd, and B–Fe pairs, respectively. Radius ratios below 1.00 give the lower 6-fold coordination as in the case of Pd–Si and Fe–B glasses.

Another interesting experimental finding in conformity with Polk's model is that the metal-metalloid distance is less than the sum of their atomic radii in all cases investigated. The results are summarized in Table 4.3 for some binary and ternary metal-metalloid glasses. It is noted that the measured metal-metalloid distance is more closely approximated by the sum of the covalent radii of the atom pair. The rather regular coordination around the metalloid and its "shortened" distance from its nearest metal neighbors are most strongly indicative of chemical ordering about the metalloid atom in the metal-metalloid glasses. Also given in Table 4.3 are the metal-metal distances in some metal-metal and rare-earth-metal glasses recently investigated. The observed separations are again shorter than the sum of the respective atomic radii but agree better with the sum of the covalent radii.

Finally the Ge–Nb distance in superconducting thin films of Nb_3Ge was determined by *Brown* et al. [4.83] with EXAFS to be 2.66 Å, which is substantially shortened from the A15 crystalline value of 2.87 Å, but is closer to the sum of the covalent radius (Table 4.3). Correspondingly the mean coordi-

Table 4.3. Distance of separation, r_{X-M}, between unlike atoms in metal-metalloid and metal-metal glasses determined by EXAFS and comparison with atomic and covalent radii addivity

Glass system	Atom type	Radius [Å]		r_{X-M} [Å] (exptal)	Expt type	Ref.
		Atomic[a]	Covalent[b]			
$Pd_{80}Ge_{20}$	Pd	1.37	1.28			
	Ge	1.37	1.22			
		2.74	2.50	2.49 ± 0.1	EXAFS	[4.85]
$Pd_{80}Si_{20}$	Pd	1.37	1.28			
	Si	1.32	1.11			
		2.69	2.39	2.40	Neutron	[4.99]
$Fe_{40}Ni_{40}P_{20}$	Fe	1.26	1.17			
	P	1.28	1.06			
		2.54	2.23	2.26	EXAFS	[4.82]
$Fe_{40}Ni_{40}P_{20}$	Ni	1.24	1.15			
	P	1.28	1.06			
		2.52	2.21	2.21	EXAFS	[4.82]
$Fe_{40}Ni_{40}B_{20}$	Ni	1.24	1.15			
	B	0.98	0.82			
		2.22	1.97	2.00	EXAFS	[4.82]
$Nb_{75}Ge_{25}$	Nb	1.46	1.34			
	Ge	1.37	1.22			
		2.83	2.56	2.66	EXAFS	[4.83]
$Ni_{63}Nb_{37}$	Ni	1.24	1.15			
	Nb	1.46	1.34			
		2.70	2.49	2.54	EXAFS	[4.107]
$DyFe_2$	Fe	1.26	1.17			
	Dy	1.80	1.59			
		3.06	2.76	2.64 ± 0.15	EXAFS	[4.84]
$TbFe_2$	Fe	1.26	1.17			
	Tb	1.80	1.59			
		3.06	2.76	2.78 ± 0.15	EXAFS	[4.84]

[a] Atomic radii from [4.110].
[b] Covalent radii from [4.111].

nation of the Ge sites is also reduced from 12 to 8 ± 2, which may be qualitatively understood in terms of the increase in Nb–Nb repulsion as the Ge–Nb distance is shortened. These findings suggest an increase in covalency in the Ge–Nb bond in going from the crystalline to the glassy state.

c) Metal Environment

The metal environment in metallic glasses can also be studied with EXAFS. With higher Z elements such as Pd, the K edge of Pd occurs at 24.36 keV. Such high-energy x-rays are difficult to come by at Stanford Synchrotron Radiation

Table 4.4. Local environments in crystalline Fe_3P: $S_4^2(I\bar{4})$, $Z=8$ (all atoms in general position, i.e., C_1 sites)

	Z_{Fe-P}	r_{Fe-P} [Å]	Δr_{Fe-P} [Å]	Z_{Fe-Fe}	r_{Fe-Fe} [Å]	Δr_{Fe-Fe} [Å]
Fe I	2	2.32–2.38	0.06	12	2.41–2.92	0.51
Fe II	3	2.26–2.41	0.15	10	2.52–2.83	0.32
Fe III	4	2.30–2.34	0.04	10	2.53–2.99	0.46
	Z_{P-Fe}	r_{P-Fe} [Å]	Δr_{P-Fe} [Å]			
P	9	2.26–2.41	0.15			

Δr denotes the spread of distances between a given pair of atoms.

Laboratory, which, till the fall of 1979, operated on a parasitic mode. Another difficulty is with the complexity of the local environment about the metal atom in these glasses. In a binary M–X glass, the possible nearest-neighbor pairs about the metal M are M–M and M–X pairs with the latter occurring at low value in r space. RDFs derived from diffraction experiments often exhibit a lower shoulder on the first peak, and only in favorable cases like Pd–Si [4.100] and using high-resolution techniques (i.e., high Q experiments) are the M–X and M–M peaks resolved. Referring back to Fig. 4.15 which shows a Fourier transform of the normalized $\chi \cdot k$ for Ni in a $Fe_{40}Ni_{40}B_{20}$ glass, the Fourier peaks at $\sim$2.0 Å (uncorrected for phase shift) comprise the contributions from all Ni–Ni, Ni–Fe, and Ni–B pairs, which may have to be extracted with careful computer simulation and curve fitting.

Crystallization studies show that the major crystallized product from transition metal-metalloid glass is an Fe_3P-type tetragonal phase. Therefore, it may be instructive to compare the nearest-neighbor environment of amorphous and crystalline metal-metalloid alloys by noting in some detail the local structure of constituent atoms in the Fe_3P structure. Using the refined crystallographic data of *Rundqvist* [4.103] for Fe_3P we have generated, by means of a computer, the distance of separation of nearest neighbors about each crystallographic nonequivalent atom in tetragonal Fe_3P. The information is summarized in Table 4.4. There are three nonequivalent Fe sites: Fe(I), Fe(II), and Fe(III) having 2, 3, and 4 nearest neighbor P, respectively. The corresponding magnetic moments measured by *Lisher* et al. [4.104] using neutron diffraction are 2.12, 1.84, and 1.25 μB, indicative of an increase in the quenching of the Fe moment with increased number of P neighbors at the Fe sites. The spread of Fe–P distances Δr_{Fe-P} for Fe(I) and Fe(III) is small, 0.05 Å, and for Fe(II) is 0.15 Å. However, the spread of Fe–Fe distances is large, varying from 0.51 to 0.32 Å. The P atom has 9 Fe nearest neighbors, found usually in the glassy state, with Δr_{P-Fe} values ranging from 0.05 to 0.15 Å. Strictly speaking from point group symmetry considerations, since *all* atoms have C_1 site symmetry (i.e., no symmetry) there is no short-range order in Fe_3P in the sense that instead of a single value there exists a distribution of bond angles, bond lengths, and coordination numbers about the constituent atoms in the structure. The lack of SRO particularly at the Fe sites in tetragonal Fe_3P may also be inherent with the structure of amorphous metallic alloys, at least with the transition-metal-metalloid glasses in contrast with the well-defined SRO in covalent network-type glasses like SiO_2 or the chalcogenide glasses. The existence of a distribution of local environments about a given constituent atom may be a distinguishing structural characteristic of metallic glasses and may account for the frequently observed nonintegral coordination numbers.

Amorphous $Fe_{75}B_{25}$ is a good model system for studying the metal environment in metal-metalloid glasses, since it can be annealed at low temperature to crystallize [4.105] homogeneously into a single Fe_3B phase which is isostructural to tetragonal Fe_3P [4.106] discussed above. Thus using the crystal as a model compound, one can hope to gain insight into the structure around the metal by measuring its EXAFS spectrum. The analysis is now underway in our laboratory.

An interesting but preliminary EXAFS study has recently been reported by *Stern* et al. [4.84] in a series of amorphous RFe_2 compounds where R = Dy, Tb. The structure about each constituent element was determined individually from their EXAFS since the K edge of Fe is at 7.1 keV and is far away from the $L_{2,3}$ edges of Dy, Tb at >8 keV. It was found that the Fe environment remains approximately the same as the crystal, but the rare-earth environment changes drastically in going from the crystal to the amorphous state. In the crystalline state, the Dy–Fe and Tb–Fe distances were determined by x-ray diffraction to be 3.036 and 3.046 Å, respectively. In the glass state, these distances decreases to 2.64 ± 0.15 and 2.7 ± 0.15 Å, respectively, as measured by EXAFS. As shown in Table 4.3, the R–Fe distances in the glass state are closer to the sum of the covalent radii. Furthermore, there is a decrease in the rare-earth coordination from the crystal value of 12 in both cases to 7.1 ± 1 and 8.4 ± 1.8 in the Dy and Tb glasses, respectively. In addition, the L_3 edge of the rare earth exhibits a negative edge shift of ~1.5 eV with respect to that of the corresponding crystal, indicative of a less ionized initial state of the rare earth in the glass. The combined findings again strongly suggest an increase in covalency in going from the crystalline to the glassy state of matter.

4.4 Concluding Remarks

It is clear that EXAFS by virture of its being an atom specific and local structure tool contributes much to our understanding of the metalloid environment in metal-metalloid glasses. In addition to qualitative verification of the various aspects of the Polk model, the technique also renders quantitative definitions of the metalloid coordination, their avoidance with each other as nearest neighbor, and accurate determination of its separation from the metal, not previously achieved with conventional diffraction techniques. The ~10%

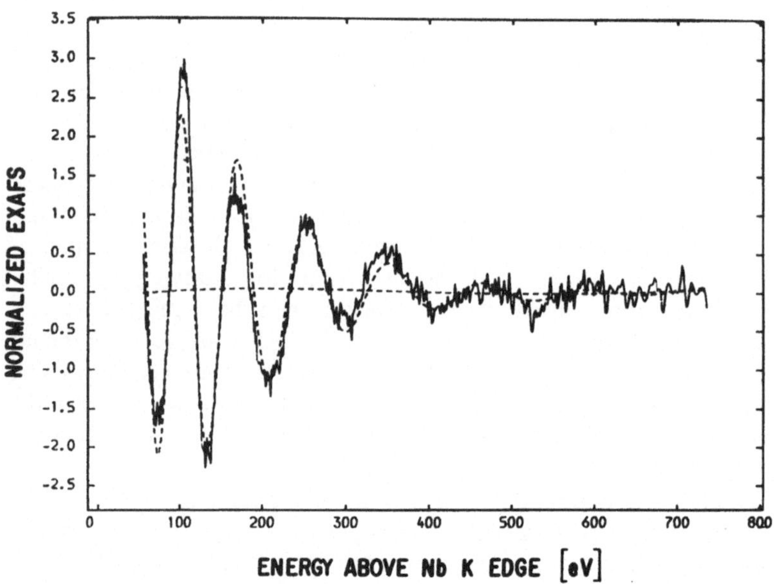

Fig. 4.19. Experimental (solid line) and fitted (dashed line) EXAFS spectra above the *K* edge of Nb in $Nb_{37.6}Ni_{62.4}$ glass (after [4.109])

decrease of the metal-metalloid separation from additivity of atomic radii indicates that there are appreciable electron interactions between the metal and metalloid constituents in the glass. The details of these chemical interactions are being studied experimentally by electron spectroscopy [4.107] and theoretically by molecular orbital cluster calculations [4.108].

Our present picture of the structure of metal-metalloid glasses may be described as a bulk DRP of metal atoms which have other metals as well as metalloid as nearest neighbors. The more structured part of the glass, however, is centered around the metalloid, which has only metals as nearest neighbor at distances reduced from the sum of radii of the atom constituents. In short, the structure of metal-metalloid glass may simply be ascribed as a DRP of metal atoms with chemical ordering around the mutually exclusive metalloid atoms.

A natural extension of EXAFS studies would be to the metal-metal glasses for which there is no adequate structural model [4.70]. By proper choice of systems, the EXAFS of both metals can be measured and their individual local structure determined so that the nearest-neighbor environment in the system is completely defined. Furthermore, structural variation with concentration may also be studied systematically since metal-metal glasses, we have seen, have large ranges of composition of glass formation. In fact some such results have been obtained by *Pettifer* et al. [4.104] in a $Nb_{37.6}Ni_{62.4}$ glass. The preliminary analysis is interesting. In Fig. 4.19 the experimental Nb *K* edge EXAFS is shown in energy space. Computer simulation using known scattering parameters showed that putting Nb as nearest neighbor will not yield a curve that fits the experimental spectrum. Nb–Nb avoidance is needed to render a good fit shown as a dashed line in Fig. 4.19. An Nb–Ni distance of 2.54 Å was determined. This again, as seen in Table 4.3, is shorter than the sum of the atom radii but is closer to the sum of the covalent radii.

In conclusion, the power of EXAFS in local structure determination cannot be overemphasized. In combination with high-resolution diffraction techniques such as pulsed neutron diffraction [4.100] and computer simulation studies like that of *Boudreaux* [4.101], the structural scientist can look forward to an exciting and profitable journey to understanding the structure of amorphous metals in particular, and the structure of the glassy state of matter in general.

Acknowledgments. The author would like to thank F. W. Lytle for his encouragement and collaboration in various aspects of EXAFS experimentation and data analysis. Discussions with R. P. Messmer on EXAFS theory and structure of metallic glasses and critical comments from E. A. Stern, G. S. Cargill III, and P. H. Gaskell on the manuscript are deeply appreciated. Thanks are also due to R. F. Pettifer for his permission to quote the Ni–Nb glass data prior to publication. The author is grateful for the experimental opportunity at Stanford Synchrotron Radiation Laboratory which is supported by NSF Grant No. DMR 73-0-7692 in cooperation with the Stanford Linear Accelerator Center and the Department of Energy. This work is supported by the US Department of Energy Division of Basic Energy Sciences, Materials Science Programs, under Contract DE-ACO2-79ER10382.

References

4.1 F.W.Lytle: In *Physics of Non-Crystalline Solids*, ed. by J.A.Prins (North-Holland, Amsterdam 1965) pp. 12–25
4.2 J.A.Bearden, A.F.Burr: Rev. Mod. Phys. **39**, 125 (1967)
4.3 E.A.Stern: Phys. Rev. B **10**, 3027 (1974)
4.4 F.W.Lytle, D.E.Sayers, E.A.Stern: Phys. Rev. B**11**, 4825 (1975)
4.5 E.A.Stern, D.E.Sayers, F.W.Lytle: Phys. Rev. B**11**, 4836 (1975)
4.6a A.D.Baer, R.Gaxiola, A.Golde, F.Johnson, B.Salsburg, H.Winick, M.Baldwin, N.Dean, J.Harris, E.Hoyt, B.Humphrey, J.Jurow, R.Melen, J.Miljan, G.Warren: IEEE Trans. NS-**22**, 1794 (1975)
4.6b S.Doniach, I.Lindau, W.R.Spicer, H.Winick: J. Vac. Sci. Technol. **12**, 1123 (1975)
4.7 P.Eisenberger, B.M.Kincaid: Science **200**, 1441 (1978)
4.8 E.A.Stern: Contemp. Phys. **19**, 289 (1978)
4.9 K.Hodgson, S.Doniach: Chem. Eng. News., p. 26 (Aug. 1978)
4.10 D.R.Sandstrom, F.W.Lytle: Annu. Rev. Phys. Chem. **30**, 215 (1979)
4.11 T.M.Hayes: J. Non-Cryst. Solids **31**, 57 (1978)
4.12 R.F.Pettifer: 4th European Physical Society Conf. (1979), Chap. 7, p. 55
4.13 F.W.Lytle, G.H.Via, J.H.Sinfelt: In *Synchrotron Radiation Research*, ed. by S.Doniach, H.Winick (Plenum, New York 1979)
4.14 R.G.Shulman, P.Eisenberg, B.M.Kincaid: Annu. Rev. Biophys. Bioeng. **7**, 559 (1978)
4.15 S.I.Chan, V.W.Hu, R.C.Gamble: J. Mol. Struct. **45**, 239 (1978)
4.16 H.Fricke: Phys. Rev. **16**, 202 (1920)
4.17 G.Hertz: Phys. Z. **21**, 630 (1920); Z. Phys. **3**, 19 (1920)
4.18 R.De L.Kronig: Z. Phys. **70**, 317 (1921); **75**, 191 (1932)
4.19 T.Hayashi: Sci. Rep. Tohoku Univ. **33**, 123 (1949); **33**, 183 (1949)
4.20a L.V.Azaroff: Rev. Mod. Phys. **35**, 1012 (1963)
4.20b W.L.Schaich: Phys. Rev. B**8**, 4028 (1973)

4.21 J.D.Hanawalt: Phys. Rev. **37**, 715 (1931)
4.22 B.M.Kincaid, P.Eisenberger: Phys. Rev. Lett. **34**, 1361 (1975)
4.23 B.M.Kincaid: Ph. D. Thesis, Stanford University (1975)
4.24 R.A.Van Nordstrand: In *Non-Crystalline Solids*, ed. by V.D.Frechette (Wiley, New York 1960) p. 108
4.25 W.F.Nelson, I.Siegel, R.W.Wagner: Phys. Rev. **127**, 2025 (1962)
4.26 E.W.White, H.A.McKinstry: Adv. x-Ray Anal. **9**, 376 (1966)
4.27 D.S.Urch: Rev. **25**, 343 (1971)
4.28 D.E.Sayers, F.W.Lytle, E.A.Stern: Adv. x-Ray Anal. **13**, 248 (1970)
4.29 D.E.Sayers, E.A.Stern, F.W.Lytle: Phys. Rev. Lett. **27**, 1204 (1971)
4.30 H.Bethe, E.Salpeter: *Quantum Mechanics of One- and Two-Electron Systems* (Springer, Berlin 1957) Sects. 59, 69
4.31 B.W.Holland et al.: J. Phys. C**11**, 633 (1978)
4.32 C.A.Ashley, S.Doniach: Phys. Rev. B**11**, 1279 (1975)
4.33 P.A.Lee, J.B.Pendry: Phys. Rev. B**11**, 2795 (1975)
4.34 P.A.Lee, G.Beni: Phys. Rev. B**15**, 2862 (1977)
4.35 R.F.Pettifer, P.W.McMillan: Philos. Mag. **35**, 871 (1977)
4.36 J.Schwinger: Phys. Rev. **75**, 1912 (1949)
4.37 M.L.Perlman, E.M.Rowe, R.E.Watson: Phys. Today **30** (1974)
4.38 F.C.Brown: Solid State Phys. **29**, 1 (1974)
4.39 E.M.Rowe, R.Watson: Sci. Am. **236**, 32 (1977)
4.40 R.Watson, M.L.Perlman: Science **199**, 1293 (1978)
4.41 F.R.Elder, R.V.Langmuir, H.Pollock: Phys. Rev. **74**, 52 (1948)
4.42 R.F.Pettifer: Ph. D. Thesis, University of Warick (1978)
4.43 S.H.Hunter: Ph. D. Thesis, Stanford University (1977)
4.44 P.Eisenberger, B.M.Kincaid: Chem. Phys. Lett. **36**, 134 (1975)
4.45 A.L.Robinson: Science **205**, 1367 (1979)
4.46 G.S.Knapp, H.Chen, T.E.Klippert: Rev. Sci. Instrum. **49**, 1658 (1978)
4.47 G.G.Cohen, D.A.Fischer, J.Colbert, N.J.Shevchik: To be published in Rev. Sci. Instrum.
4.48 E.Rowe: *Synchrotron Radiation Technology and Application*, Topics in Current Physics, Vol. 10, ed. by C.Kunz (Springer, Berlin, Heidelberg, New York 1979) Chap. 2
4.49 F.W.Lytle: Private communication
4.50 G.D.Mahan: Phys. Rev. **163**, 612, 882 (1967)
J.J.Hopfield: Comments Solid State Phys. **2**, 40 (1969)
4.51 D.A.Papaconstantopoulos: Phys. Rev. Lett. **31**, 1050 (1973)
4.52 L.D.Landau, E.M.Lifshitz: *Non-Relativistic Quantum Mechanics* (Addison-Wesley, Cambridge, Mass. 1958) pp. 121–127
4.53 B.K.Teo, P.A.Lee: J. Am. Chem. Soc. **101**, 2815 (1979)
4.54 C.Bingham, M.D.Godfrey, J.W.Turkey: IEEE Trans. AE-**15**, 58 (1967)
4.55 D.E.Sayers, E.A.Stern, F.W.Lytle: Phys. Rev. Lett. **35**, 584 (1975)
4.56 S.M.Heald, E.A.Stern: Phys. Rev. B**17**, 4069 (1979)
4.57 P.H.Citrin, P.Eisenberger, B.M.Kincaid: Phys. Rev. Lett. **36**, 1346 (1976)
4.58 B.K.Teo, P.Eisenberger, J.Reed, J.K.Barton, S.J.Lippard: J. Am. Chem. Soc. **100**, 3225 (1978)
4.59 S.P.Cramer, K.O.Hodgson: J. Am. Chem. Soc. **100**, 2748 (1978)
4.60 W.Klement, R.H.Willens, P.Duwez: Nature **187**, 869 (1960)
4.61 W.L.Johnson: J. Appl. Phys. **50**, 1559 (1979)
4.62 J.J.Gilman: Phys. Today **28**, 46 (1975)
4.63 P.Chaudhari, D.Turnbull: Science **199**, 11 (1978)
4.64 C.G.Granqvist, T.Claeson: Z. Phys. **20**, 241 (1975)
4.65 A.Calka, M.Madhava, D.E.Polk, B.C.Giessen, H.Matyja, J.Vander Sande: Scr. Metall. **11**, 65 (1977)
4.66 R.C.Ruhl, B.C.Giessen, M.Cohen, N.J.Grant: Acta Metall. **15**, 1693 (1967)
4.67 T.W.Barbee, Jr., W.A.Holmes, D.L.Keith, M.K.Pyzyna: Thin Solid Films **45**, 591 (1977)
4.68 S.Takayama: J. Mater. Sci. **11**, 164 (1976)

4.69 I.W.Donald, H.A.Davies: J. Non-Cryst. Solids **30**, 72 (1978)
4.70 G.S.Cargill, III: Solid State Phys. **30**, 227 (1975)
4.71 Y.Wasada, H.Okazaki, T.Masumoto: J. Mater. Sci. **12**, 1927 (1977)
4.72 M.H.Cohen, D.Turnbull: Nature **203**, 964 (1964)
4.73 D.Turnbull, M.H.Cohen: J. Chem. Phys. **52**, 3038 (1970)
4.74 J.D.Bernal: Nature **185**, 68 (1960); Proc. R. Soc. A**280**, 299 (1964)
4.75 J.L.Finney: Proc. R. Soc. A**319**, 479 (1970)
4.76 G.S.Cargill, III: J. Appl. Phys. **41**, 2248 (1970)
4.77 D.E.Polk: Private communication
4.78 E.J.W.Whittaker: J. Non-Cryst. Solids **28**, 293 (1978)
4.79 D.E.Polk: Acta Metall. **20**, 485 (1972)
4.80 J.L.Finney: Nature **266**, 309 (1977)
4.81 J.Wong, F.W.Lytle, J.L.Walter, F.E.Luborsky: Bull. Am. Phys. Soc. **23**, 467 (1978)
4.82 J.Wong, F.W.Lytle, R.B.Greegor, H.H.Liebermann, J.L.Walter, F.E.Luborsky: Proc. of the 3rd International Conf. of Rapidly Quenched Metals, Sussex University, Vol. II (1978) p. 345
4.83 G.S.Brown, L.R.Testardi, J.H.Wernick, A.B.Hallak, T.H.Geballe: Solid State Commun. **23**, 875 (1977)
4.84 E.A.Stern, S.Rinaldi, E.Callen, S.Heald, B.Bunker: J. Magn. Magn. Mater. **7**, 188 (1978)
4.85 T.M.Hayes, J.W.Allen, J.Tanc, B.C.Giessen, J.J.Hausen: Phys. Rev. Lett. **40**, 1282 (1978)
4.86 H.S.Chen, D.W.Polk: J. Non-Cryst. Solids **15**, 174 (1974)
4.87 T.Masumoto, R.Maddin: Mater. Sci. Eng. **19**, 1 (1975)
4.88 L.A.Davis: J. Mater. Sci. **10**, 1557 (1975)
4.89 E.M.Gyorgy, H.J.Leamy, R.C.Sherwood, H.S.Chen: AIP Conf. Proc. **29**, 198 (1976)
4.90 F.E.Luborsky, J.L.Walter: J. Appl. Phys. **47**, 3648 (1976)
4.91 J.L.Walter, F.Bacon, F.E.Luborsky: Mater. Sci. Eng. **24**, 259 (1976)
4.92 S.Rundqvist: Acta Chem. Scand. **12**, 658 (1958); **13**, 425 (1959); **61**, 1, 242, 992 (1962)
4.93 Y.Waseda, H.Okazaki, T.Masumoto: In *Proc. of Intern. Conf. on the Structure of Non-Crystalline Materials, Cambridge*, 1976 (Taylor and Francis, London 1977) p. 95
4.94 Y.Waseda, S.Tamaki: Z. Phys. B**23**, 315 (1976)
4.95 Y.Egami, R.S.Williams, Y.Waseda: Proc. of the 3rd International Conf. on Rapidly Quenched Metals, Sussex University (1978), Vol. II, p. 318; also cf Chap. 3 in this book
4.96 K.Suzuki, T.Fukumaga, M.Misawa, T.Masumoto: Mater. Sci. Eng. **23**, 215 (1976)
4.97 J.F.Sadoc, J.Dixmier: In *The Structure of Non-Crystalline Materials*, ed. by P.H.Gaskell (Taylor and Francis, London 1979) p. 85
4.98 J.Beltry, J.F.Sadoc: J. Phys. F: Metal. Phys. **5**, L110 (1975)
4.99 J.Wong, H.H.Liebermann, R.P.Messmer: To be published in J. Mater. Sci.
4.100 T.Fukunaga, M.Misawa, K.Fukamichi, T.Masumoto, K.Suzuki: Proc. of the 3rd International Conf. on Rapidly Quenched Metal, Sussex Univ., 1978, Vol. II (1978) p. 325
4.101 D.S.Boudreaux: Phys. Rev. B**18**, 4039 (1978)
4.102 J.Wong: To be published in J. Mater. Sci.
4.103 S.Rundqvist: Ark. Kemi. **20**, 67 (1962)
4.104 E.Lisher, C.Wilkinson, T.Ericsson, L.Häggstrom, L.Lundgren, R.Wappling: Proc. Int. Conf. Magnetism (Moscow) IV, 1973, p. 581
4.105 U.Herold, U.Köster: Proc. of the 3rd Intern. Conf. on Rapidly Quenched Metals, Sussex University, 1978, Vol. I (1978) p. 281
4.106 J.L.Walter, S.F.Bartram, R.R.Russell: Met. Trans. (1979)
4.107 P.Oelhafer, M.Liard, H.J.Guntherodt, K.Bernesheim, H.D.Polaschegg: To be published in Phys. Rev.
4.108 R.P.Messmmer: To be published in Phys. Rev.
4.109 R.F.Pettifer, J.Bordas, I.Donald, B.G.Lewis, H.A.Davies: To be published
4.110 R.C.Evans: *An Introduction to Crystal Chemistry*, 2nd ed. (Cambridge, London 1964) p. 87
4.111 *Sargent-Welch Periodic Table of the Elements* (1968) Catalog S–18806

5. Brillouin Light Scattering from Metallic Glasses

A. P. Malozemoff

With 5 Figures

After an introduction to this rather novel experimental technique for studying metals, a summary is given of the surface phonon velocities, g factors, surface and bulk magnetizations, and spin wave stiffness determined on metallic glasses. It is concluded that although Brillouin scattering is not unique in providing this kind of information, it could be a convenient tool for such materials-oriented studies in the future.

5.1 Overview

The first studies of metallic glasses by Brillouin light scattering [5.1–5] began in 1978, following initial experiments on polycrystalline and liquid metals, and other opaque materials [5.6–12]. A number of theoretical papers on light scattering from opaque materials have also appeared recently [5.13–18]. Although experimental results on metallic glasses are still limited in number, it is perhaps appropriate at this early stage to review the principles of the technique, which is a novel one for studying metals, to summarize the results obtained so far on metallic glasses, and to assess the future of Brillouin scattering for such materials-oriented studies.

In a typical light scattering experiment, a laser beam is incident on a sample, and scattered light is collected at a given angle as shown schematically in Fig. 5.1. The scattered light is analyzed for polarization and frequency. In addition to unwanted elastic scattering arising from defects, there often occurs weak inelastic scattering whose frequency shift is found to correspond to the frequency of various excitations in the sample. For practical purposes Brillouin and Raman scattering may be distinguished by the experimental technique used to analyze the frequency. In Raman scattering a grating monochromator is usually used, permitting resolution of frequency shifts down to a hundred gigahertz but not less because of the problem of limited discrimination against the elastically scattered light. In Brillouin scattering, a Fabry-Perot interferometer is usually used, giving higher contrast, so that excitations down to the low gigahertz range can be conveniently resolved.

The plates of the interferometer are scanned piezoelectrically, and the output of the interferometer is measured photometrically. A typical spectrum on an amorphous FeB ribbon is shown in Fig. 5.2. Large peaks of elastic

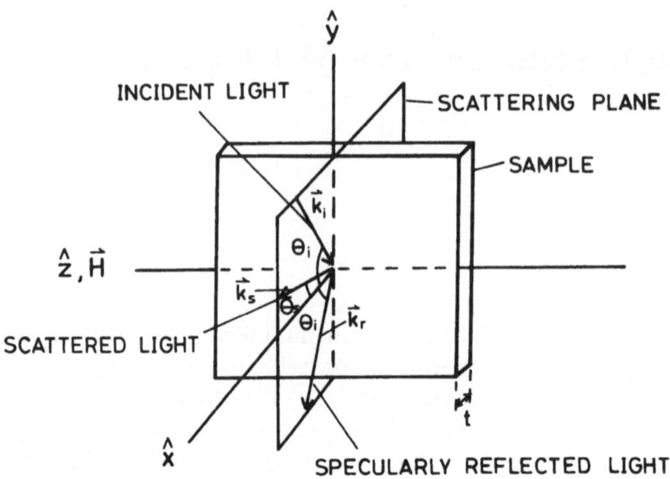

Fig. 5.1. Brillouin scattering geometry (after [5.2])

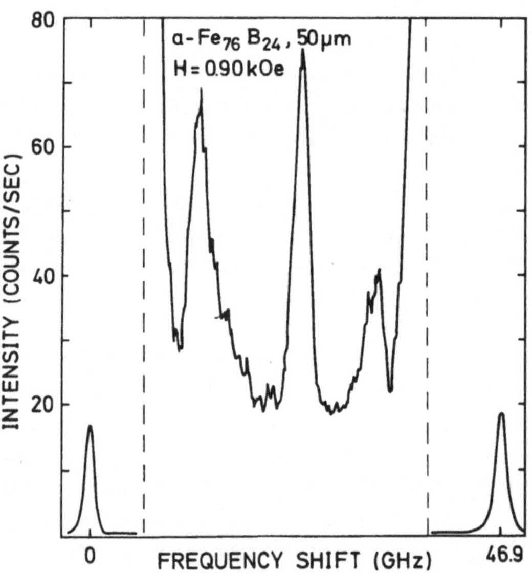

Fig. 5.2. Brillouin spectrum of magnons in a quenched ribbon of a-$Fe_{76}B_{24}$. In terms of Fig. 5.1, analyzer is perpendicular to scattering plane, $\theta_i = 64°$, $\theta_s = 0°$, λ...514.5 nm. The peaks on the sides represent elastic scattering attenuated by 10^6, and the dashed lines represent the change of scale (after [5.2])

scattering are observed at regular intervals of plate spacing corresponding to half the wavelength of the incident light. Two such peaks are shown in Fig. 5.2. Between these elastic peaks appear much weaker inelastic peaks, some downshifted in frequency, corresponding to emission of a quantum of energy and called Stokes peaks, and others upshifted in frequency, corresponding to absorption of a quantum of energy and called anti-Stokes peaks. Because there is no additional microwave or other excitation in this experiment, the absorbed quanta are presumably excited *thermally* in the sample. Because of the phenomenon of stimulated emission, Stokes and anti-Stokes peaks can be expected to be close to equal in size as long as the thermal energy kT at the ambient temperature T is much larger than the energy $\hbar\omega$ of the excitation. An exception are surface magnons (Damon-Eshbach modes), which appear only as Stokes or anti-Stokes but not both, as will be described further below. The frequency of the excitation can be calibrated in terms of the so-called free spectral range, which is the frequency shift corresponding to the spacing between two elastic peaks and which can be adjusted by varying the initial spacing of the interferometer plates.

Light scattering experiments on metals require greater experimental contrast than on transparent materials (semiconductors or insulators), for the ratio of unwanted elastic scattering from surface irregularities to the desired inelastic scattering is increased because the scattering volume for inelastic scattering is limited to a thin surface layer by the large absorption coefficient. In practice such experiments have only become possible since the recent development of a high-contrast 5-pass Fabry-Perot interferometer system by *Sandercock* [5.19]. But once these experimental difficulties were overcome, some interesting new features emerged in the study of metals which had never been seen in transparent materials. In particular, because scattering is from a thin surface layer, surface excitations, whose properties are of great interest, scatter as strongly as bulk excitations.

Another difference between metals and transparent materials is that the conventional wave vector conservation rule

$$\boldsymbol{k}_{\text{incident}} = \boldsymbol{k}_{\text{scattered}} \pm \boldsymbol{K}_{\text{excitation}}, \qquad \text{(bulk)} \tag{5.1}$$

is modified. Here $\pm$ correspond, respectively, to emission and absorption of an excitation quantum with wave vector $\boldsymbol{K}$, and the incident and scattered wave vectors lie along the incident and scattered beams in Fig. 5.1. Instead, for surface excitations which have no propagating wave vector component perpendicular to the surface, the wave vector conservation rule involves only those components of the incident and scattered light which are parallel to the surface,

$$\boldsymbol{k}_{\|\text{incident}} = \boldsymbol{k}_{\|\text{scattered}} \pm \boldsymbol{K}_{\text{excitation}}. \qquad \text{(surface)}. \tag{5.2}$$

If the light scattering geometry is held fixed but only the sample surface is rotated (e.g., around the axis $\hat{z}$ in Fig. 5.1), the surface-parallel components change and therefore the wave vector of the observed excitation changes. For bulk excitations in an isotropic medium this is not the case, and so, rotating the sample and observing whether or not the excitation frequency changes offers one test to distinguish between surface and bulk excitations.

Turning now to the wave vector selection rule for bulk excitations in metals, we find that wave vector components parallel to the surface are conserved as in (5.2), but because of the short light penetration depth, described by an absorption coefficient α [i.e., light attenuated as $\exp(-\alpha x)$] one can expect the wave vector of the light within the material to be smeared out. Taking the Fourier transform of $\exp(-\alpha x)$, one finds a wave vector distribution of the form $(\alpha^2+k^2)^{-1}$, so that in effect all components are present with equal density up to wave vectors of order α. Typically α is $10^6\,\mathrm{cm}^{-1}$ at 514.5 nm in a-FeB, which is an order of magnitude larger than the light wave vector $10^5\,\mathrm{cm}^{-1}$, and thus light can couple to a much larger section of the Brillouin zone than in transparent materials. In particular, bulk magnons, whose energy E increases quadratically with wave vector above a magnetostatic or anisotropy gap E_0, have a one-dimensional density of states of the form $(E-E_0)^{-1/2}$. Therefore, if the matrix element for the magneto-optical coupling is independent of energy, one can expect an asymmetric light scattering peak with this form, that is, with a tail to higher energies, and such peaks are indeed observed, as shown in Fig. 5.2.

The question arises, does the Brillouin scattering experiment on amorphous materials differ in any qualitative way from the same experiment on crystalline materials? Experiment shows that it does not. For example very similar phonon and magnon spectra, both in shape and overall intensity, are observed in amorphous FeB ("a-FeB") and in polycrystalline Fe [5.1, 2, 10, 11]. The reason for the similarity is that the low-energy excitations observed in Brillouin scattering are continuum excitations. That is, their wavelengths are long compared to atomic dimensions, and therefore they are not sensitive to atomistic structure. By contrast Raman scattering probes higher energy excitations which *are* sensitive to atomic structure. A well-known result in the Raman spectra of amorphous semiconductors is that very broad peaks are observed. These peaks can be interpreted crudely as densities of phonon states which are all visible because of the breakdown of the wave vector selection rule in the amorphous structure. Presumably similar results can be expected in amorphous metals, with the added complication that the signals will be weak because of the small scattering volume. Indeed we have attempted to detect Raman scattering from amorphous $Fe_{80}B_{20}$ and $Fe_{40}Ni_{40}P_{14}B_6$ ribbons but have observed no distinguishable structure above the experimental noise level [5.20].

5.2 Experimental Results and Discussion

5.2.1 Surface Phonons

When the electric vector of the incident and scattered light is polarized in the scattering plane (see Fig. 5.1), one pair of Stokes-anti-Stokes peaks of equal intensity is observed by Brillouin scattering in all metallic glasses studied so far

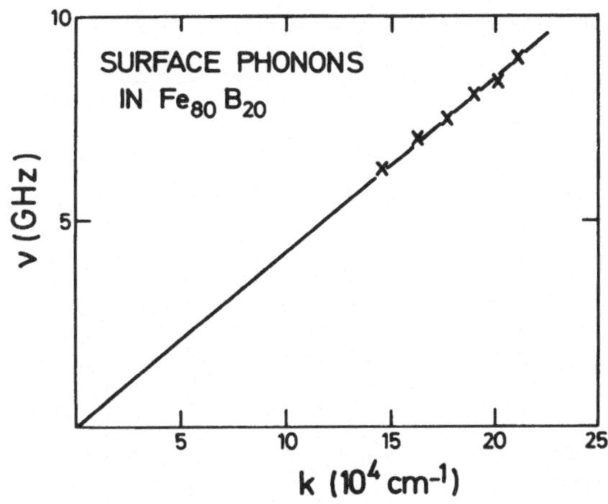

Fig. 5.3. Frequency shift of surface phonon scattering as a function of wave vector, as determined from (5.2) and varied by rotating the sample about $\hat{z}$ of Fig. 5.1 [5.41]

[5.1–3]. In this polarization the peaks are independent of applied field (this was checked particularly carefully for a-$Fe_{1-x}B_x$ where unusually large magnetoelastic effects are known to occur). Furthermore, the frequency is found to shift as the sample is rotated, as shown in Fig. 5.3 [the wave vector was determined from (5.2)]. These results identify the peaks as surface phonons (Rayleigh waves), whose properties are well known from classical calculations. The propagation velocity V_s of the surface phonons is given by the slope of the ω vs k data of Fig. 5.3, and according to classical calculations it should be equal to

$$V_s = \frac{0.87 + 1.12\sigma}{1+\sigma} \sqrt{\frac{E}{2\varrho(1+\sigma)}}, \tag{5.3}$$

where E is Young's modulus, σ is Poisson's ratio, and ϱ is the density. Given that (5.3) is weakly sensitive to σ, and σ usually varies only between 0.3 and 0.4, the sound velocity can be used to determine Young's modulus provided that the density is known.

A compilation of experimental room temperature surface-wave velocities from Brillouin scattering is given in Table 5.1. These results can be compared to predictions using (5.3) and measurements of Young's modulus by more conventional vibrating reed or ultrasonic experiments [5.22–27]. In these latter experiments, which use much lower frequencies, E has been found to depend strongly on the magnetic state of the sample, decreasing in the demagnetized state (E_D) by an amount $\Delta E = E_s - E_D$ from its value in the saturated state (E_s). $\Delta E/E_D$ can be as large as 70% in annealed $Fe_{82}B_{18}$ [5.24]. The effect is known to arise from magnetostrictive strains associated with domain wall movement. For purposes of comparison with Brillouin scattering, it would seem that E_s rather than E_D should be used in (5.3) because at the gigahertz frequencies of the Brillouin surface phonons, domain walls cannot respond. This view is

Table 5.1. Observed and calculated surface phonon velocities V_s in metallic glasses, ϱ = density, σ = Poisson ratio, E = saturated Young's modulus

Sample	ϱ [gm/cm^3]	σ	E [$\times 10^{12}$ dyn/cm^2]	$V_{s\text{-calc}}$ [$\times 10^5$ cm/s]	$V_{s\text{-obs}}$ [$\times 10^5$ cm/s]	Ref.
$Fe_xCo_{75-x}Si_{15}B_{10}$ $x=0$ to 6					2.5 ± 0.07	[5.3]
$Fe_{40}Ni_{40}P_{14}B_6$	7.7	0.37	1.34	2.35	2.38 ± 0.03	[5.1, 23]
$Fe_{20}Ni_{60}B_{20}$	7.94	$\sim$0.33	1.53	2.51	2.44	[5.3, 26]
$Fe_{40}Ni_{40}B_{20}$	7.72	0.341	1.60	2.60	2.57	[5.3, 26]
$Fe_{60}Ni_{20}B_{20}$	7.55	0.365	1.67	2.67	2.70	[5.3, 26]
$Fe_{76}Mo_4B_{20}$					2.47	[5.3]
$Fe_{86}B_{14}$					2.47 ± 0.1	[5.3]
$Fe_{83}B_{17}$	7.55	0.096	2.06	3.15[a]	2.6 ± 0.06[a]	[5.1, 24, 27]
$Fe_{80}B_{20}$	7.40	0.3	1.69	2.75	2.64 ± 0.04	[5.1, 3, 23]
$Fe_{76}B_{24}$					2.87	[5.3]
$Fe_{72}B_{28}$					3.0	[5.3]

[a] Calculated V_s based on ϱ, σ, and E measured on Tohoku sample, V_s measured on Allied Chemical sample

supported by the observation that there is no difference in the Brillouin surface phonon velocity in the demagnetized and saturated states.

Table 5.1 shows values for V_s calculated from (5.3). There is reasonable agreement except in the $Fe_{1-x}B_x$ series, which is of particular interest because of its unusually large magnetoelastic interactions. At the time of this writing there seem to be severe discrepancies in measured elastic constants of samples prepared by several groups – Allied Chemical (AC), Bell Laboratories, and Tohoku University (TU). For AC–$Fe_{80}B_{20}$, $E_s = 1.69 \times 10^{12}$ dynes/cm^2, $\sigma = 0.3$ [5.23]. These results are relatively insensitive to annealing [5.25] and give moderately good agreement with the Brillouin result on the same sample. If $\sigma = 0.3$ is assumed for the entire series, the Brillouin results on the AC samples give a Young's modulus which increases strongly with boron content, in good agreement the Bell results [5.26]. By contrast for TU–$Fe_{80}B_{20}$, E_s is 2.07×10^{12} dynes/cm^2 and very sensitive to annealing (10% change after 2 h at 200 °C) [5.24]. For TU $Fe_{83}B_{17}$, σ is only 0.1, a suspiciously low value [5.27], although anomalously low values of ν have also been reported in annealed amorphous $Fe_{40}Ni_{38}Mo_4B_{18}$ [5.25]. The strong annealing dependence of E_s in TU $Fe_{1-x}B_x$ points to an extraordinary degree of structural relaxation, which is all the more remarkable because magnetic parameters like moment and Curie temperature change relatively less with annealing [5.28]. Calculation of V_s using parameters for annealed TU $Fe_{83}B_{17}$ gives a value far above the Brillouin result on AC $Fe_{83}B_{17}$. This discrepancy is clearly related to differences in AC and TU samples. The origin of the difference is not yet known, but it suggests the existence of very different as-grown amorphous states in the $Fe_{1-x}B_x$ series [5.42].

The above data were all on quenched ribbons. Surface phonons have also been observed in thin sputtered films. However, if the film thickness is comparable to the penetration depth of the surface phonon, which in turn is comparable to its wavelength, the velocity will be affected by the elastic characteristics of the substrate. In $Fe_{80}B_{20}$ films deposited on silicon, the velocity was found to increase below about 2000 Å for wavelengths of order 3000 Å [5.2]. Thus larger thicknesses are required to obtain reliable material parameters for films.

In summary, results so far indicate that Brillouin scattering can be reliably used to measure sound velocity and hence the saturated Young's modulus in metals.

5.2.2 Bulk and Surface Magnons: Thick Materials

When the electric vector of the incident light is polarized in the scattering plane but the scattered light is analyzed perpendicular to the scattering plane, a rather complex spectrum is observed depending on the thickness of the sample and applied magnetic field (assumed to lie perpendicular to the scattering plane as in Fig. 5.1. For sufficiently thick materials (e.g., greater than several thousand angstroms for a-$Fe_{1-x}B_x$) only three peaks are observed as shown in Fig. 5.2 for a-$Fe_{80}B_{20}$: one Stokes-anti-Stokes pair of asymmetrically shaped peaks, each with a tail to the high-energy side, and a single symmetrically shaped but asymmetrically positioned peak whose position shifts from the Stokes to the anti-Stokes side by reversing the polarity of the magnetic field.

This latter peak can be identified as a surface magnon, whose remarkable nonreciprocal properties were first predicted by *Damon* and *Eshbach* [5.29] and first observed by Brillouin scattering by *Grünberg* and *Metawe* [5.7]. The angular frequency ω of this mode is given by [5.29, 30]

$$\omega=\gamma(H+2\pi M_s+\zeta Dk_y^2), \tag{5.4}$$

where γ is the gyromagnetic ratio $g\mu_B/\hbar$, g the g factor, μ_B the Bohr magneton, $\hbar$ Planck's constant divided by 2π, M_s the magnetization, H the applied field in the sample plane (along $\hat{z}$ in Fig. 5.1), D the spin wave stiffness, k_y the wave vector in the sample plane and perpendicular to H (along $\hat{y}$ in Fig. 5.1), and ζ a constant (2 according to an analytic theory and 1.76 according to numerical calculations). This formula is valid only when the field is perpendicular to k_y in the sample plane. If the field is turned towards $\hat{y}$, the surface mode is expected to shift towards the bulk modes and merge with them above a critical angle [5.17, 29]. This effect has been beautifully confirmed in experiments on polycrystalline iron films [5.11].

If the conventional geometry of Fig. 5.1 is used and the Brillouin frequency shift of the surface mode is plotted versus applied field strength, a straight line is obtained in accord with (5.4), from whose slope γ can be determined and from

Table 5.2. Room temperature magnetization and g factor of metallic glasses determined from Brillouin scattering off surface and bulk magnons. Samples prepared by rapid quenching (Q) or sputtering (S)

Sample	$4\pi M$ [kOe]	g	Ref.
$Fe_{40}Ni_{40}P_{14}B_6(Q)$	7.5	2.05 ± 0.04	[5.1]
$Fe_{86}B_{14}(Q)$	14.5	2.10 ± 0.03	[5.3]
$Fe_{80}B_{20}(Q)$	15.5	2.09 ± 0.03	[5.3]
$Fe_{72}B_{28}(Q)$	14	2.10 ± 0.03	[5.3]
$Fe_{78}B_{22}(S)$	15.5	2.09 ± 0.03	[5.2]
$Co_{85}B_{15}(S)$	13.5	2.15 ± 0.04	[5.2]
$Co_{78}B_{22}(S)$	10.5	2.19 ± 0.04	[5.2]
$Co_{70}Fe_5Si_{15}B_{10}(Q)$	8.3		[5.3]

whose intercept M_s can be determined. For all materials studied so far, $2\pi M_s$ is several kilogauss while $\zeta D k_y^2$ is at most 50 Oe and can be neglected, considering experimental accuracy of a few percent in $4\pi M_s$. A summary of room temperature values for g and M_s is given in Table 5.2. The g factors of a-$Fe_{1-x}B_x$ and a-$Co_{1-x}B_x$ are close to the values for crystalline Fe (2.10) and Co (2.2). A g value of 2.07 for a-$Fe_{40}Ni_{40}P_{14}B_6$ has also been measured by ferromagnetic resonance [5.32], in reasonable agreement with the value in Table 5.2. Magnetizations in Table 5.2 are also in agreement with standard magnetic measurements, where available.

The asymmetrically shaped peaks in Fig. 5.2 can be identified as bulk magnons on the basis of the following arguments. As mentioned in Sect. 5.1, when the light penetration depth is small compared to its wavelength, the light couples to a one-dimensional density of states, which for bulk magnons leads to an asymmetric peak with a tail to higher energies as observed. Furthermore a plot of the peak position versus field fits the expected dispersion relation [5.30, 31]

$$\omega = \gamma[(H + Dk_\perp^2)(H + Dk_\perp^2 + 4\pi M_b)]^{1/2}, \tag{5.5}$$

where $k_\perp = k_x^2 + k_y^2$ (see Fig. 5.1) and M_b is the bulk magnetization. There are several points to note about (5.5). First, this formula is independent of film thickness, assuming free or fully pinned boundary conditions at the surfaces and provided only that there is no component of wave vector k_z along the applied field direction [5.30, 31]. Second, we distinguish bulk magnetization M_b in (5.5) from surface magnetization M_s in (5.4) because of the possibility of a surface layer of changed magnetization. Such effects have been seen in oxidized polycrystalline iron films, where M_s and M_b extracted from fits to (5.4) and (5.5) were noticeably different but were brought into agreement by polishing the sample [5.10, 11]. In the amorphous materials listed in Table 5.1, no discrepancy has been observed, indicating good surface quality, and no change has occurred on polishing [5.1, 2]. This was fortunate because a proper theory to treat the case of different M_s and M_b is lacking at present.

A third point is that experimental fits to (5.5) for the samples of Table 5.2 give values for $Dk_\perp^2$ ranging rather haphazardly from 100 to 500 Oe. This is puzzling because if there is no pinning at the surface, the peak of the bulk magnon spectrum should occur at $k_x=0$, and estimates of Dk_y^2 give less than 50 Oe. This discrepancy may arise from surface pinning in the following way [5.2]. If the exchange field $Dk_\perp^2$ of a given mode is less than the surface pinning field H_p, the spins at the surface are pinned, and since the Kerr magneto-optic interaction is proportional to the square of the magnetization precession amplitude M_x and is furthermore limited to the surface, the intensity of this mode will be small. Only for $Dk_\perp^2$ greater than H_p will the full intensity appear, and this implies a shift of the peak of the bulk magnon spectrum by a frequency corresponding to $Dk_\perp^2 \approx H_p$ in (5.5).

5.2.3 Bulk and Surface Magnons: Thin Films

As the thickness of a sample is reduced, its Brillouin magnon spectrum changes in a characteristic way [5.2, 4, 5]. The field dependence of the surface mode, when fit to (5.4), appears to give reduced magnetization and increased g factor [5.2]. In fact the true magnetization and g factor are unchanged but the Damon-Eshbach theory predicts such deviations from (5.4) when $k_y t < 1$, where t is the film thickness [5.43].

For the bulk modes, the asymmetric peak breaks up into a substructure which becomes more widely spread apart the thinner the film. A spectrum of a sputtered film of a-$Fe_{80}B_{20}$, 106 nm thick, is shown in Fig. 5.4, where four anti-Stokes bulk magnons appear on the right and three Stokes magnons on the left (the fourth being masked by the surface magnon). These magnons presumably correspond to individual standing spin wave modes with different quantized wave vector components k_x normal to the plane of the film. If the boundary conditions are fully pinned or unpinned, k_x is quantized according to

$$k_x = n\pi/t\,, \tag{5.6}$$

where n is a positive integer. An interesting case arises if there is a nonzero component of wave vector in the film plane $k_y \neq 0$, as is the case in the typical scattering geometry (Fig. 5.1). In this case, theory predicts that even for unpinned boundary conditions, a mode with $n=0$, that is a uniform mode through the thickness of the film, does not exist [5.30].

The field dependence of the standing spin waves such as those of Fig. 5.4 has been measured for a series of a-$Fe_{80}B_{20}$ films ranging in thickness from 45 nm on up, and the data have been fitted to (5.5) and (5.6) [5.4, 5]. Good fits were achieved for films with a thickness of up to ~100 nm, giving values for D of $(1.2 \pm 0.2) \times 10^{-9}$ Oe cm^2. A possible explanation for the failure of fits at greater

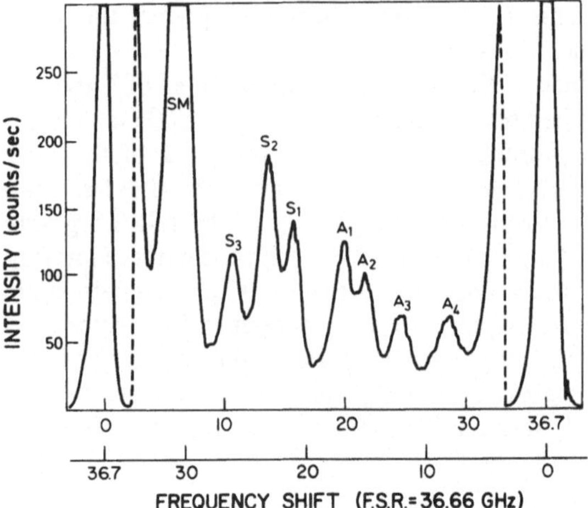

Fig. 5.4. Brillouin spectrum of magnons in a sputtered film of a-$Fe_{80}B_{20}$ 100 nm thick, with a magnetic field in the sample plane. SM is the Stokes surface magnon while A_1-A_4 and S_1-S_3 label the anti-Stokes and Stokes bulk magnons (standing spin waves). The flanking peaks represent elastic scattering attenuated by 10^5 (after [5.4])

thicknesses is that surface pinning fields strongly affect the frequency and intensity of modes whose exchange fields are comparable to or less than the pinning field. For a measured D of 1.2×10^{-9} Oe cm^2, and a thickness of 100 nm, this implies a pinning field of order 100 Oe, within the range observed in quenched ribbons as mentioned above [5.44].

The result $D=(1.2\pm0.2)\times10^{-9}$ Oe cm^2 for a-$Fe_{80}B_{20}$, also confirmed in ferromagnetic resonance results like those in Fig. 5.5, is surprisingly high compared to values of $D=0.7\times10^{-9}$ Oe cm^2 deduced from low-temperature magnetization measurements using standard Bloch theory [5.4, 5, 33]. It should be noted that the Brillouin and ferromagnetic resonance results correspond to long wavelength spin waves while much shorter wavelength spin waves dominate low-temperature magnetization measurements performed at and above 4.2 K. This raises the interesting possibility of anomalies in the spin wave dispersion of amorphous ferromagnets as a function of wave vector. Yet not all amorphous ferromagnets show this kind of discrepancy [5.34–37]. Recently theoretical proposals have been made for anomalies when there are correlated exchange or anisotropy fluctuations [5.38] or when the spins have a canted structure [5.39]. Both possibilities may be uniquely present in a-FeB because of its Invar properties [5.40], which are generally believed to be related to strong magnetic inhomogeneities [5.45].

5.3 Conclusion

In this conclusion, we assess the potential of Brillouin scattering as a tool for studies of materials such as the metallic glasses.

From the above sections, it is clear that Brillouin scattering permits determination of both elastic and magnetic parameters of a material with the

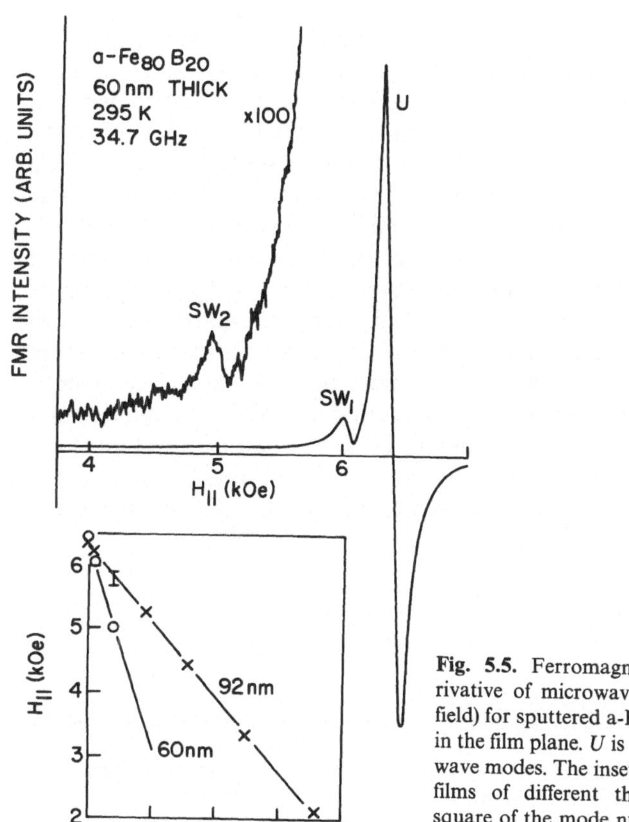

Fig. 5.5. Ferromagnetic resonance spectrum (derivative of microwave absorption as a function of field) for sputtered a-$Fe_{80}B_{20}$, 60 nm thick, with field in the film plane. U is the uniform mode, SW_i the spin wave modes. The inset shows mode positions for two films of different thicknesses plotted versus the square of the mode number n (after [5.5])

same apparatus. From the surface phonon velocity, Young's modulus can be obtained if the density and Poisson's ratio are known. This method may be uniquely convenient for samples which are too small for vibrating reed or ultrasonic experiments. As far as magnetic parameters are concerned, Brillouin scattering should be compared to ferromagnetic resonance (FMR), which also permits determination of the g factor, magnetization and, in thin films, spin wave stiffness. The differences between the two techniques are well illustrated in a comparison of Figs. 5.4 and 5.5. At the present state of development of the technology, FMR has a far higher resolution and signal-to-noise ratio, and it can be conveniently carried to low temperatures, while Brillouin signals from thermal magnons fall in intensity in proportion to temperature. Nevertheless Brillouin scattering has some essential advantages which could be capitalized on if the experimental techniques are further developed. First of all, it is like a frequency-variable spectrometer – all frequencies and fields can be measured simultaneously, whereas in practice FMR is usually conducted at fixed frequencies, varying field and sample orientation. Thus g factors can be determined directly from the field dependence of mode frequency in Brillouin scattering, and use of the Damon-Eshbach mode, which is linear in field, enhances the accuracy of the measurement. Similarly, D can be determined with greater absolute accuracy by direct observation of the field dependence of the standing spin waves, which is very difficult in FMR. A second point is that a standard FMR experiment requires some sort of nonuniformity in the sample to excite the standing spin waves. When nonuniformity is small, as in Fig. 5.5, the intensity of these modes is also small. By contrast these modes scatter light with approximately the same intensity, as shown in Fig. 5.4. Thus in principle Brillouin scattering has an advantage in ideally uniform materials. A third point is that by virtue of the comparison of surface and bulk modes, Brillouin scattering permits a check of magnetization uniformity in the surface layer.

In summary the information Brillouin scattering provides is not unique and can be obtained by more classic methods. However, if the sensitivity and resolution of the technique continue to improve, Brillouin scattering could become a very convenient tool for characterizing metallic glasses and other metals.

References

5.1 P.H.Chang, A.P.Malozemoff, M.Grimsditch, W.Senn, G.Winterling: Solid State Commun. **27**, 617 (1978)
5.2 A.P.Malozemoff, M.Grimsditch, J.Aboaf, A.Brunsch: J. Appl. Phys. **50**, 5885 (1979)
5.3 A.P.Malozemoff, P.H.Chang, M.Grimsditch: J. Appl. Phys. **50**, 5896 (1979)
5.4 M.Grimsditch, A.P.Malozemoff, A.Brunsch: Phys. Rev. Lett. **43**, 711 (1979)
5.5 M.Grimsditch, A.P.Malozemoff, A.Brunsch, G.Suran: J. Magn. Magn. Mater. **15–18**, 769 (1980)
5.6 J.G.Dill, E.M.Brody: Phys. Rev. B**14**, 5218 (1976)
5.7 P.Grünberg, F.Metawe: Phys. Rev. Lett. **39**, 1561 (1977)
5.8 F.Metawe, P.Grünberg: J. Magn. Magn. Mater. **9**, 80 (1978)
5.9 J.R.Sandercock: Solid State Commun. **26**, 547 (1978)
5.10 J.R.Sandercock, W.Wettling: IEEE Trans. M-**14**, 442 (1978)
5.11 J.R.Sandercock, W.Wettling: J. Appl. Phys. **50**, 7784 (1979)
5.12 P.Grünberg: J. Magn. Magn. Mater. **15–18**, 766 (1980)
5.12a A.S.Borovik-Romanov, N.M.Kreines: J. Magn. Magn. Mater. **15–18**, 760 (1980)
5.12b R.E.Camley, M.Grimsditch: Phys. Rev. B**11** (1980)
5.12c M.Grimsditch, P.Grünberg: *Spin Wave Stiffness of Iron Measured by Brillouin Scattering* (to be published)
5.13 B.I.Bennett, A.A.Maradudin, L.R.Swanson: In *Proceedings of the Second International Conference of Light Scattering in Solids*, ed. by M.Balkanski (Flammarion, Paris 1971) p. 443; Ann. Phys. (Paris) **71**, 357 (1972)
5.14 R.Loudon: Phys. Rev. Lett. **40**, 581 (1978)
5.15 M.G.Cottam: J. Phys. C**11**, 165 (1978)
5.16 R.E.Camley, D.L.Mills: Solid State Commun. **28**, 321 (1978)
5.17 R.E.Camley, D.L.Mills: Phys. Rev. B**18**, 482 (1978)
5.18 N.L.Rowell, G.I.Stegeman: Phys. Rev. B**18**, 2598 (1978)
5.18a M.G.Cottam: J. Phys. C**12**, 1709 (1979)
5.18b R.E.Camley: J. Appl. Phys. **50**, 5272 (1979)

5.18c R.C. Moul, M.G. Cottam: J. Phys. C**12**, 5191 (1979)
5.18d M.G. Cottam, R.C. Moul: J. Magn. Magn. Mater. **15–18**, 1085 (1980)
5.18e R.E. Camley: Phys. Lett. **45**, 283 (1980)
5.18f R.E. Camley, T.S. Rahman, D.L. Mills: *Theory of Light Scattering by the Spin Wave Excitations of Thin Ferromagnetic Films.* Technical Report No. 80.39 (University of California, Irvine 1980)
5.19 J.R. Sandercock: In *Proceedings of the Second International Conference on Light Scattering in Solids,* ed. by M. Blakanski (Flammarion, Paris 1971) p. 9
5.20 G. Güntherodt, S. Hüfner, A.P. Malozemoff: Private communication
5.21 I.A. Viktorov: *Rayleigh and Lamb Waves* (Plenum, New York 1967)
5.22 B.S. Berry, W.C. Pritchet: AIP Conf. Proc. **34**, 292 (1977)
5.23 C.P. Chou, L.A. Davis, M.C. Narasimhan: Scr. Metall. **11**, 417 (1977)
C.P. Chou, L.A. Davis, R. Hasegawa: J. Appl. Phys. **50**, 3334 (1979)
5.24 M. Kikuchi, K. Fukamichi, T. Masumoto, T. Jagielinski, K.I. Arai, N. Tsuya: Phys. Status. Solidi. (a) **48**, 175 (1978)
5.25 G. Hausch, E. Török: Phys. Status. Solidi. (a) **50**, 159 (1978)
E. Török, G. Hausch: J. Magn. Magn. Mater. **10**, 303 (1979)
5.26 H.S. Chen, J.T. Krause: Scr. Metall. **11**, 761 (1977)
5.27 K. Fukamichi, M. Kikuchi, H. Hiroyoski, T. Masumoto: *Rapidly Quenched Metals III*, Vol. 2 (The Metals Society, London, 1978) p. 117
5.28 K. Fukamichi: Private communication
5.29 R.W. Damon, J.R. Eshbach: J. Phys. Chem. Solids **19**, 308 (1961)
5.30 R.E. DeWames, T. Wolfram: J. Appl. Phys. **41**, 987 (1970)
T. Wolfram, R.R. DeWames: Phys. Rev. B**1**, 4358 (1970)
5.31 M. Sparks: Phys. Rev. B**1**, 383 (1970)
5.32 S.M. Bhagat, S. Haraldson, O. Beckman: J. Phys. Chem. Solids **38**, 593 (1977)
5.33 R. Hasegawa, R. Ray: Phys. Rev. B**20**, 211 (1979)
5.34 R.J. Birgeneau, J.A. Tarvin, G. Shirane, E.M. Gyorgy, R.C. Sherwood, H.S. Chen, C.L. Chien: Phys. Rev. B**18**, 2192 (1978)
5.35 J.R. McColl, D. Murphy, G.S. Cargill, III, T. Mizoguchi: AIP Conf. Proc. **19**, 172 (1978)
5.36 J.J. Rhyne, J.W. Lynn, F.E. Luborsky, J.L. Walter: J. Appl. Phys. **50**, 1583 (1979)
5.37 G. Suran, R.J. Gambino: J. Appl. Phys. **50**, 7671 (1979)
5.38 V.A. Ignatchenko, R.S. Iskhakov: Sov. Phys. JETP **47**, 725 (1978); **45**, 526 (1977)
5.39 M.A. Continentino, N. Rivier: J. Phys. F**9**, L145 (1979)
5.40 K. Fukamichi, M. Kikuchi, S. Arakawa, T. Masumoto: Solid State Commun. **23**, 955 (1977)
5.41 P.H. Chang, A.P. Malozemoff, M. Grimsditch: Private communication
5.42 G. Suran: Private communication

Notes Added in Proof

5.42

Recent measurements by *M. Grimsditch* (private communication) on samples of TU–$Fe_{1-x}B_x$ (x from 0.12 to 0.22) provided by *M. Kikuchi* have given $V_s = (2.37 + 2.3x) \times 10^5$ cm/s, both for as-grown and annealed surfaces. These values are 0.1 to 0.2×10^5 cm/s higher than the results for AC–$Fe_{1-x}B_x$, but they are far below the value calculated in Table 5.1 from measured elastic constants of TU–$Fe_{83}B_{17}$. This result shows that the TU samples *are* elastically different from the AC samples, but not by as much as the earlier constant measurements had indicated. The reliability and flexibility of the Brillouin scattering method is demonstrated here; in comparison, the conventional elastic constant measurements are difficult, being limited by sample size and brittleness, which explains why so few results have been reported so far.

5.43

Recently a more explicit theory of this effect has been given [5.18c] and applied to measurements on thin films of iron [5.12c].

5.44

Recent calculations [5.12b, 5.18f] have shown that the unusual intensity distribution of the peaks in Fig. 5.4 can be explained very nicely using a model *without* pinning. In particular, the peculiar reduction in intensity of the S_1 peak compared to S_2 appears to arise from the orthogonality requirement between S_1 and the surface mode. The surface mode has a maximum precession amplitude at the surface; so the surface amplitude and hence intensity of S_1 is reduced, while A_1 is unaffected because the surface mode is on the Stokes side. These results may not be inconsistent with the conclusions in the text about a pinning field of order 100 Oe, because according to [5.18f] it takes much larger pinning fields to significantly change the intensity ratios of S_i and A_i peaks.

5.45

Further support for these conclusions has come from recent Brillouin scattering measurements on thin films of iron [5.12c] in which similar standing spin wave spectra were analyzed. The resulting D agreed well with the results of low temperature magnetization measurements. This result confirms the reliability of the Brillouin scattering technique for determining D and indicates the discrepancy in amorphous FeB is not simply a measurement error.

6. Theory of the Structure, Stability, and Dynamics of Simple-Metal Glasses

J. Hafner

With 22 Figures

Metallic glasses formed by simple metals only are the ideal testing ground for a microscopic theory of the amorphous state. Pseudopotential methods for constructing the interatomic forces in binary alloys of arbitrary structure are reviewed. The knowledge of the interatomic potentials allows a microscopic calculation of the structure and of the thermodynamic properties of the crystalline, liquid, and amorphous intermetallic phases. Interrelations between the interatomic potentials, the amorphous structure, the phase diagram, and the glass-forming ability are discussed. Various theoretical attempts to calculate the dynamical properties of amorphous metals are presented. The correlations between the structural and the electronic properties are briefly reviewed.

6.1 Background

When twenty-five years ago amorphous metals were first produced by condensation from the vapor onto a cooled substrate [6.1], they were considered merely as laboratory curiosities. Six years later *Klement* and co-workers [6.2] demonstrated that it is possible to produce amorphous metallic alloys by rapid quenching from the liquid – quite in the same way as conventional window glass is produced by quenching from the melt. The investigation of the physical and chemical properties of these "metallic glasses" proved their technological usefulness, which is due to a unique combination of metallic and glassy behavior. In the last ten years many metallic alloys have been produced in an amorphous state [6.3], and it is now widely believed that nearly all metallic liquids (except perhaps pure liquid metals) would undergo a transition to a glassy state, provided that crystallization could be bypassed. Whether this is possible depends – at a given cooling rate – on thermodynamic conditions that favor the disordered (liquid or amorphous) relative to the crystalline state and on kinetic conditions that inhibit nucleation to a crystalline structure. We shall not be concerned with the kinetic factors of glass formation; our main interest will be in the thermodynamic conditions. It is usual to speak of the "stability" of a metallic glass. This is not completely correct: glasses are not in a thermodynamically stable state, they are only metastable with respect to the thermodynamic ground state. Geometrically, the problem might be described in terms

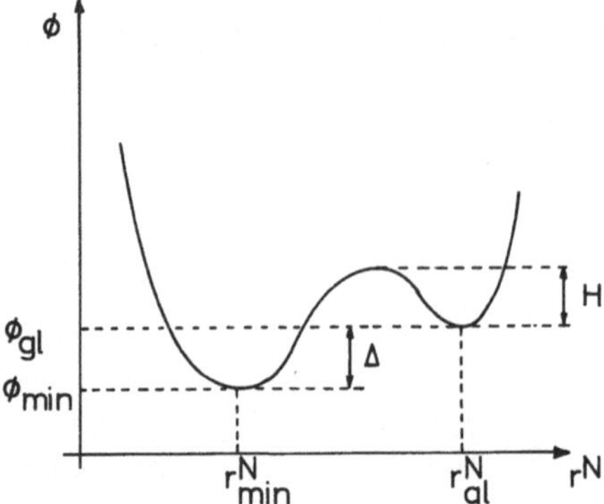

Fig. 6.1. Schematic sketch of potential energy ϕ in the configuration space C_N

of a configuration space C_N of $3N$-6 variables, N being the number of atoms in the system (or the number of atoms in the model we use to describe the system). Any given structure (i.e., any collection of $3N$ atomic coordinates) corresponds to a single point in C_N fixed by a $3N$-6 dimensional vector $\boldsymbol{r}_N$. With each point $\boldsymbol{r}_N$ we associate a potential energy function $\phi(\boldsymbol{r}_N)$, the stable structure being represented by the point $\boldsymbol{r}_N^{\min}$ corresponding to the absolute minimum of ϕ. Possible metastable configurations correspond to local minima of the potential energy function. Schematically, we might picture the situation in the form of one-dimensional plot of C_N (Fig. 6.1): we want to calculate the structure $\boldsymbol{r}_N^{\min}$ and the energy $\phi^{\min}$ of the stable crystalline state (which is not necessarily given by a single crystalline phase, but may be a two-phase mixture) and the structure $\boldsymbol{r}_N^{\mathrm{gl}}$ and the energy ϕ^{gl} of the metastable glassy state. The stability of this state depends on the driving thermodynamic potential $\Delta = \phi^{\min} - \phi^{\mathrm{gl}}$ and on the height of the potential barrier H separating the amorphous and the crystalline states. The one-dimensional projection of C_N gives a formal description of the transformation from the amorphous to the crystalline structure.

Of course it is quite impossible to search the entire configuration space for possible minima in the configurational energy. We must have an idea where to start. Two very important correlations are particularly useful to facilitate our task. Ever since the different microcrystalline models have been definitely superseded by the homogeneously disordered models (for a discussion of the "microcrystalline" versus "homogeneously disordered" controversy, see [6.4]), the correlation between the glassy and a very stable liquid state has served as a starting point to the various attempts of model building. After the very considerable refinements introduced in the last years, the packing models of the glassy state allow for a quite accurate reproduction of the diffraction data, but they bring no really satisfying explanation for the stability of the glassy state. The reason is that hard-packing models contain no and soft-packing models (using interatomic potentials of the Lennard-Jones or Morse type) only very little alloy chemistry. Evidently the most urgent need is for a realistic description of the interatomic interactions. Here the correlation between glass formation and the formation of certain classes of stoichiometric crystalline compounds may be very helpful. It is clear, however, that the microscopic theory of the structure and stability of crystalline intermetallic compounds is not so much further advanced than the theory of the amorphous structure. Only in a few special cases where some progress has been recently achieved is a quantum mechanical calculation of the bonding forces possible. It was a very happy coincidence that just at the time when the first such calculations (on topologically close-packed structures formed by simple metal alloys [6.5]) were published, the production of the first transition metal free metallic glasses was reported [6.6]. This opened the way to microscopic theory of glass formation [6.7, 8].

The correlation between glass formation and the formation of certain types of stoichiometric crystalline compounds is twofold: *Giessen* and *Wagner* [6.9] have pointed out that the "formation of amorphous phases is promoted if the competing crystalline phases have complex structures with high coordination numbers and several widely differing atomic positions". Examples are the μ and σ phases and the C16–$CuAl_2$-type and $CuMg_2$-type phases. For a description of the different crystal structures mentioned in this chapter we refer the reader to [6.10] and to the Structure Reports. On the other side it has been shown by *Polk* and *Giessen* [6.11] that the existence of several types of very stable (usually close-packed) intermetallic compounds in one part of the phase diagram is in some sense complementary to the glass formation in another concentration range. If we pursue this correlation somewhat further, we find that the metallic glasses can be tentatively classified in three main groups: the first contains the now classic transition metal-metalloid (T–M) glasses [6.3]; examples are Fe–B, Pd–Si, etc. In the extreme case we can include the Rb–O and Cs–O glasses [6.12]. Typical crystal structures belonging to this class are the Fe_3C-(cementite), Fe_2P-, and MnP-type phases (for a review of the structures of the transition metal carbides, phosphides, etc., see [6.14]). A common feature of these lattices is the metalloid site coordination by transition metals only, in the form of slightly distorted trigonal prisms. Again, the "anticluster" structures of the Rb and Cs suboxides with their octahedral coordination of the oxygen by alkali atoms fit losely into this scheme [6.13]. Furthermore these structural units persist over a wide range of concentrations, suggesting that they may have considerable stability. A structural model based on the random packing of such trigonal prismatic units instead of single atoms has very recently been proposed by *Gaskell* [6.15].

The simple-metal glasses [6.16] (S–S, e.g., Mg–Zn, Ca–Mg, Ca–Al, Mg–Cu), the simple metal-transition metal [6.17] (S–T, e.g., Ti–Be, Zr–Be), the simple metal-rare earth [6.18] (S–R, e.g., La–Al, La–Ga), and the transition metal-rare earth glasses [6.4, 19] (T–R, e.g., Gd–Co, Tb–Fe) form a second group. Their phase diagrams are characterized by the formation of highly stable AB_2 Laves phases or closely related *Frank-Kasper* [6.20] ($CaCu_5$, Th_2Mn_{17}, $BaCd_{11}$, $NaZn_{13}$) phases [6.10] at majority concentrations of the smaller B atoms. Glasses are formed at majority concentrations of the larger A atoms (though the glass-forming region may be extended to include the Laves phase by using high-speed quenching techniques, e.g., for T–R glasses). In the glass-forming region there are usually no or only relatively unstable compounds. In most cases their crystal structures are so complex that they have not been resolved as yet; among the few exceptions are the Mg_2Cu and the $CuAl_2$ phases. Unifying features of the stable structures are the principle of close packing and the coordination in the form of certain types of polyhedra. The smaller majority atoms have coordination number (CN) 12, icosahedral for the Laves phases, not or partically not icosahedral for the other structures. The CN12 polyhedra are linked by interpenetrating larger coordination polyhedra (CN16 Friauf polyhedra for the Laves phases and even larger polyhedra – CN18, CN20, CN22 for the other structures). The quasi-omnipresence of icosahedral coordination is at least indicative for possible local motifs in the amorphous structures.

The glasses formed by transition metals only [6.21] (T–T, e.g., Nb–Ni, Cu–Zr) constitute a third group characterized by complex tetrahedrally close-packed Frank-Kasper structures such as the μ and the σ phases. Icosahedral coordination (CN12) around the smaller atoms alternates with CN14, CN15 (Frank-Kasper polyhedra) around the larger atoms. Contrary to the second group the glass-forming region overlaps with the often rather broad homogeneity range of the crystalline compounds. While the crystalline phases of the former group are strictly ordered and characterized by a size ratio $R_A/R_B \geqq 1.15$, the μ and σ phases show a tendency towards substitutional disorder with a size ratio $R_A/R_B \leqq 1.15$. The formation of the μ and σ phases requires the existence of a well-defined stacking sequence of the coordination polyhedra over fairly large distances ($\gtrsim 30$ Å), which will easily be disturbed at high cooling rates. *Wang* [6.22] has recently proposed a structural model for amorphous intertransition metal alloys based on a random stacking of the typical coordination polyhedra (icosahedra and Frank-Kasper polyhedra).

Finally, the composition diagrams of all three classes of glass-forming alloys show very limited terminal solid solubilities, indicating a positive enthalpy of formation for solid solutions. The enthalpy of formation of liquid solutions on the other side is usually negative. Some prototype phase diagrams are shown in Fig. 6.2 [6.23].

In this chapter we review and extend a microscopic theory of the structure, of the stability [6.7, 8], and of the elementary excitations (electronic and dynamic [6.24]) of simple-metal glasses. Recent advances in the theory of pseudopotentials for binary alloys [6.5, 25] allow a more realistic treatment of electronic (charge transfer and screening) effects in the alloy pseudopotential and hence a more reliable calculation of the interatomic interactions. The calculation of the ground state properties (relative structural energies, enthalpies, and volumes of formation) of crystalline intermetallic compounds and solid solutions in some glass-forming simple-metal systems provides a first check of the usefulness and reliability of these effective interatomic pair potentials. The atomic structure of the disordered (liquid or amorphous) phases

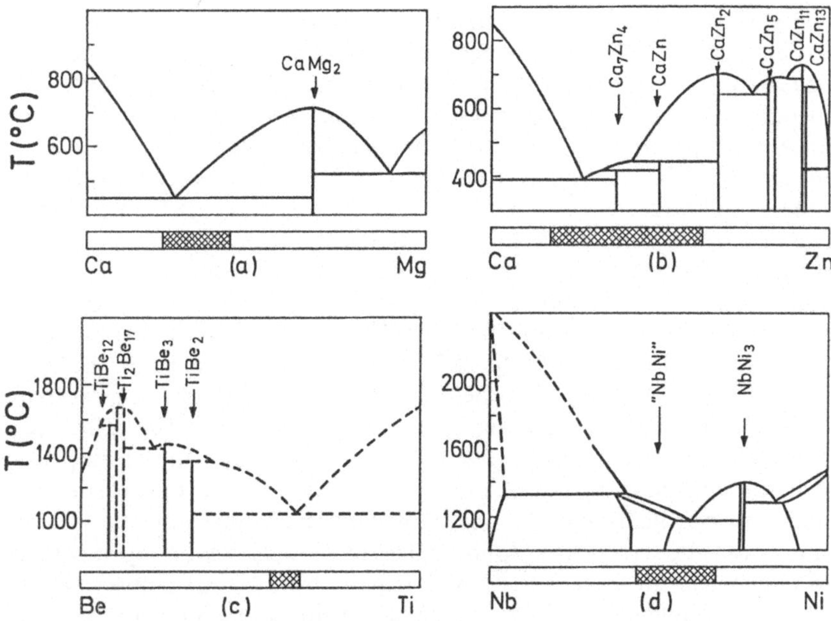

Fig. 6.2a–d. Some typical phase diagrams of glass-forming systems [6.23], the hatched area indicates the width of the glass-forming region

may be calculated either using a fully numerical procedure (molecular dynamics, Monte Carlo or cluster relaxation) or using one of the thermodynamic perturbation theories (Gibbs-Bogoljubov variational method [6.26, 27] or Weeks-Chandler-Andersen (WCA) method [6.28]). In our treatment it turns out that structures (crystalline, liquid, or amorphous) are stabilized when the atoms lie at the minima of the oscillatory interatomic pair potentials. Several models have been proposed to explain the relative stability of metallic glasses [6.29–32]. Now, their predictions can be checked against our microscopic calculations. We shall discuss the role of the traditional alloy-chemical factors (size ratio, valence electron concentration, and electronegativity difference) in the stabilization of amorphous metallic alloys.

The use of the thermodynamic calculations enables us to calculate a phase diagram from first principles in at least semiquantitative agreement with experiment. The interrelation between the formation of eutectic phase diagrams, the high stability of Frank-Kasper phases, and the formation of amorphous alloys is readily understood. We can discuss the coexistence of certain types of crystal structures with metallic glasses; we can even get some idea of the nature of the free energy barrier against nucleation to the crystalline state. Technically, the application of our method is restricted to nontransition metals, but the principle found here is of such striking simplicity that it offers some hope for a better understanding of all metallic glasses.

Finally we shall briefly discuss the dynamical and electronic properties of simple-metal glasses. The vibrational spectra are of particular interest because of the unusual low-temperature thermodynamic [6.18, 33], acoustic [6.34], and thermoconductivity [6.35] properties of amorphous substances. Different theoretical attempts to investigate the dynamical properties – the equation-of-motion method, the molecular dynamics, and the momentum-expansion techniques – will be reviewed. The rather unusual electronic transport properties of amorphous metallic alloys (high resistivities and low, often even negative, temperature coefficients of the resistivity) are thought to be characteristic of the disordered state. There is still an open controversy whether the diffraction model of the electronic transport (the Faber-Ziman theory for liquid metals as extended to metallic glasses [6.36–38]) or other models such as the $s-d$ scattering model of *Mott* [6.39] or the pseudo-Kondo model of *Cochrane* et al. [6.40] are the appropriate description of these phenomena. The diffraction model seems to be particularly appealing because of its conceptual simplicity and its ability to explain the interrelation between the transport properties and the stability of metallic glasses. However, the numerical application to systems containing transition metals is doubtful [6.41] (a weak scattering theory is applied to a strong scattering situation). So again it will be instructive to apply the theory to simple-metal alloys where the justification for using a weak scattering assumption is beyond doubt.

6.2 Interatomic Interactions in Binary Alloys

To begin with, let us sketch the path to be followed for an ab initio calculation of a phase diagram:

1) Calculate the self-consistent atomic potentials of the components.
2) Assume composition.
3) Assume possible atomic structure.
4) Assume lattice constants.
5) Superpose atomic potentials, calculate the self-consistent band structure and the total energy.
6) Repeat 4) and 5) for different values of the lattice constants to determine their equilibrium values.
7) Repeat 3) to 6) for other possible structures.
8) Repeat 2) to 7) for other concentrations.

As a final result of this procedure we obtain the enthalpies of the possible alloy phases at $T=0$ K and at a given pressure as a function of concentration. The stable phases and the boundaries of their homogeneity ranges are given by the well-known common tangent construction. Beyond the determination of the stable phases, we can get at least an indication on the possible formation of metastable phases, e.g., if the system is constrained against phase separation. The procedure given above is restricted to $T=0$ K; for finite temperatures we have to add the thermal entropy which is the sum of a vibrational and a configurational part. In principle this necessitates a complete knowledge of the vibrational and the defect structure of the alloy. For configurationally disordered systems there is an additional complication: starting from an initial guess of the interatomic forces, we have to calculate the atomic structure. Knowing the atomic structure, we have to calculate the electronic structure. This gives an improved estimate of the interatomic interactions and the whole cycle has to be iterated to self-consistency. It is quite clear that in general the realization of this program is a tremendous computational task. However, for simple metals pseudopotential theory [6.42, 43] allows some substantial simplifications. Within second-order perturbation and linear screening theories, the total energy calculation is reduced to the computation of lattice sums (in either direct or reciprocal space) over a set of structure-independent effective interatomic pair potentials.

6.2.1 Pseudopotentials for Binary Systems

Pseudopotential theory has enormously contributed to the advances in the microscopic theory of the binding, the atomic structure, the phase transitions, the vibrational, and the electronic properties of pure simple metals. The progress in the application of this scheme to alloys has been comparatively slow. This is related to the fact that the pseudopotential is often considered as a purely atomic property which is – except for conduction-electron screening effects – independent of the atom surroundings. In reality the pseudopotential is a collective property which describes the scattering of electrons by an atom in a given effective medium (the average potential of all other ions and electrons in the metal). If this effective medium changes (e.g., by alloying), the pdeudopotential changes too. These alloying effects are easily incorporated in a pseudopotential theory based on an expansion of the conduction electron states in terms of orthogonalized plane waves (OPWs) [6.25]. In the alloy, an OPW is constructed by projecting a plane wave onto the subspace orthogonal to the core states ψ_{nl}^{A} and ψ_{nl}^{B} of the components A and B. Inserting this expansion in the one-particle Schrödinger equation, one obtains a generalized *Phillips* and *Kleinman* [6.44] equation for the pseudo-wave-function, i.e., the smooth part of the conduction electron state. Within this general class of pseudopotentials, an optimized pseudopotential may be selected using *Cohen* and *Heine*'s criterion [6.45] of the smoothest pseudo-wave-function. The remaining ingredients of the method are a local density approximation for the core-core and core-valence electron exchange and correlation interactions (for any details concerning the construction of the crystal potential and the application of the resulting pseudopotential to calculate the properties of the pure metals see [6.46]) and the use of the *Vashishta-Singwi* function [6.47] for many-body effects in the conduction electron screening function. The resulting *electron-ion* pseudopotential may be factorized in terms of form factors w_A, w_B

describing the scattering properties and structure factors S_A, S_B describing the spatial arrangement of the ions. The charge density distribution of a one-OPW state consists of a negative homogeneous part and strictly localized positive charge distributions centered at the ionic sites, the so-called orthogonalization holes arising from the orthogonality constraint of the Pauli principle. The charge distribution in the metal is described by a superposition of neutral "pseudoatoms" [6.48], consisting of the ionic core, the orthogonalization hole, and the valence electron screening charge. The screening charge density $n_i(r)$ of an atom i is related to the form factor w_i via a generalized Poisson equation [6.49]. The pseudoatoms i and j interact via an effective central pair potential $V_{ij}(|\boldsymbol{r}_i - \boldsymbol{r}_j|)$ which is composed by the Coulomb repulsion between ions with the effective valences Z_i^*, Z_j^* (effective valence Z^* = nominal valence Z + orthogonalization hole n^{oh}) plus an indirect interaction via the conduction electrons ([6.42, 49] and Sect. 6.2.2).

The method is well documented in several papers [6.5, 7, 8, 25, 27]; only a few specific points should be made here. If we alloy metals with appreciably different valence electron densities, e.g., Ca (atomic volume $\Omega = 43.50\,\text{Å}^3$, valence $Z = 2$, electron density $Z/\Omega = 0.046\,e/\text{Å}^3$) and Al ($\Omega = 16.51\,\text{Å}^3$, $Z = 3$, $Z/\Omega = 0.182\,e/\text{Å}^3$), the one which "sees" an increased electron density must now form a bigger orthogonalization hole, while around the other ion it is reduced. The Ca orthogonalization hole increases from $-0.353\,e$/at. in the pure metal to $-0.835\,e$/at. in a $CaAl_2$ alloy, while the Al orthogonalization hole is reduced from -0.241 to $-0.161\,e$/at. This is equivalent to an electron transfer in the direction that is expected from the electronegativity difference, i.e., from Ca to Al (for a discussion of the interrelation orthogonalization hole-electronegativity see [6.51]). The electrical neutrality of the pseudoatoms requires the screening charges to compensate for the change in the orthogonalization charges. In order to maintain the electroneutrality even on a local level, the additional screening charge of the electropositive ion (the one with the larger orthogonalization hole) is concentrated in the core region. This is accompanied by an inward shift of the Friedel oscillations in the screening charge (Fig. 6.3). Around the electronegative ion, the effect is just the inverse. Thus the screening effect yields an electron transfer to the electropositive ion – contrary to what has to be expected from the electronegativity difference. An early discussion of these competing electron transfer effects arising from the electronegativity and electroneutrality principles has been given by *Pauling* [6.52].

6.2.2 Effective Pair Potentials

The interatomic potentials follow very closely the redistribution of the screening charge. The change in the short-range part is equivalent to a compression of the more electropositive ion and a slight expansion of the electronegative ion. This is an effect which is well known in the metallurgical literature as the

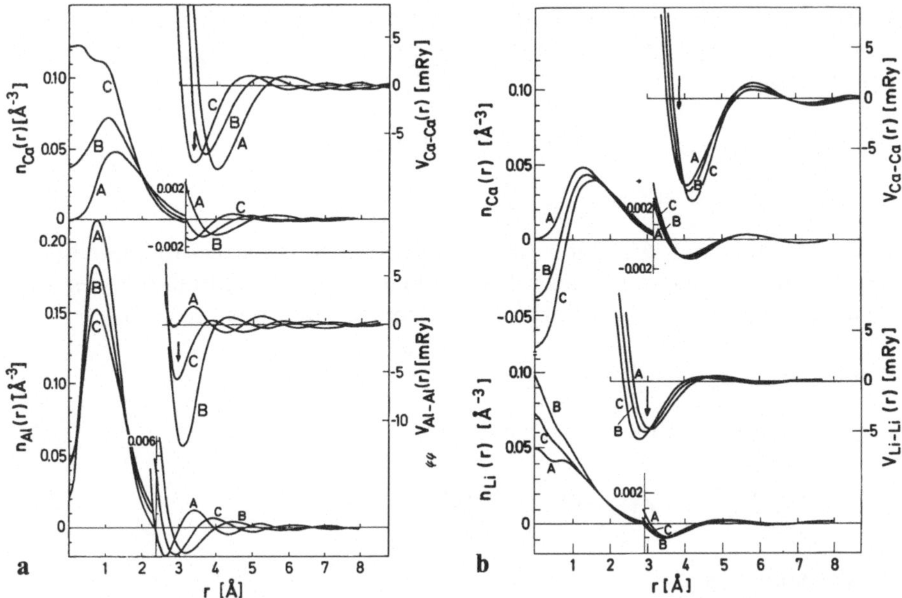

Fig. 6.3a, b. Pseudoatom (screening) charge density $n_i(r)$ and effective interatomic potentials $V_{ij}(r)$. (**a**) For Ca and Al atoms, (**b**) for Ca and Li atoms in the pure metals (Curve *A*) and in alloys of the composition $Ca_{0.67}X_{0.33}$ (Curve *B*) and $Ca_{0.33}X_{0.67}$ (Curve *C*), demonstrating the "chemical compression" of the electropositive atom in the alloy. The vertical arrows indicate the interatomic distances in the Laves phases $CaAl_2$ and $CaLi_2$, respectively [6.7, 8]

"chemical compression" [6.53]. The effects discussed here exist in all alloys, but they are certainly more prominent in alloys with large differences in electronegativity and electron density. In two recent papers it has been demonstrated how such effects contribute to stabilize the tetrahedrally close-packed structures of the stoichiometric alkali–alkali compounds (Na_2K, Na_2Cs, K_2Cs – hexagonal Laves phases, K_7Cs_6 – hexagonal stacking variant of the μ phase structure) [6.5]. The effects are even more important in heterovalent alloys such as the series $CaLi_2$–$CaMg_2$–$CaAl_2$ (cf. Sect. 6.3).

The configurational (structure-dependent) part of the internal energy per atom of the crystalline alloy is given by the lattice sums $\left(\sum_{i(A)}, \sum_{i(B)}\right.$ extend over the *A* or *B* sublattices, *N* is the number of atoms in the crystal [6.5, 25]$\left.\right)$

$$E_{st}=(2N)^{-1}\left[\sum_{i(A)}\sum_{j(A)}{}' V_{AA}(|\boldsymbol{r}_i-\boldsymbol{r}_j|)+\sum_{i(B)}\sum_{j(B)}{}' V_{BB}(|\boldsymbol{r}_i-\boldsymbol{r}_j|)\right.$$
$$\left.+2\sum_{i(A)}\sum_{j(B)}{}' V_{AB}(|\boldsymbol{r}_i-\boldsymbol{r}_j|)\right]. \qquad (6.1)$$

Using the partial pair distribution functions $g_{lm}(r)$, we can write the configurational energy of an amorphous or liquid phase as (c_A and c_B are the concentrations)

$$E_{st}=2\pi[c_A^2\int V_{AA}(r)g_{AA}(r)r^2dr+c_B^2\int V_{BB}(r)g_{BB}(r)r^2dr$$
$$+2c_Ac_B\int V_{AB}(r)g_{AB}(r)r^2dr]\,. \tag{6.2}$$

Equations (6.1) and (6.2) allow an interpretation of the structural stability in terms of pair interactions and a first estimate of the structural energies. For their exact calculation, a reciprocal space representation is preferable,

$$\bar{E}_{st}=E_{bs}+E_{es}\,; \tag{6.3}$$

here E_{bs} is the band structure energy, given by

$$E_{bs}=\sum_{i,j}\sum_{\boldsymbol{q}}{}' S_i^*(\boldsymbol{q})S_j(\boldsymbol{q})F_{ij}(q)\,,\qquad i,j=A,B \tag{6.4}$$

for crystalline, or by

$$E_{bs}=\sum_{i,j}\frac{\Omega_a}{2\pi^2}(c_ic_j)^{1/2}\int F_{ij}(q)S_{ij}(q)q^2dq,\quad i,j=A,B \tag{6.5}$$

for disordered structures. Here Ω_a is the average atomic volume, the $S_{ij}(q)$ are the Ashcroft-Langreth partial static structure factors [6.54], and the $F_{ij}(q)$ are the energy wave number characteristics, i.e., the Fourier transform of the indirect ion-electron-ion interactions. The electrostatic energy E_{es} is given by

$$E_{es}=\frac{1}{r_a}\sum_{i,j}Z_i^*Z_j^*\alpha_{ij},\quad i,j=A,B\,. \tag{6.6}$$

r_a is the radius of a sphere with the average atomic volume; the α_{ij} are the Ewald coefficients. For crystalline lattices they can be computed using the well-known Ewald-Fuchs method [6.55]; for disordered structures they are given in terms of integrals over the partial static structure factors [6.56],

$$\alpha_{ij}=(c_ic_j)^{1/2}\frac{2r_a}{\pi}\int_0^\infty[S_{ij}(q)-\delta_{ij}]dq\,. \tag{6.7}$$

6.3 Crystalline Alloys

In Sect. 6.1 we have pointed out that the constitution diagrams of many glass-forming systems are characterized by the formation of tetrahedrally close-packed compounds of the Frank-Kasper type at one side of the diagram (at majority concentrations of the smaller atoms) and glass formation at the other side (at majority concentrations of the larger atoms).

6.3.1 Tetrahedrally Close-Packed Frank-Kasper Compounds and Related Structures

Topologically close-packed structures are usually thought to be determined by the size ratio of the constituent atoms. For the AB_2 Laves phases, the ideal size ratio derived from the requirement that both $A-A$ and $B-B$ contacts are formed is $R_A/R_B=1.225$. In practice, the size ratio deduced from the Goldschmidt radii of the pure metals does not scale with the observed stability of the compounds. A striking example is the series of $CaLi_2$–$CaMg_2$–$CaAl_2$ Laves phases: $CaAl_2$ with its highly nonideal size ratio ($R_{Ca}/R_{Al}=1.40$) is an extremely stable, congruently melting compound with a large negative enthalpy of formation; $CaLi_2$ with a nearly ideal size ratio ($R_{Ca}/R_{Li}=1.26$) is relatively unstable and decomposes peritectically at low temperature. Glasses are formed in the Ca–Mg and Ca–Al systems [6.16], but apparently not in the Ca–Li system.

For the pure metals, the position of the first attractive minimum in the pair potential agrees well with the interatomic distance in a close-packed structure (two times the Goldschmidt radius). Now, the position of this minimum changes on alloying (cf. Sect. 6.2.1) and it is certainly a more sensible measure of the atomic size than any rigid atomic radius. The effective size ratios deduced from the interatomic potentials appropriate for the composition of the Laves phase are now (see Fig. 6.3) $R_{Ca}/R_{Al}=1.15$, $R_{Ca}/R_{Mg}=1.21$, $R_{Ca}/R_{Li}=1.37$, in striking agreement with the observed stability of the compounds. For the $CaAl_2$ and $CaMg_2$ phases, the interatomic distances fit exactly into the minima of the pair potentials (Figs. 6.3a and 6.4), i.e., the calculated potential energy function allows for nearly strain-free $A-A$, $A-B$, and $B-B$ bonds. For the $CaLi_2$ phase on the other side, the $A-A$ bonds are considerably strained (Fig. 6.3b). This pair potential argument provides a very elegant explanation of "Simon's rule" [6.57]. This rule – whose validity is now established for a wide class of stoichiometric intermetallic AB_n compounds including the Laves phases – expresses the additivity of the interatomic distances,

$$d_{ij}=F_{ij}(R_A+nR_B); \tag{6.8}$$

the interatomic distances d_{ij} are related to the concentration weighted average of the atomic radii via the geometrical coefficients F_{ij} [6.58]. If R_A and R_B are sufficiently different, (6.8) predicts a negative volume of formation. This rule is the counterpart of Vegard's rule for mixed crystals, which predicts the additivity of the atomic volumes. However, Simon's rule alone does not yield a strain-free configuration: for larger size ratios, the contacts between the larger atoms

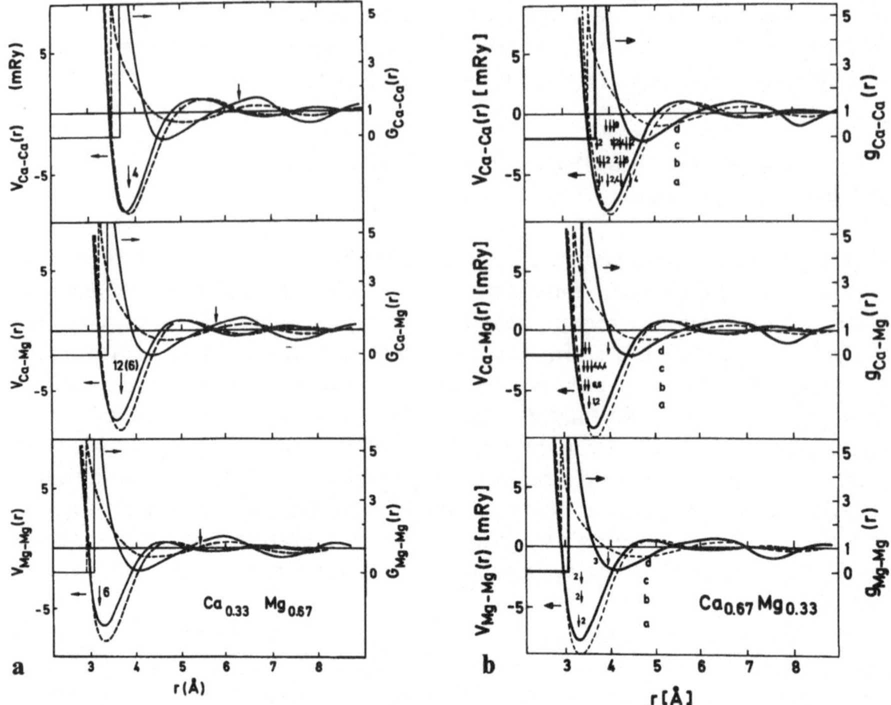

Fig. 6.4a, b. Interatomic pair potentials $V_{ij}(r)$ and partial pair distribution functions $g_{ij}(r)$. (**a**) For $Ca_{0.33}Mg_{0.67}$, (**b**) for $Ca_{0.67}Mg_{0.33}$ in the amorphous (solid lines, $T=25\,°C$) and in the liquid (dashed lines, $T=900\,°C$) states. The vertical arrows indicate the interatomic distances in the $MgZn_2$-type Laves phase (**a**) and in different hypothetical crystalline structures (**b**): *a* $CuAl_2$ type, *b* $NiMg_2$ type, *c* $CuMg_2$ type, *d* $TiNi_2$ type [6.8]

appear to be considerably compressed. In our picture, the individual atomic radii are modified, but – at least in the case of weak chemical bonding – their average value $(R_A+nR_B)/(n+1)$ is left approximately constant. In the case of strong chemical bonding (large differences in electronegativity and/or electron density), the "chemical compression" effect is superposed to the volume contraction predicted by Simon's rule. This explains why for $CaAl_2$ the observed volume contraction is larger ($\Delta\Omega=-15.5\%$) than predicted by Simon's rule ($\Delta\Omega=-7.5\%$), while there is even a slight volume expansion ($\Delta\Omega=2.1\%$) for $CaLi_2$, compared to $\Delta\Omega=-3.8\%$ predicted by Simon's rule.

For heterovalent alloys, a full ab initio calculation of the thermochemical ground state properties is complicated by the fact that the electronic effects described above require a reconsideration of the structure-independent first-order contribution to the binding energy. This will be left to a subsequent publication. For the homovalent alloys, there is an excellent agreement between the calculated thermochemical data and experiment, e.g., for $CaMg_2$:

$\Delta H(\text{th.}) = -3.6$ kcal/g·at., $\Delta H(\text{exp.}) = -3.2$ kcal/g·at.; $\Delta\Omega(\text{th.}) = -9.2\%$, $\Delta\Omega(\text{exp.}) = -5.6\%$.

One can go even one step further: the free parameters of the Laves phase structure (i.e., those not determined by space group symmetry) may be calculated by minimizing the energy in the configuration space of the atomic positions. This calculation shows that the electronic effects discussed above are vital in stabilizing the tetrahedral networks of the A and the B sublattices [6.5, 8]. The structural energy differences between the different stacking variants of the Laves phases are also correctly predicted by our theory [6.8]: in accordance with experiment $CaAl_2$ is shown to be cubic ($MgCu_2$ type), while $MgZn_2$, $CaMg_2$, and $CaLi_2$ are hexagonal ($MgZn_2$ type). Since the nearest-neighbor configuration is identical for all Laves phase variants, this shows that our potentials are very realistic even for larger distances.

The pair potential argument is not restricted to the relatively simple Laves phases, but appears to be applicable to a large variety of tetrahedrally close-packed structures. A first example was the explanation of the K_7Cs_6 structure; another example is given in Fig. 6.5: it is shown that the interatomic distances of the $CaZn_2$ ($CeCu_2$-type), $CaZn_5$ ($CaCu_5$-type), $CaZn_{11}$ ($BaCd_{11}$-type), and $CaZn_{13}$ ($NaZn_{13}$-type) compounds fit very well into the minima of the pair potentials [6.8]. Ca–Zn is a very interesting system because of its extremely wide glass-forming region.

6.3.2 Mixed Crystals

As pointed out in Sect. 6.1 glass formation is restricted to systems with a very low solid solubility. From our pair potential argument it is clear that due to the size difference a substitutionally disordered arrangement of the atoms on any possible periodic lattice will be energetically very unfavorable. Indeed we calculate quite large positive enthalpies of formation and slightly positive volumes of formation for all possible solid solutions. It is interesting to mention that the structural energy differences between possible polymorphs of Ca-rich disordered alloys are extremely small. This implies that the potential energy surface in the configuration space is very flat, showing several local minima with nearly equal depth. This may be a further hint to correlations between the thermal stability of amorphous solids on one side [6.59] and the thermal anomalies of noncrystalline solids and polymorphism on the other side [6.60].

6.3.3 Intermetallic Compounds in the Glass-Forming Composition Range

The formation of a metallic glass clearly requires that possible intermetallic compounds with a composition falling in the glass-forming range have a higher or at least only slightly lower free energy than the glass itself. Only a very few

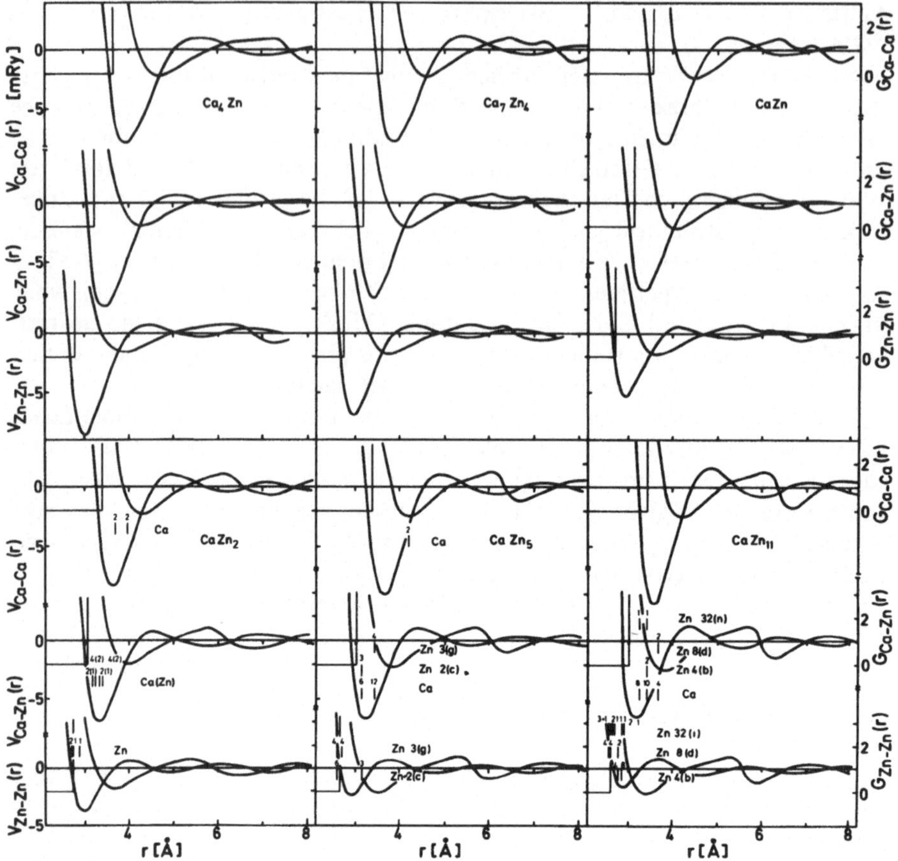

Fig. 6.5. Interatomic pair potentials $V_{ij}(r)$ and partial pair distribution functions $g_{ij}(r)$ for the supercooled liquid (~glassy) state at $T=25\,°C$ for the Ca–Zn system. The vertical bars indicate the interatomic distances in the crystalline $CaZn_2$ ($CeCu_2$-type), $CaZn_5$ ($CaCu_5$-type), and $CaZn_{11}$ ($BaCd_{11}$-type) compounds. The notation for the atomic positions is that of the International Tables for x-Ray Crystallography [6.8]

simple-metal glasses form stable intermetallic compounds of this composition. Their crystal structures as well as those of the metastable crystalline phases formed during the recrystallization process are usually very complex and have not been resolved. One of the few examples is found in the Mg–Cu system: $MgCu_2$ is a cubic Laves phase; the glass-forming region comprises a stable phase Mg_2Cu with an orthorhombic structure. On heating, the glass undergoes a glass transition and then recrystallizes in the Mg_2Cu structure [6.61]. Another possible example is found in the Cu–Zr system, where the glass-forming range includes a compound Zr_2Cu, which has been tentatively described as $MoSi_2$ type [6.23].

The selection of possible crystalline structures to be compared with the amorphous structure is of course always somewhat arbitrary, but following these examples and on the basis of size ratio considerations we choose the Mg_2Cu, the Mg_2Ni, the $CuAl_2$ structures (both closely related to the Mg_2Cu type), and the $MoSi_2$, the AlB_2, and the $TiNi_2$ structures (both related to the $MoSi_2$ type). The relative stability of A_2B phases (Ca_2Mg, Ca_2Al, and Mg_2Zn) with these crystal structures was estimated using the real space formulation of the configurational energy ((6.1, 2), for details see [6.8]). The interatomic distances are compared with the pair potentials at the example of Ca_2Mg (Fig. 6.4b). It is evident that the structures of the $MoSi_2$ family (only the case of the $TiNi_2$ type is shown) cannot be stabilized by pair potentials alone because of their rather anisotropic coordination. Their configurational energies are distinctly higher than those of the amorphous phases (see Sects. 6.5 and 6.6). On the other hand, the interatomic distances of the $CuAl_2$, the Mg_2Cu, and the Mg_2Ni lattices fit quite well into the minima of the pair potentials. The estimate of the structural energy differences predicts that the most stable crystalline structures are Mg_2Ni for $CaAl_2$ and Mg_2Cu for Ca_2Mg and Mg_2Zn. However, the configurational energy of the amorphous phase is nearly the same. The energy difference ΔE relative to the most stable crystalline compound is $\Delta E \sim -280$ cal/g·at. for a-Ca_2Al, $\Delta E \sim 100$ cal/g·at. for a-Ca_2Mg, and $\Delta E \sim 200$ cal/g·at. for Mg_2Zn. Considering that the amorphous phase has a higher entropy of formation, we find that the amorphous alloys have a lower free energy than any of the crystalline phases considered here. A detailed comparison shows that the number of $A-A$ and $A-B$ contacts is larger in the crystalline than in the amorphous A_2B compounds, whereas the number of $B-B$ neighbors is much smaller. Correspondingly, the $A-B$ interactions prefer the crystalline state, the $A-A$ interaction energy is approximately equal in both states, whereas the $B-B$ interactions prefer the amorphous state. Since due to electronic screening effects (cf. Figs. 6.3–5), the $B-B$ pair potential has a very deep minimum, this last contribution dominates.

6.4 Liquid Alloys

The pseudopotential method allows a formulation of the internal energy in terms of effective pair potentials and partial structure factors or partial pair correlation functions (6.2–7). In principle, the pair correlation functions may be calculated from the pair potentials by solving one of the integral equations of the theory of liquids (see, e.g., [6.62]). For binary systems and realistic pair potentials, this is a very hard computational task. A much simpler way to get a reasonably accurate value of the free energy is to use thermodynamic perturbation theory, i.e., to lowest order

$$F \cong F_0 + \langle V \rangle_0 + \ldots, \tag{6.9}$$

where F_0 is the free energy of a reference system and $\langle V \rangle_0$ is the expectation value of the perturbation. It is well known that hard spheres are a very suitable reference system for liquid metals and alloys. Moreover, closed-form expressions for the partial structure factors [6.54] and the thermodynamic quantities of hard sphere mixtures are furnished by the Percus-Yevick theory [6.63–65]. The Carnahan-Starling version of the hard sphere equation of state has been used in our calculations. The attractive parts of the interatomic potentials and the softness of their repulsive parts may be taken into account by extending the perturbation expansion beyond the lowest order [6.66]. For pure metals this yields very interesting results for more complicated liquid structures, e.g., for liquid Ga and liquid Sn [6.67]. For liquid alloys however, we have to stick to the lowest order perturbation expression: a very convenient criterion for choosing an optimal hard sphere reference system is provided by the Gibbs-Bogoljubov inequality [6.68]. It states that the right-hand side of (6.9) constitutes an upper bound to the exact free energy. Hence the hard sphere diameters σ_A and σ_B will be chosen to minimize $(F_0 + \langle V \rangle_0)$. The resulting σ_A, σ_B obey very well the relation [6.27, 50, 54]

$$V_{ij}(\sigma_{ij}) - V_{\min} = \tfrac{3}{2} k_B T, \qquad i,j = A, B, \tag{6.10}$$

i.e., the difference between the pair potential at σ_{ij} and its minimum value is just equal to the mean kinetic energy per particle. The obvious meaning of (6.10) is that σ_A, σ_B, and $\sigma_{AB} = 0.5(\sigma_A + \sigma_B)$ are average collision distances. Moreover (6.10) tells us that the hard sphere diameters (and further the partial structure factors) are determined mainly by the short-range part of the potential, i.e., by the depth of the attractive minimum and the steepness of the repulsive part.

Details of the application of this method to simple metals and alloys have been presented elsewhere; the agreement with experiment is very gratifying for both structures and thermodynamic quantities [6.8, 27]. Here we shall present only the results for the free enthalpy ΔG, the enthalpy ΔH, the entropy ΔS, and the volume $\Delta\Omega$ of formation of liquid Ca–Mg alloys (Fig. 6.6). From the ΔG and ΔH curves we see that there is no particular stabilization of the melt at the eutectic composition ($\sim$27 at.% Mg), the maximum stability (largest ΔG) being achieved at a concentration of $\sim$67 at.% Mg (experimentally at $\sim$54 at.% Mg). Our calculation shows that the asymmetry of the ΔG curve is due to packing effects, as evidenced by the concentration dependence of $\Delta\Omega$ and of the packing fraction η. For purely geometrical reasons it is much easier to achieve a high packing fraction at a majority concentration of the smaller atoms. It appears to be unnecessary to assume the formation of $CaMg_2$ "associates" [6.69] to explain the concentration dependence of ΔG.

Again it is very instructive to look at the interrelation between the interatomic potentials $V_{ij}(r)$ and the interatomic distances, as given by the pair correlation functions $g_{ij}(r)$ (Figs. 6.3–5). From (6.2) it is clear that if the maxima of $g_{ij}(r)$ fall into the minima of $V_{ij}(r)$, we will get a very low configurational energy. For the Ca–Mg and Ca–Zn systems, this "constructive interference"

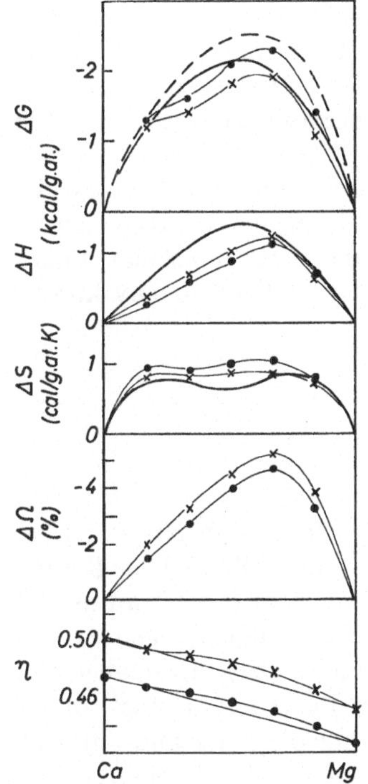

Fig. 6.6. Free enthalpy ΔG, enthalpy ΔH, entropy ΔS, and volume $\Delta\Omega$ of formation, and the packing fraction η for liquid Ca–Mg alloys. The points and crosses represent the calculated values at $T=900\,^{\circ}\mathrm{C}$ (●) and $T=700\,^{\circ}\mathrm{C}$ (×). The solid line represents the experimental results of *Sommer* et al. [6.69] ($T=737\,^{\circ}\mathrm{C}$), and the dashed line the ΔG (for $T=923\,^{\circ}\mathrm{C}$) quoted by *Hultgren* et al. [6.70]. (After [6.8])

between $V_{ij}(r)$ and $g_{ij}(r)$ is realized over the entire composition range. We note that the physical mechanism for the stabilization of the liquid alloy is identical to that for the stabilization of the Laves phase.

However it would be misleading to think that the periodicity in the oscillations of the interatomic potentials forces a similar periodicity upon the atomic structure, i.e., upon the pair distribution functions of the liquid (or, as we shall see, of the amorphous) alloy. The simple relation (6.10) demonstrates very clearly that the atomic structure is determined mainly by the short-range part of the interatomic potentials. The constructive interference between $V_{ij}(r)$ and $g_{ij}(r)$ leads to a very low electronic total energy and contributes to an additional stabilization of the disordered phase.

6.5 The Structure of Amorphous Alloys

In the early stage the theory of the atomic structure of metallic glasses has been characterized by the "microcrystalline" versus "homogenously disordered" controversy. After the microcrystalline hypothesis had been definitely ruled out on the basis of diffraction experiments [6.4], there was a widespread agreement that the *Bernal* dense random packing of hard sphere model (DRPHS) [6.71] is the obvious starting point for structural modelling. When improved neutron and x-ray scattering data became available [6.72–75], it turned out that there is at least some degree of local order which cannot be described in terms of a random packing of individual atoms. Further evidence for some structure in the chemical environment comes from extended x-ray absorption fine structure (EXAFS) data [6.76] and from hyperfine field measurements [6.77]. This is the motivation for the construction of models based on the random spatial repetition of larger local units (one should avoid the term "molecular units", which could be misleading) in a way that cannot be related to any known crystalline structure [6.15, 22].

6.5.1 Structural Models

a) Sphere Packing

Cargill [6.4] has compared the pair correlation function of an $Ni_{76}P_{24}$ glass with that obtained by *Finney* [6.78] (laboratory-built DRPHS) or *Bennett* [6.79] (computer-generated DRPHS) and pointed out that there is an overall agreement. The use of a single-sphere DRPHS for composite glasses was justified by the Polk hypothesis [6.80] which assumes that the smaller and softer metalloid atoms are located in the large cavities (Bernal holes) of the DRPHS network. An attempt was made to predict the observed concentration dependence of the stability of T–M glasses from an analysis of the number and size of these holes. It turned out however that the observed size of the metalloid atoms is not always compatible with the size of the available Bernal holes [6.15].

Hence the most obvious improvement of the DRPHS model is the step from a single-sphere to a binary DRPHS model. The first such model has been reported by *Sadoc* et al. [6.81] for T–M glasses; binary DRPHS structures for T–R glasses have been generated by *Cargill* and *Kirkpatrick* [6.82]. *Von Heimendahl* [6.83] and *Barker* et al. [6.84] independently made the important step of relaxing the Bernal structure under the action of soft pair interactions of a Lennard-Jones type. A further possibility to improve the DRPHS model is to force a certain degree of "tetrahedral perfection" on nearest-neighbor sites [6.82]. Some justification for this procedure is found in the frequent occurrence of tetrahedrally close-packed compounds in glass-forming systems. Recent work of *Cargill* [6.19] on Gd–Co glasses combines all three types of refinements and produces very accurate x-ray diffraction patterns. However, the work of *Boudreaux* [6.85] on relaxed binary DRPHS models for T–M glasses should serve as a warning: it shows excellent agreement for the composite (x-ray weighted) pair correlation function, but the agreement for the partial pair correlation functions is quite disappointing (though it must be admitted that the experimental partial pair correlation functions [6.86] should be considered with some reserve). This shows that we need more than a single scattering experiment to assess the validity of a given model structure.

Another problem arises from the use of finite clusters in the relaxation calculations, typically some thousand atoms are relaxed. In a cluster of that size more than 30% of the atoms are located at the surface. Their near-neighbor statistics and the local number density are quite different from those of the bulk and the unbalancing of the interatomic forces at the surface will introduce considerable distortions in the cluster. One way out of the dilemma is to use periodic boundary conditions [6.24] as is usually done in Monte Carlo or molecular dynamics calculations on disordered systems. Even if it has to be admitted that the periodic replication of the cluster introduces some elements of lattice symmetry in the results, the use of periodic boundary conditions is certainly an improvement over earlier calculations using free clusters as long as the interaction range is small compared to the dimensions of the cluster.

Finally it should perhaps be emphasized that the relaxation procedure generally does not maintain the topology of the original DRPHS model. It appears that the improvement achieved by relaxing the DRPHS model is mainly due to the change in the local structure and coordination [6.85, 87]. A topological analysis of the relaxed structure appears to be necessary before drawing any conclusions [6.88].

It is worth to mention another useful aspect of hard sphere systems: as shown in Sect. 6.4 they allow one to formulate certain approximate equations of state, which can serve as a starting point for the calculation of realistic systems. *Weeks* [6.89] has proposed that the analytical expressions for the partial structure factors of hard sphere mixtures (calculated in the Percus-Yevick or mean spherical approximations) should be a reasonable first approximation to the structure of metallic glasses.

b) Designed Models

The DRPHS models start from the similarity of the pair correlation functions in the liquid and in the amorphous states. While this remains obviously true for the long- and medium-range structure, the more recent diffraction experiments support the point of view that the short-range order in the amorphous phase may be similar to the crystalline structure. Thus, instead of building models by a random packing of individual atoms, one might begin with larger local units having the desired coordination and a topology borrowed from the crystalline structure. The model is built by a random stacking of those local units.

A model based on trigonal prismatic units was proposed by *Gaskell* for T–M glasses [6.15, 90]. The topology of the local units resembles the Fe_3C (cementite) structure. The model was relaxed using Lennard-Jones potentials, with some additional constraints conserving the type of coordination. The calculated interference function reproduces the measured neutron diffraction data very well.

Wang [6.22] proposed a model for the short-range structure of amorphous intertransition metals which is based on the random packing of interpenetrating Frank-Kasper coordination polyhedra. The model has several appealing features, since the same packing principle – tetrahedral close packing – is thought to apply to both the amorphous and the crystalline phases. However, it appears to be difficult to build in three dimensions, since it is impossible to pack polytetrahedral units together without serious stacking misfits at the boundaries of the local ordered units [6.91]. This is a difficulty that Wang's model shares with earlier models based on icosahedral packing [6.91–94]. The construction of these models is motivated mainly by three facts: i) For atoms interacting via attractive pair potentials it is established both theoretically and experimentally that the equilibrium structure of small microclusters is not identical to the bulk structure, but is rather based on the 13-atom icosahedron. ii) Structures produced by slow molecular dynamic cooling of droplets of Lennard-Jones fluids show icosahedral packing [6.93, 94]. iii) Icosahedral coordination is a very frequent local motif in the crystalline structures characterizing the glass-forming binary metallic systems. Though *Briant* and *Burton* [6.94] have shown that the interference functions of assemblies of the smaller of these microclusters possess the features characteristic for metallic glasses, it is still far from established that they really exist in amorphous bulk phases. An investigation of a realistic packing model in terms of the local coordination appears to be desirable.

6.5.2 The Structure of Simple-Metal Glasses

All structural models discussed above suffer from the fact that the interatomic potentials used to relax the starting model are known to be unrealistic. The only substances for which realistic pair potentials are known are the simple metals. Using the pair potentials described in Sect. 6.2, *von Heimendahl* [6.7, 24] has performed a cluster relaxation calculation for the $Mg_{0.7}Zn_{0.3}$ glass. The relaxation was started from a rhombic dodecahedron containing 800 atoms which was cut out of the center of Finney's DRPHS model [6.78]. Periodic boundary conditions were applied. The particular choice of the geometry was motivated by the fact it approximates a sphere most closely. The distribution of the two different kinds of atoms over the available atomic positions was assumed to be random. The relaxation procedure was essentially equivalent to the steepest descent algorithm (in the configuration space of all 2400 coordinates) described by *Hoare* [6.95]: the total force on atom i of type A in the position $\boldsymbol{r}_i$ is the sum over the pair interactions

$$\boldsymbol{F}_A(\boldsymbol{r}_i) = -\sum_{j(A)\neq i} \vec{\nabla} V_{AA}(|\boldsymbol{r}_j - \boldsymbol{r}_i|) - \sum_{j(B)} \vec{\nabla} V_{AB}(|\boldsymbol{r}_j - \boldsymbol{r}_i|); \tag{6.11}$$

an analogous expression holds for the forces on the B atoms. If the atom i is displaced by a vector $\delta\boldsymbol{r}_i$ from its position $\boldsymbol{r}_i$ while all other atoms are kept fixed at their positions, the force $\boldsymbol{F}_A(\boldsymbol{r}_i)$ changes and in a linear approximation we have

$$\boldsymbol{F}_A(\boldsymbol{r}_i+\delta\boldsymbol{r}_i)=\boldsymbol{F}_A(\boldsymbol{r}_i)+(\delta\boldsymbol{r}_i\cdot\vec{\nabla})\boldsymbol{F}_A(\boldsymbol{r}_i). \tag{6.12}$$

With the same constraint, the force-free position of atom i is given by

$$\boldsymbol{F}_A(\boldsymbol{r}_i+\delta\boldsymbol{r}_i)=0 \tag{6.13}$$

which yields the 3×3 system of linear equations for the displacement vector $\delta\boldsymbol{r}_i$,

$$-\boldsymbol{F}_A(\boldsymbol{r}_i)=(\delta\boldsymbol{r}_i\cdot\vec{\nabla})\boldsymbol{F}_A(\boldsymbol{r}_i). \tag{6.14}$$

In each relaxation run the displacement vectors $\delta\boldsymbol{r}_i$ of all atoms are calculated and stored. Then new positions $\boldsymbol{r}_i'$ are calculated by adding a fraction f of the $\delta\boldsymbol{r}_i$ to $\boldsymbol{r}_i$

$$\boldsymbol{r}_i'=\boldsymbol{r}_i+f\cdot\delta\boldsymbol{r}_i. \tag{6.15}$$

Starting from the new positions the procedure is repeated until the last set of $\delta\boldsymbol{r}_i$ is negligible. Interatomic interactions over a radius of $r=6.68$ Å have been taken into account. This distance corresponds to the position of the minimum after the split second peak in the total pair distribution function. In the average each particle interacts with 57 neighbors. The interaction radius is still comparatively small compared to the smallest radius of the cluster ($r=14.65$ Å).

The distorting forces on all atoms together form a multidimensional force $\boldsymbol{F}(\boldsymbol{r}^N)=\vec{\nabla}[\phi(\boldsymbol{r}^N)]$, ϕ being the configurational potential energy E_{st} [see (6.1)]. Following the direction of $\boldsymbol{F}(\boldsymbol{r}^N)$ by stepwise computation, each atom will move continuously, leading to a unique minimum $\boldsymbol{r}^N_{min}$ where the gradient $\vec{\nabla}[\phi(\boldsymbol{r}^N)]$ vanishes. In this way each point $\boldsymbol{r}^N$ (i.e., each starting configuration) is uniquely associated with a minimum in $\phi(\boldsymbol{r}^N)$ via a steepest descent path, but it is not sure that there is not another minimum with an even lower potential energy which is connected with $\boldsymbol{r}^N$ only via a nonsteepest descent line (e.g., crossing a saddle point of the potential energy surface). Using the present relaxation algorithm, such a minimum will be accessible only from a different starting configuration.

In Fig. 6.7 the total and two partial pair distribution functions (PDFs) are shown together with the corresponding interatomic pair potentials. For large distances, the PDFs approach the total and the two partial densities, respectively. Due to the smaller number of particles, the Zn PDF shows more fluctuations than the two others. The positions of the first peaks in the PDFs agree well with the first minima in the pair potentials. This one-to-one correspondence extends over the three to four nearest-neighbor shells; for larger distances pair potentials and PDFs go gradually "out of phase". This will be an important element in the discussion of the stability of the amorphous state.

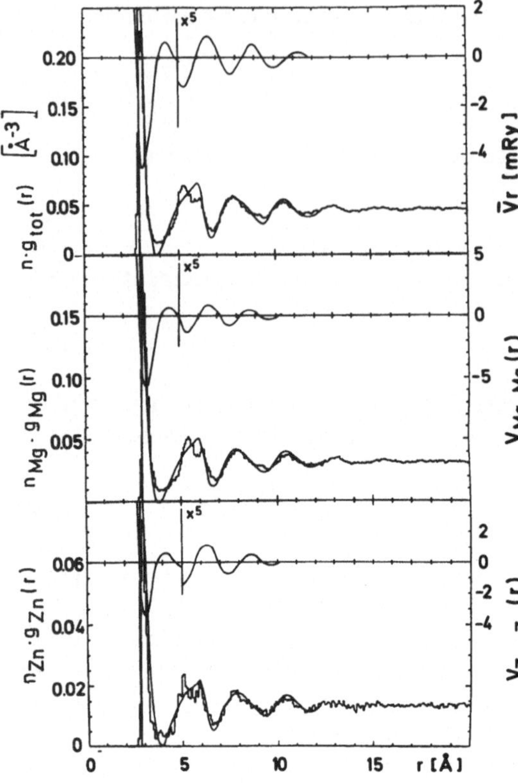

Fig. 6.7. Total and partial pair distribution functions $g_{ij}(r)$ for the $Mg_{0.7}Zn_{0.3}$ glass. The histogram shows the result of the cluster relaxation calculation, the continuous line the result of the thermodynamic variational technique ($T = 25\,°C$). The average and the partial interatomic potentials $V_{ij}(r)$ are shown on a displaced scale (right hand) [6.7]

The splitting of the second peak in the partial PDFs is very distinct. It is somewhat smeared out in the total PDF. The second peaks in both the total and the partial structure factors are split in the same way as for the PDFs, which is surprising since it is not a simple consequence of the Fourier transform. Both features suggest a relatively high degree of topological (but not substitutional) short-range order.

Neutron and x-ray diffraction experiments have been performed by *Rudin* [6.96]. They are not sufficient to resolve the partial structure factors. The composite interference function $S(q)$ may be expressed in terms of the *Ashcroft-Langreth* [6.54], or the *Bhatia-Thornton* [6.97] structure factors as (see, e.g., *Wagner* [6.98] and *Chieux* [6.98a])

$$\begin{aligned} S(q) = &\frac{1}{\langle f(q)\rangle^2}[c_A f_A(q)^2 S_{AA}(q) + c_B f_B(q)^2 S_{BB}(q) \\ &+ 2(c_A c_B)^{1/2} f_A(q) f_B(q) S_{AB}(q) + \langle f(q)\rangle^2 - \langle f(q)^2\rangle] \\ = &\frac{1}{\langle f(q)\rangle^2}[\langle f(q)\rangle^2 S_{NN}(q) + 2\langle f(q)\rangle \Delta f(q) S_{NC}(q) \\ &+ \Delta f(q)^2 S_{CC}(q) - c_A c_B \Delta f(q)^2], \end{aligned} \tag{6.16}$$

where $f_A(q)$ and $f_B(q)$ are the atomic scattering form factors. For the case of $Mg_{0.7}Zn_{0.3}$ we get

$$\begin{aligned} S^X(q) &= 0.33\,S_{\mathrm{MgMg}}(q) + 0.89\,S_{\mathrm{ZnZn}}(q) + 1.09\,S_{\mathrm{ZnMg}}(q) - 0.22 \\ &= S_{NN}(q) - 2.07\,S_{NC}(q) + 1.07\,S_{CC}(q) - 0.22 \end{aligned} \tag{6.17}$$

for x-ray scattering and

$$\begin{aligned} S^N(q) &= 0.66\,S_{\mathrm{MgMg}}(q) + 0.34\,S_{\mathrm{ZnZn}}(q) + 0.95\,S_{\mathrm{ZnMg}}(q) - 0.002 \\ &= S_{NN}(q) - 0.18\,S_{NC}(q) + 0.008\,S_{CC}(q) - 0.002 \end{aligned} \tag{6.18}$$

for neutron scattering. Equation (6.18) is exact, while in (6.17) we have approximated the q-dependent x-ray form factors by their values for $q=0$. In this approximation (Warren-Krutter-Morningstar approximation [6.98]), we can express the composite pair correlation functions in terms of the partial pair correlation functions as

$$\begin{aligned} g(r) &= \frac{1}{\langle f\rangle^2}\left[c_A^2 f_A^2 g_{AA}(r) + c_B^2 f_B^2 g_{BB}(r) + 2c_A c_B f_A f_B g_{AB}(r)\right] \\ &= g_{NN}(r) + \frac{2\langle f\rangle \Delta f}{\langle f\rangle^2} g_{NC}(r) + \frac{c_A c_B \Delta f^2}{\langle f\rangle^2} g_{CC}(r) \end{aligned} \tag{6.19}$$

and specifically for x-rays

$$\begin{aligned} g^X(r) &= 0.23\,g_{\mathrm{MgMg}}(r) + 0.27\,g_{\mathrm{ZnZn}}(r) + 0.50\,g_{\mathrm{MgZn}}(r) \\ &= g_{NN}(r) - 2.07\,g_{NC}(r) + 0.22\,g_{CC}(r) \end{aligned} \tag{6.20}$$

and for neutrons

$$\begin{aligned} g^N(r) &= 0.46\,g_{\mathrm{MgMg}}(r) + 0.10\,g_{\mathrm{ZnZn}}(r) + 0.44\,g_{\mathrm{MgZn}}(r) \\ &= g_{NN}(r) - 0.18\,g_{NC}(r) + 0.002\,g_{CC}(r)\,. \end{aligned} \tag{6.21}$$

The pair correlation functions resulting from the static cluster relaxation calculation were folded with a gaussian broadening function to simulate thermal disorder and to obtain nearest-neighbor peak widths similar to those found experimentally. The theoretical and experimental results for the composite interference functions and the composite radial distribution functions are shown in Figs. 6.8 and 6.9. In making the comparison we have to remember that this is a true first-principles calculation. Hence we should not expect the same kind of agreement as in other modelling studies where atomic sizes, "tetrahedron perfection", potential parameters, etc., are optimized in order to reproduce the measured diffraction data. Nevertheless, the agreement of the calculated neutron interference function and radial distribution function is

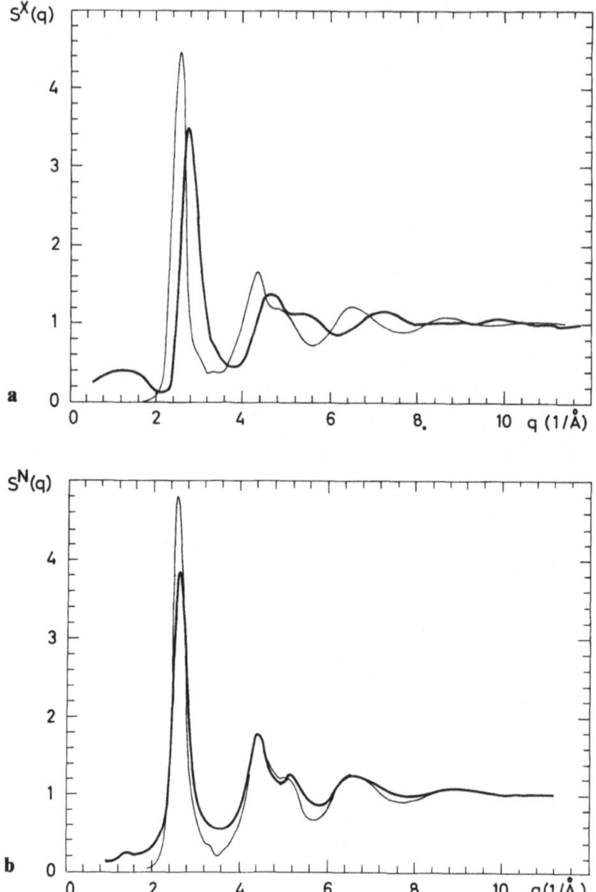

Fig. 6.8a, b. Composite interference functions for amorphous $Mg_{0.7}Zn_{0.3}$. **(a)** For neutron diffraction, **(b)** for x-ray diffraction. The thick lines show the experimental result of *Rudin* [6.96], the thin lines the theoretical model

excellent; every single structural detail (see the structure in the higher order peaks of the reduced radial distribution function, Fig. 6.10) is reproduced. The oscillations are only less strongly damped and there is a slight "phase shift" in the calculated distribution function. Both features point to the fact that though the overall topology seems to be correctly described by the model, the mean interatomic distances are slightly overestimated and the size difference is slightly underestimated. This is corroborated by analyzing the x-ray results, where the agreement with experiment is much poorer. Considering that the main difference between the x-ray and the neutron results is the increased weight of the Zn–Zn correlations (6.17–21), we are led to the conclusion that our calculations yield Mg–Mg distances in good agreement with experiment,

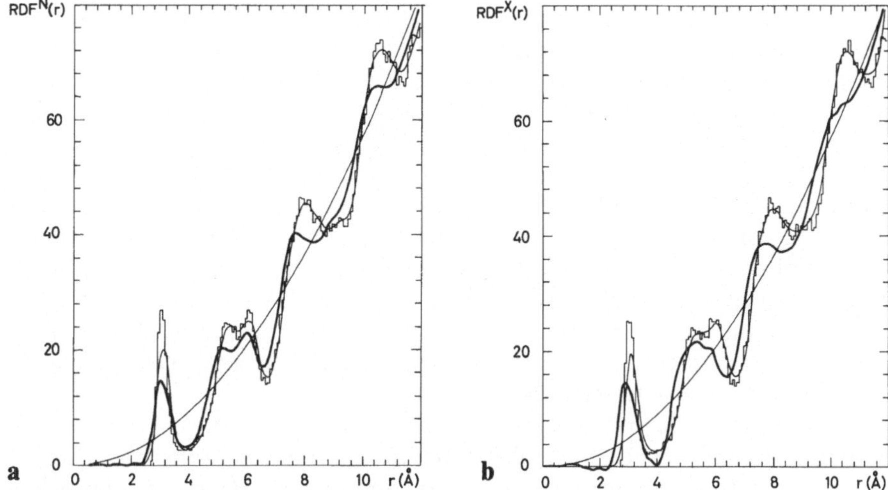

Fig. 6.9a, b. Composite radial distribution functions RDF(r) for amorphous $Mg_{0.7}Zn_{0.3}$. **(a)** Neutron weighted, **(b)** x-ray weighted. The thick lines show the experimental RDF(r) of *Rudin* [6.96], the histogram the result of the static cluster relaxation calculation; this RDF has been folded with a gaussian broadening function (thin line)

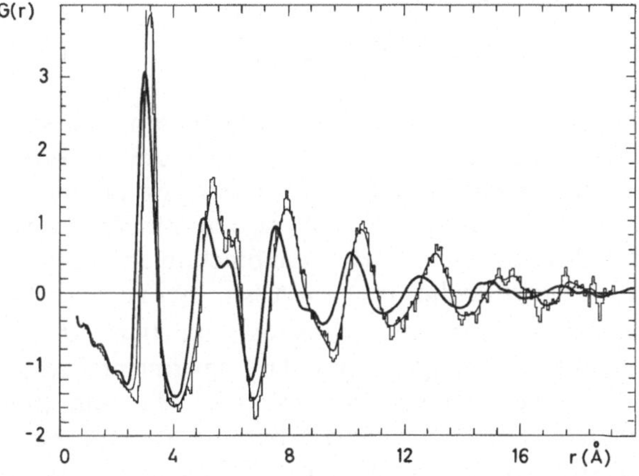

Fig. 6.10. Composite (neutron-weighted) reduced radial distribution function $G(r)$ for amorphous $Mg_{0.7}Zn_{0.3}$ (cf. Fig. 6.9)

whereas the Zn–Zn distances seem to be overestimated by about 0.12–0.15 Å, possibly due to the neglect of $s-d$ hybridization (cf. Sect. 6.3.2 and [6.8]). Another feature which is not reproduced by the model is the rather strong prepeak in the x-ray interference function. According to the most recent experiments of *Rudin* [6.96] the prepeak is definitely an intrinsic feature of the structure of amorphous $Mg_{0.7}Zn_{0.3}$. In order to understand the effect of the prepeak, we Fourier transformed $S^X(q)$ once more, this time with the prepeak removed as sketched in Fig. 6.8a. We find that the effect of the prepeak upon the distribution functions is rather weak; it modifies only the first and second peaks while beyond $r=8$ Å the distribution function is left unchanged. Due to the prepeak, the intensity of the first peak in $g^X(r)$ is reduced by $\sim 4\%$, that of the second peak is increased by $\sim 2\%$. Very recently, a similar but much stronger prepeak and similar changes in the distribution functions of amorphous $Ti_{0.34}Cu_{0.66}$ have been found by *Sakata* et al. [6.74] using neutron diffraction. In the case of Mg–Zn the prepeak is extremely weak in the neutron interference function; hence we conclude that it is almost entirely due to Zn–Zn correlations. On this evidence we can associate the prepeak with a weak tendency to chemical ordering between Zn and Mg: the Zn atoms try to avoid homocoordination on nearest-neighbor sites. This is in agreement with packing considerations: packing efficiency requires that direct contacts between the smaller minority atoms are to be avoided.

An analysis of the partial coordination numbers in the model cluster before and after relaxation shows that the statistical distribution of the two different kinds of atoms is not modified by the relaxation process. Hence an appropriate description of the prepeak requires that some degree of chemical ordering is incorporated in the starting structure.

Some interesting topological details may be found by considering the difference

$$\begin{aligned} g^X(r)-g^N(r) &= -0.89\,g_{NC}(r)+0.22\,g_{CC}(r) \\ &= -0.23\,g_{\mathrm{MgMg}}(r)+0.17\,g_{\mathrm{ZnZn}}(r)+0.06\,g_{\mathrm{MgZn}}(r)\,. \end{aligned} \tag{6.22}$$

Since $g_{CC}(r)$ is very small, this difference is due mainly to the number density-concentration correlation $g_{NC}(r)$. We see (Fig. 6.11) that the theoretical result qualitatively reproduces the experimental result, apart from a few additional features: there is a strong negative peak ar $r\sim 1.4$ average atomic diameters. The interrelation between details in the pair distribution function and possible local configurations in DRPHS networks has been discussed by *Finney* [6.78] and by *Bennett* [6.79]. Contributions to the correlation functions at $\sqrt{2}$ diameters arise from local octahedral configurations. The experimental result suggests that a certain number of such configurations exists for Mg atoms (but they are not more frequent than in DRPHS structures), but they are suppressed for Zn atoms. At the present stage any further detailed discussion of possible topological details of the amorphous $Mg_{0.7}Zn_{0.3}$ structure would be purely speculative.

It is also interesting to note that the main features of the structure (i.e., the intensity, position, and shape of the oscillations in the distribution functions) are quite well reproduced by an analytical hard sphere calculation in the Percus-Yevick approximation. The hard sphere diameters are determined via the Gibbs-Bogoljubov variational method for a hypothetical supercooled liquid alloy at room temperature (Fig. 6.7). This result is important for two reasons: i) it demonstrates that except for some degree of topological short-

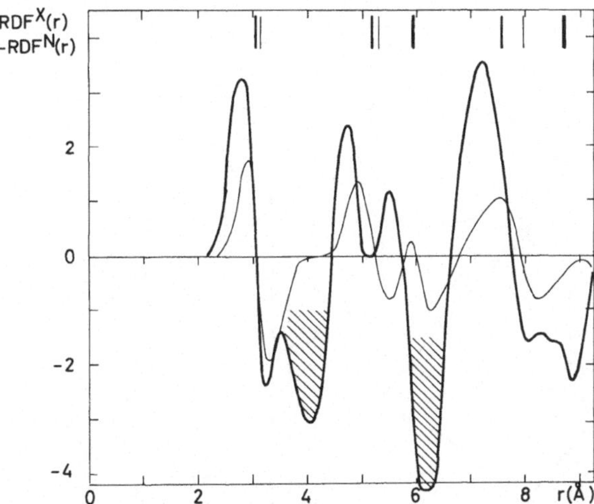

Fig. 6.11. Difference between the composite radial distribution function for x-ray and neutron scattering. Thick line, experiment; thin line, theory. The hatched areas emphasize the main qualitative differences

range order (which is manifest through the splitting of the higher order peaks as discussed above), the structure of the glass is identical to that of the supercooled melt and ii) it allows one to study the interrelation between the PDFs and the pair potentials without the expense of a full cluster relaxation calculation. In Figs. 6.4 and 6.5 it is shown that for Ca–Mg and Ca–Zn the pair potentials and the PDFs are well "in phase" over the whole range of nonnegligible interatomic forces. For these systems, this is true for all concentrations, while for Ca–Al the relation holds only for Ca-rich alloys. As the Al content is increased the maxima of the PDFs move out of the minima of the pair potentials for second and higher neighbors [6.8]. Hence it is clear that the same physical mechanism – optimal adaption of the interatomic distances to the minima of the interatomic potentials – determines the atomic structure and the stability of crystalline, liquid, and amorphous phases.

6.6 The Stability of Amorphous Alloys

In the absence of any possibility of predicting realistic quantitative values for the free enthalpy of formation of the supercooled liquid, the existing glass formation models [6.29–32] are necessarily mainly speculative. Now the *Gibbs-Bogoljubov* method allows us to make at least an estimate of the thermodynamic properties of the supercooled liquid state. The results for Ca–Mg at room temperature are shown in Fig. 6.12. They are similar to the results for the stable liquid alloy (Sect. 6.4): ΔH and $\Delta\Omega$ are assymmetric due

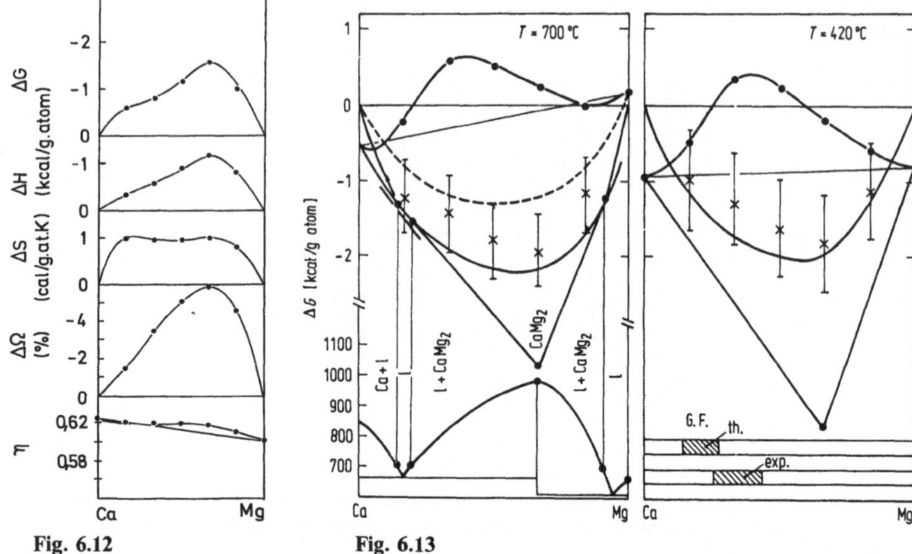

Fig. 6.12 **Fig. 6.13**

Fig. 6.12. Free enthalpy ΔG, enthalpy ΔH, entropy ΔS, volume $\Delta\Omega$ of formation, and packing fraction η of supercooled liquid ($\sim$glassy) Ca–Mg alloys at $T = 25\,°C$ (relative to the supercooled pure metals) [6.8]

Fig. 6.13. Free enthalpy of formation ΔG (relative to the pure liquid, or supercooled liquid metals) of Ca–Mg alloys at different temperatures: ● solid solution, × liquid solution, ⊙ Laves phase. The error bars represent the estimated uncertainty, the solid line is a tentative interpolation. The common tangent construction for the construction of the phase diagram and of the glass-forming region is indicated [6.8]

to packing effects, ΔS deviates substantially from the ideal (parabolic) form. The main point is that there is no minimum in ΔG in the eutectic region. It is interesting to note that the ΔG curve is even concave at $\sim$35 at.% Mg. If this is not a spurious effect it means that the supercooled liquid tends to phase separate into liquids with $\sim$15 at.% and $\sim$50 at.% Mg. *St. Amand* and *Giessen* [6.16] noted that "in the Ca–Mg system, heating may result in phase separation of a single glass into two glasses". This point deserves further investigation.

The calculated equilibrium density of the supercooled $Ca_{0.67}Mg_{0.33}$ melt ($\varrho = 1.609\ g/cm^3$) is only 2% higher than the experimental density of the amorphous $Ca_{0.7}Mg_{0.3}$ alloy ($\varrho = 1.579\ g/cm^3$ [6.99]). As pointed out by *Chen* and *Jackson* [6.100], the density of an amorphous metal depends on its thermal history. The thermal expansion coefficient changes discontinuously at the glass transition temperature, which in turn depends on the cooling rate. Thus our calculated density is rather an upper bound (though apparently a very realistic one) to the density of the amorphous alloy.

If we supplement the thermodynamic calculations for the liquid phase with a similar thermodynamic variational calculation for the solid state [6.101, 102], we can calculate temperature-dependent $\Delta G(c)$ curves for both the solid and liquid phases, which is all we need for a first-principles construction of the phase diagram. Such $\Delta G(c, T)$ diagrams for Ca–Mg (based on an even simpler estimate of the solid-state thermodynamic properties [6.8]) are shown in Fig. 6.13. At a given temperature the limits of stability of each phase are determined by the well-known common tangent construction. In this way it is possible to derive a fully realistic phase diagram. The melting temperature of the Laves phase is somewhat overestimated, since the theoretical ΔG is lower than the experimental one for the Laves phase and vice versa for the liquid phase (Sects. 6.3.1 and 6.4). The picture demonstrates very clearly in which way the eutectic composition and temperature are determined by the relative magnitudes of ΔG for the competing liquid and crystalline phases.

A similar $\Delta G(c, T)$ plot for a temperature below the lowest eutectic shows that there is a limited concentration range in which ΔG of the supercooled liquid is very close to the straight line representing the stable ($Ca+CaMg_2$) two-phase mixture (in interpolating the ΔG curve we ignore for the moment the slight inflection at ~35 at.% Mg). Since the enthalpies of formation for possible crystalline phases of this composition are even higher (Sect. 6.3.3), the supercooled liquid has a lower free enthalpy than any single crystalline phase of this composition. If the system is constrained against phase separation (this is now a kinetic condition), glass formation will occur with great ease.

That the constitutional conditions resulting in eutectic formation tend to enhance the glass-forming tendency has been the basis of most glass-forming models. They only differ in that these conditions are attributed either to the stabilization of the liquid relative to the crystalline mixture (which in turn is explained by packing considerations as in the Polk model [6.29] or by electronic effects as in the Nagel-Tauc model [6.30]) or to a destabilization of the crystalline state at glass-forming compositions [6.31]. Our quantitative approach demonstrates that there is some truth in all of these models.

The energetics of phase stabilization and destabilization may be understood in terms of the pair potential approach. The cornerstone of our theory is the existence of an oscillating effective pair potential which is attractive for all neighbors within the range of nonnegligible interatomic forces. For disordered structures, this is expressed by the "constructive interference" between pair potentials and pair distribution functions (PDFs). Inspired by the well-known asymptotic behavior of the oscillations in the pair potentials

$$V_{ij}(r) \to A\frac{\cos(2k_F r)}{r^3} \tag{6.23}$$

and in the pair distribution functions

$$[g_{ij}(r)-1] \to B\frac{\sin(Q_{ij}^p r)}{r} \tag{6.24}$$

[k_F is the Fermi wave vector and the Q^p_{ij} are the wave vectors where the partial structure factors $S_{ij}(q)$ have their first peak], *Beck* and *Oberle* [6.103] suggested a pair potential interpretation of the *Nagel-Tauc* rule [6.30] which says

$$Q^p_{ij} \approx Q^p \approx 2k_F . \tag{6.25}$$

Nagel and *Tauc*'s original explanation of this rule (which is indeed met by a large number of amorphous alloys) is built in analogy to the Mott-Jones formulation of the Hume-Rothery rules for crystalline alloys: Hume-Rothery's critical valence electron concentrations (VECs) correspond to the case of a Jones zone plane touching the Fermi sphere, i.e., $|\boldsymbol{Q}| = 2k_F$, where $\boldsymbol{Q}$ is now a vector of the reciprocal lattice. In this case a gap opens at $\boldsymbol{Q}/2$ and produces a minimum in the density of states. In analogy to this situation *Nagel* and *Tauc* proposed that in the disordered state there should be a "pseudogap" which is manifest through a minimum in the density of states. This pseudogap must be isotropic and should lower the electronic energy compared to crystalline structures with anisotropically distributed gaps. However, recent photoemission experiments [6.104, 105] failed to confirm the postulated s-like character of the electronic density of states of $Pd_{0.8}Si_{0.2}$ near the Fermi level, and low-temperature specific heat measurements [6.106] on amorphous Pd–Si–Cu and Pd–Si alloys failed to confirm the minimum in the density of states. In this situation, *Beck* and *Oberle*'s conjecture seems to offer an alternative explanation of the *Nagel-Tauc* rule. However, this explanation considers only the VEC effect and neglects the size factor. It is well established that only alloys with a size ratio that is appreciably different from unity show a good glass-forming tendency. The Q^p_{ij} depend on the effective atomic sizes; they are related to the effective hard sphere diameters through a phenomenological relation due to *Blétry* [6.107]

$$Q^p_{ij} = \frac{1}{\sigma_{ij}}[7.64 - 4.32(\bar{\sigma}/\sigma_{ij} - 1)] \tag{6.26}$$

with the average diameter $\bar{\sigma} = c_A\sigma_A + c_B\sigma_B$. Now it is immediately clear that for different Q^p_{ij}'s, the three partial PDFs cannot oscillate in phase in the asymptotic region. On the other side, the electronic screening function is the same for all three interatomic potentials and the V_{ij}'s do oscillate in phase in the asymptotic region. Hence a "constructive interference" between $g_{ij}(r)$ and the corresponding $V_{ij}(r)$ over large distances is impossible. Anyway, the matching between the pair potentials and the PDFs is much more important for short and intermediate distances where both functions deviate quite substantially from their asymptotic forms. The existence of this relation is seen to be a more complicated function of the traditional alloy chemical factors: i) size ratio, ii) strong chemical bonding (charge transfer and screening), iii) valence electron concentration. The action of these three factors upon the stability of the glassy state is closely interconnected.

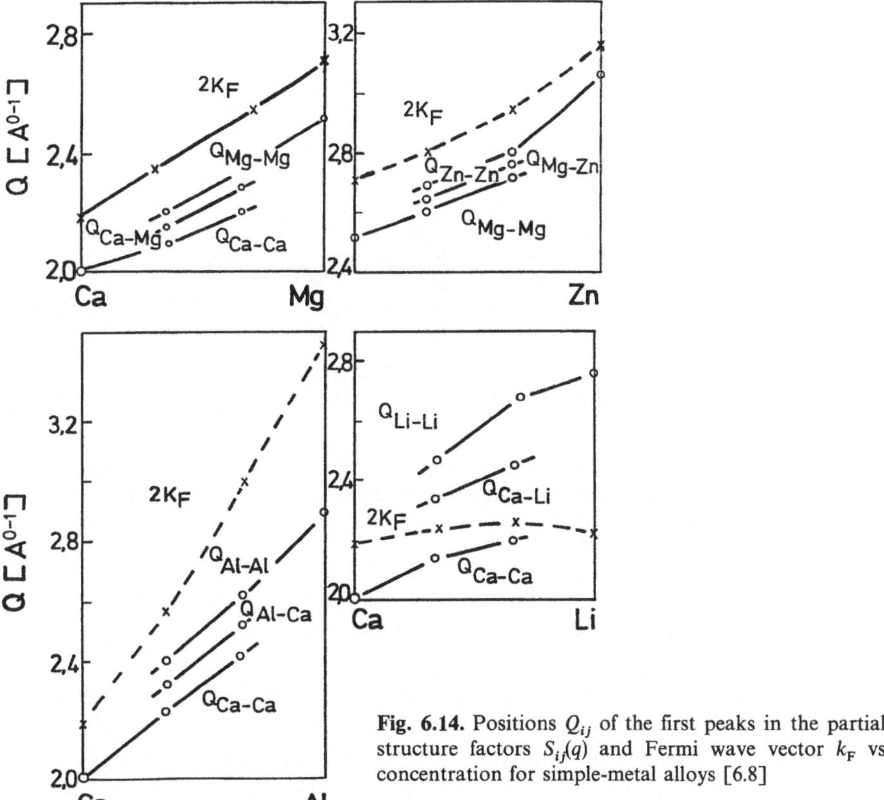

Fig. 6.14. Positions Q_{ij} of the first peaks in the partial structure factors $S_{ij}(q)$ and Fermi wave vector k_F vs concentration for simple-metal alloys [6.8]

The *Nagel-Tauc* rule for amorphous simple-metal alloys is examined in Fig. 6.14. We find that the alloy which fulfills this relation most closely is Ca–Li, the only one which does not form a glass. The reason is that the large effective size ratio caused by electronic effects (Sect. 6.2.2) prohibits any energetically favorable interrelation between the PDFs and the pair potentials for second and higher neighbors [6.8]. For the Ca-based amorphous alloys the importance of the VEC has been critically discussed by *Giessen* et al. [6.16]. They showed that the glass transition temperature (and hence the thermal stability) is lowest for the divalent alloys for which the Nagel-Tauc rule predicts the maximum stability (Fig. 6.15a). Evidently the amorphous alloys with a higher (Ca–Al, Ca–Ga) or a lower (Ca–Cu, Ca–Ag, Ca–Au, Ca–Pd) VEC are stabilized by strong chemical bonding effects arising from large differences in both electronegativity and electron density of the components. An explicit demonstration of these effects has been given here at the example of Ca–Al, and we expect the effects to be quite similar for the Ca-noble metal alloys. The VEC is important but since (counting only *s* electrons) it can vary only from 1 to 4

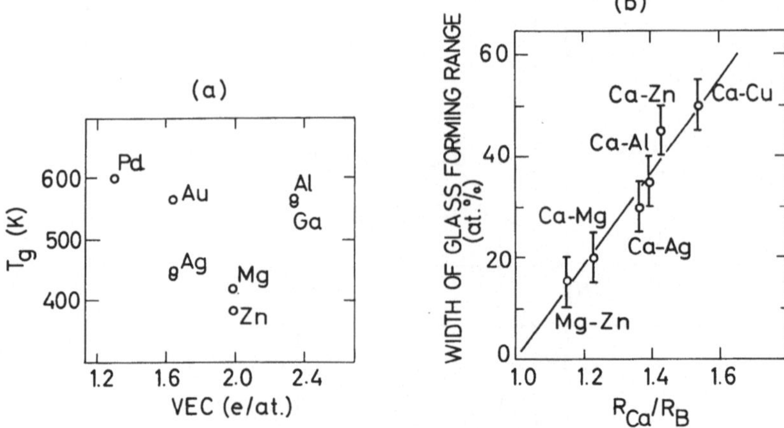

Fig. 6.15. (**a**) Glass transition temperature T_g vs valence electron concentration VEC and (**b**) width of the glass-forming range vs the ratio of the Goldschmidt radii of the components for several amorphous simple metal alloys. (After [6.16])

(and k_F varies only with the cube root of the VEC), it is not a very sharp criterion.

St. Amand and *Giessen* [6.16] stressed the importance of the size ratio. Again at the example of the Ca-based alloys they demonstrated that the width of the glass-forming region (and hence the easiness of glass formation) scales with the nominal component size ratio (Fig. 6.15b). However, the use of the nominal size ratio is misleading. Our results show convincingly that the effective size ratio of Ca–Al and Ca–Zn is much smaller (Figs. 6.3 and 6.5) and that easy glass formation requires an effective size ratio between 1.10 and 1.30. The effect of a large nominal component size ratio is rather indirect: the inspection of the phase diagrams [6.23] shows that with an increasing nominal R_A/R_B the stoichiometry of the highest melting intermetallic compound (i.e., the one with the highest stability relative to the liquid state) is shifted to higher B concentrations (e.g., from $CaMg_2$ in Ca–Mg to $CaCu_5$ in Ca–Cu or to $CaZn_{11}$ in Ca–Zn; compare Fig. 6.2). At the example of Ca–Zn Fig. 6.5 demonstrates in which way the concentration-dependent variation of the effective pair potentials (which is of course possible only if the atomic diameters in the pure metals are sufficiently different) allows for the formation of a large number of intermetallic compounds with comparable thermal stability. The $\Delta G(c)$ curve of the disordered state will be somewhat asymmetric and nonparabolic, similar to what we have seen at the example of Ca–Mg. Schematically, the $\Delta G(c)$ curve we expect for this situation is sketched in Fig. 6.16 (the estimate of the ΔG-values for the solid phases is based on the thermochemical data of [6.108]). We see that the curves for the disordered and the stable crystalline phases are very close over a large concentration range,

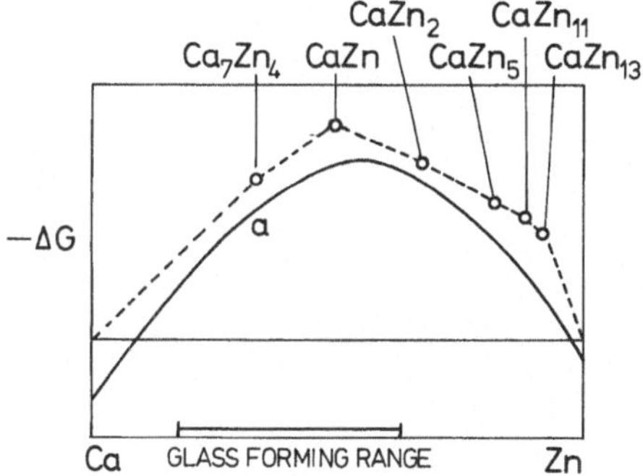

Fig. 6.16. Schematic of free enthalpy of formation ΔG for the Ca–Zn system

allowing a wide glass-forming region without however yielding necessarily a high thermal stability.

This brings us back to the interrelation between crystalline structures and glass formation. We have shown that the common principle underlying the formation of the *B*-metal-rich crystalline, the liquid, and the amorphous phases is tetrahedral close packing. This will always be more effective on a regular crystalline than on a disordered array. Thus, on the *B*-metal-rich side of the constitution diagram the configurational energy distinctly favors the crystalline state. Moreover in this concentration range the partial coordinations are similar in the crystalline and in the amorphous states [6.8]. Hence the potential energy barrier to nucleation of the crystalline structure will be comparatively low. This may be the reason why there is no glass formation in the *B*-metal-rich eutectic. On the other side the *A*-metal-rich crystalline structures are definitely not tetrahedrally close packed; their configurational energy is comparable to that of the amorphous phase (Sect. 6.3.3). Furthermore, their partial coordinations are definitely different from those of the amorphous phase [6.8]. This means that the nucleation of a crystalline structure requires a diffusive displacement of many atoms, yielding a much higher potential barrier.

We see that our simple pair potential argument, combined with quantitative microscopic calculations, brings considerable progress over the traditional, too oversimplified glass-formation models. One interesting aspect is that the theory emphasizes all the well-known alloy chemical factors (size ratio, chemical bonding, VEC) that are known to be important in the discussion of the crystalline intermetallic phases. Technically the application of the pair potential argument is restricted to simple-metal systems, but its simplicity makes its application to other systems very appealing. A nice example for the validity of the argument in more complicated systems has recently been given by *Leitz* and *Buckel* [6.109]. They showed that for quench-condensed amorphous Sb–Au films the position of the first few peaks in the pair distribution function coincides with the minima of the pair potentials if it is assumed that the first maximum in $g(r)$ coincides with the first minimum in $V(r)$ (that is where the size factor enters) and that the period of the oscillations in $V(r)$ is nearly free-

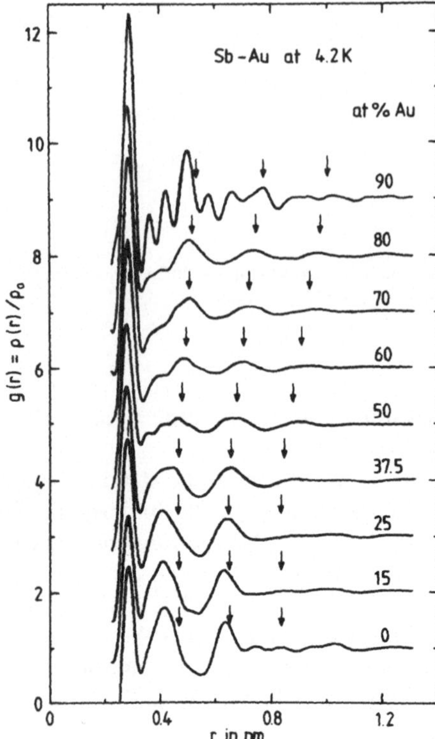

Fig. 6.17. Composite (x-ray weighted) pair distribution functions $g(r)$ for amorphous Sb–Au alloys. The arrows indicate the estimated positions of the minima in the pair potentials. (After [6.109])

electron-like, i.e., π/k_F (Fig. 6.17). k_F is calculated in a free-electron approximation (which is justified on the basis of electronic transport measurements made in [6.110]). The matching between the pair potentials and the pair distribution functions extends over the whole glass-forming region (37.5 to 80 at.% Au). It is valid even at Sb concentrations where the Nagel-Tauc rule is definitely violated.

This offers some hope that the simple pair potential argument will be helpful not only for simple-metal alloys but also for a more complete understanding of alloys containing transition metals and metalloids.

6.7 The Dynamical Properties of Amorphous Metals

The dynamical properties of amorphous substances are of particular interest because of a possible interrelation between the low-temperature properties and dynamical excitations at low energy. Considerable progress has been made in recent years in the understanding of the vibrational properties of amorphous semiconductors and oxide glasses [6.111]. Much of this has been stimulated by the wealth of experimental information offered by optical spectroscopy. Similarly exciting experimental results are still lacking for metallic glasses and the progress in the theory of their dynamical properties has been comparatively slow.

In amorphous alloys (as in all disordered materials) there is no conservation of crystal momentum to help us in the theoretical treatment of their vibrational properties. In this section we shall discuss different methods to overcome this difficulty: i) the equation-of-motion method, ii) the molecular dynamics approach, and iii) the continued fraction expansions. Finally recent theoretical results for the dynamical structure factors of amorphous $Mg_{0.7}Zn_{0.3}$ will be reviewed.

6.7.1 The Equation-of-Motion Method

The equation-of-motion method [6.112, 113] is a very efficient technique to calculate the properties of systems whose Hamiltonian is quadratic in the dynamical variables. Such a Hamiltonian may be transformed to normal coordinates, e.g., the displacements $\boldsymbol{u}_i(t)$ of the atoms of a system of N particles may be written as

$$\boldsymbol{u}_i(t)=m_i^{-1/2}\sum_k \boldsymbol{e}_{ki}[A_k\cos(\omega_k t)+B_k\sin(\omega_k t)/\omega_k], \tag{6.27}$$

where m_i denotes the mass of the atom in the equilibrium position $\boldsymbol{r}_i$. The kth normal mode is characterized by the frequency ω_k and the polarization vector $\boldsymbol{e}_{ki}$. ω_k and $\boldsymbol{e}_{ki}$ contain all the dynamical information on the system and can in principle be calculated by diagonalizing the Hamiltonian. For ordered systems this is a very advantageous procedure, since the lattice periodicity allows one to reduce the problem of diagonalizing the $3N\times 3N$ force constant matrix to the repeated diagonalization of the $3r\times 3r$ (r is the number of particles in the unit cell) dynamical matrix. For disordered systems this is impossible and the direct diagonalization of the full force constant matrix is of course very inefficient. The equation-of-motion method starts from the Newtonian equation of motion

$$\ddot{\boldsymbol{u}}_i(t)=m_i^{-1}\boldsymbol{F}_i(t), \tag{6.28}$$

the force $\boldsymbol{F}_i(t)$ on the ith particle being given by (6.11). It is clear that only nonzero force constants contribute to $\boldsymbol{F}_i$ and that the sparseness of the force constant matrix (which arises from the finite range of the interatomic forces) is fully exploited. The integration of (6.28) is done numerically after an appropriate choice of the initial conditions $\boldsymbol{u}_i(0)$ and $\dot{\boldsymbol{u}}_i(0)$. In the simplest form this is done by using the difference equation

$$\boldsymbol{u}_i(t+\Delta)=\boldsymbol{u}_i(t-\Delta)-2\boldsymbol{u}_i(t)+\Delta^2 m_i^{-1}\boldsymbol{F}_i(t)+O(\Delta^4) \tag{6.29}$$

with the time step Δ; in practice more efficient integration formulae are used [6.113]. The knowledge of the $\boldsymbol{u}_i(t)$ is more than we really need to know on the vibrations of the system. Interesting quantities such as the vibrational density of states or the dynamical structure factor may be computed directly.

a) The Vibrational Density of States

The vibrational density of states projected onto a given initial state $\boldsymbol{u}_i(0)$ of the system is given by the following integral representation [6.114]:

$$g(\omega)=\pi^{-1}\int_0^\infty \sum_i \boldsymbol{u}_i(0)\cdot\boldsymbol{u}_i(t)\cos(\omega t)dt\,. \tag{6.30}$$

Equations (6.28) and (6.30) are evaluated numerically. In an actual calculation the integration in (6.30) has to be truncated at a finite T and a damping function has to be introduced in order to eliminate unwanted oscillations [6.113, 114]. In the first application of the method to model systems representing amorphous metals, *von Heimendahl* and *Thorpe* [6.115] considered the effect of topological disorder alone, i.e., the variation of the interatomic forces with the interatomic distance was neglected. Their calculated density of states turned out to be very similar to that of crystalline close-packed metals (Fig. 6.18a), with separate peaks for longitudinal- and transverse-type models. The effect of quantitative disorder was first investigated by *Rehr* and *Alben* [6.114]. This is an aspect which is typical for amorphous metals, since metallic interatomic potentials are known to vary rapidly over the width of the nearest-neighbor peak in the pair distribution function. *Rehr* and *Alben* concluded that quantitative disorder almost completely destroys the structure in the vibrational density of states. The effect depends on the amount of quantitative disorder; it is stronger for a Lennard-Jones- than for a Morse-type potential because the force constant derived from the former shows a larger variation (Fig. 6.18).

Inelastic neutron scattering results of *Suck* and *Rudin* in [6.116] on $Cu_{0.46}Zr_{0.54}$ alloys have shown that the vibrational density of states $g(\omega)$ at low energies is higher for the amorphous than for the crystalline phase and possibly varies linearly with ω. The former is corroborated by the result of the model calculations, but the suggested linear variation at low ω is not reproduced by the calculation.

b) The Dynamical Structure Factor

For disordered binary systems, the double-differential coherent one-phonon scattering cross section for neutrons may be written as

$$\frac{d^2\sigma}{d\omega d\Omega}=N\frac{k_\mathrm{f}}{k_\mathrm{i}}\sum_{\alpha,\beta=A,B} a_\alpha a_\beta S^{(1)}_{\alpha\beta}(\boldsymbol{q},\omega)\,. \tag{6.31}$$

Here $\hbar k_\mathrm{i}$ and $\hbar k_\mathrm{f}$ are the initial and final momenta of the neutron, $\hbar q$ and $\hbar\omega$ are the momentum and energy transfers. a_α is the product of the scattering length and of the Debye-Waller factor for an atom of type α. The $S^{(1)}_{\alpha\beta}(\boldsymbol{q},\omega)$ are the one-

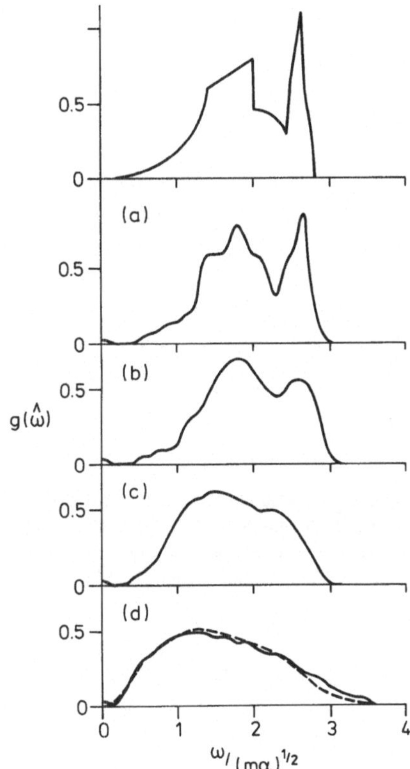

Fig. 6.18. Vibrational density of states for fcc crystal (top), and (**a**) 480 atom fcc cluster, (**b**)–(**d**) 500 atom amorphous model with (**b**) constant force constants, (**c**) a Morse potential, and (**d**) a Lennard-Jones potential, calculated using the equation-of-motion method. (After [6.114].) The dashed line in (**d**) shows the result of the molecular dynamics calculation of *Rahman* et al. [6.117]

phonon partial dynamical structure factors. In the equation-of-motion method they are calculated in the form

$$S^{(1)}_{\alpha\beta}(\boldsymbol{q},\omega)=\frac{1}{2\pi N}\sum_{i(\alpha)}\sum_{j(\beta)}\mathrm{e}^{-\mathrm{i}\boldsymbol{q}(\boldsymbol{r}_i-\boldsymbol{r}_j)}\int_{-\infty}^{\infty}\mathrm{e}^{-\mathrm{i}\omega t}[\boldsymbol{q}\cdot\boldsymbol{u}_i(t)\boldsymbol{q}\cdot\boldsymbol{u}_j(0)]\,dt\,. \tag{6.32}$$

The results of *Rehr* and *Alben* [6.114] for a normalized $\tilde{S}(\boldsymbol{q},\omega)$ for longitudinal excitations (their calculation refers to a single-atom model)

$$\tilde{S}(\boldsymbol{q},\omega)=S(\boldsymbol{q},\omega)m\omega/q^2[n(\hbar\omega/k_\mathrm{B}T)+1] \tag{6.33}$$

are plotted in Fig. 6.19, for the Morse-type interatomic potential. The dynamical structure factor does not depend dramatically on the amount of quantitative disorder. A sharp longitudinal acoustic mode is seen for small $\boldsymbol{q}$. As $\boldsymbol{q}$ increases there is a minimum in the energy in the vicinity of the first peak in the static structure factor. This shows that some effects of long-range order are preserved in the amorphous structure. At these $\boldsymbol{q}$ values the structure can statically absorb momentum, allowing low-energy excitations to contribute to the scattering [6.114].

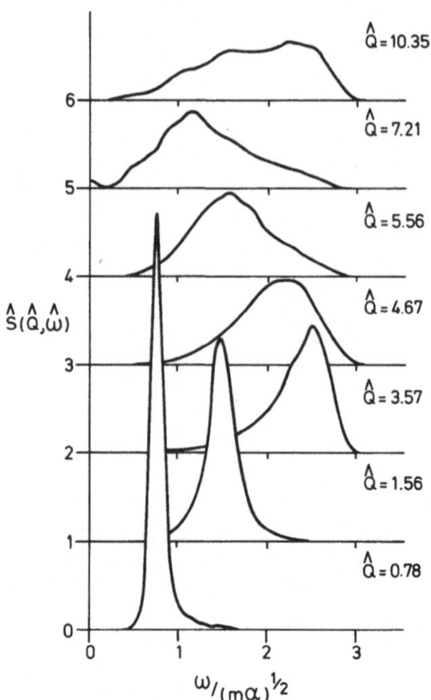

Fig. 6.19. Dynamical structure factor $S(Q, \omega)$ for the amorphous model, calculated using the equation-of-motion method and assuming a Morse potential. (After [6.114])

6.7.2 Molecular Dynamics

Rahman et al. [6.117] have studied the low-temperature properties of an amorphous system of 500 particles interacting via a Lennard-Jones potential using molecular dynamics techniques. They calculated the density of states by directly diagonalizing the dynamical matrix and by establishing the statistics of its eigenvalues. The computation was tractable only for a subsystem of 200 particles, embedded in the larger original cluster. This was done at different moments in time and in different regions of the periodicity volume of the 500-particle system. Their calculated density of states agrees well with the equation-of-motion result of *Rehr* and *Alben* for the same system (cf. the dashed line in Fig. 6.18d).

6.7.3 Continued-Fraction Techniques

The dynamical structure factor $S(\boldsymbol{q}, \omega)$ for coherent scattering is given by the Fourier transform of a density correlation function

$$S(\boldsymbol{q}, \omega) = \frac{1}{2\pi N} \sum_{i,j} \mathrm{e}^{-\mathrm{i}\boldsymbol{q}(\boldsymbol{r}_i - \boldsymbol{r}_j)} \int_{-\infty}^{\infty} \mathrm{e}^{-\mathrm{i}\omega t} \langle \mathrm{e}^{-\mathrm{i}\boldsymbol{q}\boldsymbol{u}_i(t)} \mathrm{e}^{\mathrm{i}\boldsymbol{q}\boldsymbol{u}_j(0)} \rangle dt \tag{6.34}$$

(for simplicity we quote only the formulae for the total dynamical structure factor, the angle brackets indicate thermodynamic expectation values). Such correlation functions for dynamical variables are very efficiently calculated using *Mori's* [6.118] continued-fraction technique. A relaxation shape function $F(\boldsymbol{q}, \omega)$ is defined by

$$F(\boldsymbol{q}, \omega)=\frac{1}{2\pi}\int_{-\infty}^{\infty}\frac{\sum_{i,j} \mathrm{e}^{-\mathrm{i}\boldsymbol{q}(\boldsymbol{r}_i-\boldsymbol{r}_j)}\langle \mathrm{e}^{-\mathrm{i}\boldsymbol{q}\boldsymbol{u}_i(t)}\mathrm{e}^{\mathrm{i}\boldsymbol{q}\boldsymbol{u}_j(0)}\rangle}{\sum_{i,j} \mathrm{e}^{-\mathrm{i}\boldsymbol{q}(\boldsymbol{r}_i-\boldsymbol{r}_j)}\mathrm{e}^{-\mathrm{i}\boldsymbol{q}\boldsymbol{u}_i(0)}\mathrm{e}^{\mathrm{i}\boldsymbol{q}\boldsymbol{u}_j(0)}}\mathrm{e}^{-\mathrm{i}\omega t}dt \tag{6.35}$$

so that for $\boldsymbol{q} \neq 0$ we have

$$S(\boldsymbol{q}, \omega)=S(\boldsymbol{q})\,F(\boldsymbol{q}, \omega)\,. \tag{6.36}$$

Mori's treatment leads to a continued-fraction expansion for the Laplace transform of the relaxation shape function

$$F(\boldsymbol{q}, \omega)=\pi^{-1}Re\{\tilde{F}(\boldsymbol{q}, \mathrm{i}\omega)\}\,, \tag{6.37}$$

$$\tilde{F}(\boldsymbol{q}, s)=\cfrac{1}{s+\cfrac{d_1}{s+\cfrac{d_2}{s+\cfrac{d_3}{s+\dots}}}} \tag{6.38}$$

where $\tilde{F}$ denotes the Laplace transform of F. Using (6.36) this leads to the following expansion for the dynamical structure factor [6.119, 120]

$$S(\boldsymbol{q}, \omega)=\pi^{-1}S(\boldsymbol{q})Re\cfrac{1}{s+\cfrac{d_1}{s+\cfrac{d_2}{s+\cfrac{d_3}{s+\dots}}}} \tag{6.39}$$

with $s=\mathrm{i}\omega$. The coefficients d_i are related to the frequency moments $\langle\omega^n\rangle$ of $S(\boldsymbol{q}, \omega)$

$$\langle\omega^n\rangle=\int_{-\infty}^{\infty}\omega^n S(\boldsymbol{q}, \omega)d\omega\,; \tag{6.40}$$

for the first few terms, this relation is given by

$$\begin{aligned} d_1&=\langle\omega^2\rangle/S(\boldsymbol{q}) \\ d_2&=\langle\omega^4\rangle/\langle\omega^2\rangle-\langle\omega^2\rangle/S(\boldsymbol{q}) \\ d_3&=(\langle\omega^6\rangle-\langle\omega^4\rangle^2/\langle\omega^2\rangle)/[\langle\omega^4\rangle-\langle\omega^2\rangle^2/S(\boldsymbol{q})]\,. \end{aligned} \tag{6.41}$$

To proceed further, the continued-fraction expansion has to be truncated at some stage. This is a rather critical step and we refer to the specialized literature [6.119, 120] for any further discussion. A similar expansion might be devised for the one-phonon part of $S^{(1)}(\boldsymbol{q},\omega)$. The fourth moment of the one-phonon dynamical structure factor is given by [6.119]

$$\begin{aligned}\langle\omega^4_{S^{(1)}}\rangle &= \frac{1}{q^2}\frac{k_B T}{m^2}\varrho\int g(r)[1-\cos(\boldsymbol{q}\boldsymbol{r})](\boldsymbol{q}\cdot\vec{\nabla})^2 V(r)d^3r \\ &\equiv \omega(\boldsymbol{q})^2\end{aligned} \tag{6.42}$$

and may be used to define a "phonon dispersion relation" in an approximation that is essentially equivalent to an assumption of independent phonons, similar to the theory of *Hubbard* and *Beeby* [6.121]. The method is readily generalized to binary systems [6.119]. In this case the fourth moments of the partial one-phonon structure factors (6.32) form a matrix $\langle\omega^4_{S^{(1)}_{\alpha\beta}}\rangle$; approximate "dispersion relations" might be derived by diagonalizing this matrix. This allows a classification of the vibrational modes as "acoustic" or "optical". For small $\boldsymbol{q}$ this assignment is justified by the form of the eigenvectors.

The recursion method of *Haydock* et al. [6.122] is similar in spirit to the Mori technique. It allows the calculation of the density of states without explicitly calculating the normal mode frequencies. As yet the method not been applied to the vibrational properties of amorphous metals (for an application to calculate the electronic properties, cf. *Gaspard*'s chapter in [6.116]), but recently *Meek* [6.123] used the recursion method to discuss the interrelation between the vibrational spectra and the topological structure of amorphous semiconductors.

6.7.4 The Dynamics of Amorphous $Mg_{0.7}Zn_{0.3}$

Starting from the pair potentials and the equilibrium structure described in Sects. 6.2 and 6.5, *von Heimendahl* [6.24] has calculated the one-phonon dynamical structure factor for the $Mg_{0.7}Zn_{0.3}$ glass. The phonon dispersion relations for crystalline Mg have been calculated by *Hafner* and *Eschrig* [6.46], for Zn by *Hafner* [6.124], and for the Laves phase $MgZn_2$ by *Eschrig* et al. [6.125] using the same type of potentials. Very good agreement with experiment was achieved in each case. In the amorphous calculation, full account was taken of topological, quantitative, and substitutional disorder. The results for both longitudinal and transverse excitations are shown in Fig. 6.20. The results are qualitatively similar to the model calculations of *Rehr* and *Alben*. The k dependence of the peaks is shown in Fig. 6.21. These "dispersion relations" show a clear maximum for longitudinal excitations at $k \simeq 1.21\,\text{Å}^{-1}$, i.e., at approximately half the value of the wave vector of the first peak in the static structure factor ($Q^p \cong 2.54\,\text{Å}^{-1}$). The subsequent minimum is not really identifi-

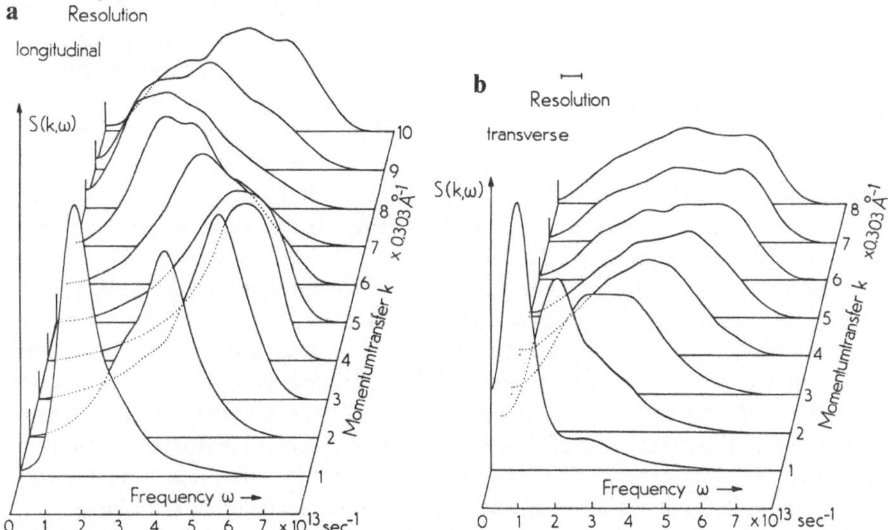

Fig. 6.20a, b. Dynamical structure factor for amorphous $Mg_{0.7}Zn_{0.3}$ for longitudinal (**a**) and transverse (**b**) excitations. (After [6.24])

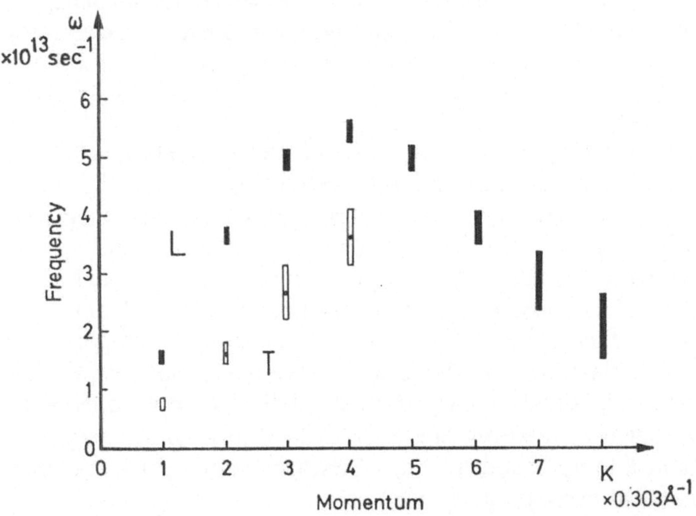

Fig. 6.21. "Phonon dispersion relations" for amorphous $Mg_{0.7}Zn_{0.3}$. (After [6.24])

able, because the peaks in the structure factor become too broad. For transverse waves the vibrational energy is well defined for small momenta, reflecting the rigidity of the amorphous solid against shear deformations. Apparently there is no peak in the dispersion relations for transverse excitations.

Because of instrumental limitations, no one-phonon peaks could be detected in the inelastic neutron scattering experiments [6.116]. At present the only experimental information comes from the Brillouin scattering experiments of *Grimsditch* and *Güntherodt* [6.126] who measured a surface sound velocity of $V_{\text{surf}} = 2.13 \times 10^5$ cm/s. If we assume a Poisson ratio of $\sigma = 0.30$ (this is approximately equal to the σ of pure Mg and Zn metals), we can deduce longitudinal and transverse sound velocities of $V_L = 4.3 \times 10^5$ cm/s, $V_T = 2.3 \times 10^5$ cm/s. From von Heimendahl's dispersion relations (Fig. 6.20) one might estimate $V_L = 5.1 \times 10^5$ cm/s and $V_T = 2.5 \times 10^5$ cm/s.

A comparison of the dispersion relations of the glass with the results for the crystalline metals and alloys shows that the maximum frequencies vary roughly linearly with concentration, independent of the structure. Anomalously low vibrational energies have been predicted by *Eschrig* et al. [6.125] for some transverse optical modes in $MgZn_2$. They are a peculiarity of the crystal structure and the mass ratio of this compound. No related features are found in the glass. The comparison of the elastic part of the dispersion relations does not reveal any softening of the long-wave part of the spectrum of the glass.

Beck and *Tomanek* [6.119] used the continued-fraction technique to calculate the dynamical structure factor, using some simplifying assumptions: i) simpler pair potentials and ii) analytical hard core structure factors in the Percus-Yevick approximation. Using this information they could calculate the fourth moment; the calculation of the sixth moment already necessitates further assumptions on the triplet correlation functions. Despite these limitations they were able to reproduce qualitatively von Heimendahl's results. They calculated explicitly partial structure factors $S_{\alpha\beta}(\boldsymbol{q}, \omega)$. Their results offer a possible explanation for the structure in von Heimendahl's dynamical structure factors as arising from acoustic- and optic-type excitations.

On the basis of inelastic neutron scattering measurements on glassy $Pd_{0.80}Si_{0.20}$, *Windsor* et al. [6.127] concluded that apart from some loss of structure the vibrational spectrum of the metallic glass resembles closely that of the crystalline material. The more recent experiments of *Suck* et al. [6.116] on glassy Cu–Zr and the theoretical results on Mg–Zn seem to support this conclusion. For an explanation of the low-temperature properties of the metallic glasses, a search for subsidiary minima of the potential energy in configuration space (such as done by *Smith* [6.128] for amorphous semiconductors) appears to be a more promising line of research than the investigation of phononlike dynamical excitations.

6.8 Electronic Properties

A discussion of the electronic properties of amorphous metals goes far beyond the scope of the present chapter. Nevertheless it seems to be important to say a few words on the interrelation between the glass-forming ability and the

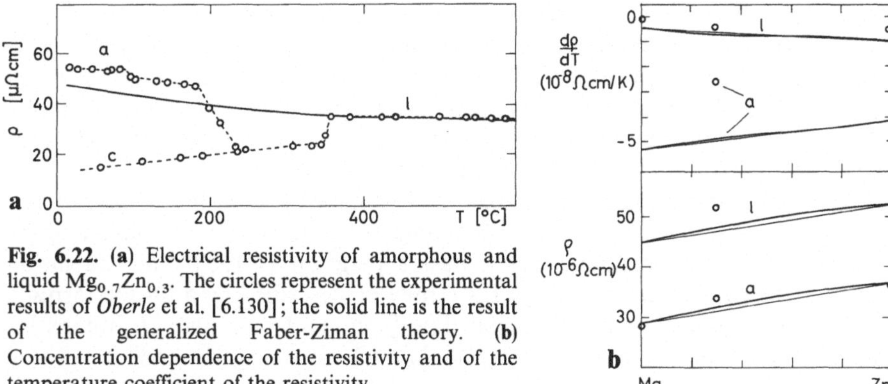

Fig. 6.22. (**a**) Electrical resistivity of amorphous and liquid $Mg_{0.7}Zn_{0.3}$. The circles represent the experimental results of *Oberle* et al. [6.130]; the solid line is the result of the generalized Faber-Ziman theory. (**b**) Concentration dependence of the resistivity and of the temperature coefficient of the resistivity

electronic transport properties. It has often been observed that a good glass-forming ability is linked to a negative temperature coefficient of the electrical resistivity (TCR) [6.129]. In principle this interrelation seems to be very easy to understand since the extension of the Faber–Ziman theory for liquid metals to amorphous alloys [6.36–38] predicts negative TCRs if the Nagel–Tauc rule $Q^p \approx 2k_F$ is satisfied. A numerical verification of this conjecture is always difficult for systems containing transition metals. In this respect, the simple-metal glasses are the ideal systems to study. In Fig. 6.22a we show the electrical resistivity of liquid and amorphous $Mg_{0.7}Zn_{0.3}$ as calculated using the pseudopotentials described in Sect. 6.2 and the partial hard sphere structure factors calculated using the thermodynamic variational technique (Sects. 6.4 and 6.5). For the liquid state the agreement between theory and experiment is perfect; the resistivity in the amorphous state is ~15% higher than in the supercooled liquid state, possibly due to an additional scattering from structural defects. The calculated TCR of the amorphous state ($d\varrho/dT = -5.3 \times 10^{-8}$ Ωcm/K) is distinctly more negative than the experimental one ($d\varrho/dT = -2.6 \times 10^{-8}$ Ωcm/K). This comes from the fact that the first peak in the hard sphere model structure factors is too sharp and demonstrates that the correlation to the supercooled melt should not be extended too far. Since we know how to calculate the partial dynamical structure factors (Sect. 6.7) we can in principle calculate the electronic transport properties down to lowest temperatures. Such a calculation is presently being done.

The concentration dependence of both the electrical resistivity and the TCR (Fig. 6.22b) is very smooth, almost linear. This shows that a correlation of the electrical transport properties of the molten alloy either with the eutectic composition or with the composition of the Laves phase does not exist. A resistivity maximum in sputtered amorphous Mg–Zn films at the composition $MgZn_2$ reported by *Hauser* and *Tauc* [6.131] seems to be due to structural defect scattering. The temperature variation of the TCR depends very sensitively on the $Q^p \leftrightarrow 2k_F$ relation and on the form of the pseudopotential form factors for these values of the wave vector. At low temperatures the largest contributions to the Faber-Ziman integral come from $q \sim Q^p$ and the TCR is negative; at higher temperatures the contributions from $q \sim 2k_F$ dominate and the TCR may become positive. This explains the resistivity minimum in molten Zn [6.132] and the temperature variation of the TCR in Mg: negative TCR for "amorphous" Mg and TCR ~ 0 at the melting point.

Similarly interesting correlations may be found between the structural and the optical properties: in the frequency range below the interband threshold the optical properties of simple-metal glasses (Mg–Zn [6.133]) and transition metal-metalloid glasses (Au–Si [6.134]) may be described by a Drude formula with an energy-dependent relaxation time $[1/\tau(\omega) = 1/\tau_0 + a(\hbar\omega)^2]$. Indeed for disordered metals and alloys with $Q^p \sim 2k_F$ such an energy dependence of $\tau(\omega)$ may be derived using a generalized Faber–Ziman formula [6.135].

6.9 Conclusions

In a recent review paper on the structure of amorphous metals, *Hoare* [6.95] has established a list of "some of the requirements that any good theory of a glass might reasonably be expected to fulfil". In a first group of sine qua non conditions he lists (we have taken some freedom to reformulate and rearrange his conditions)

1a) Reproduction of the main essentials of the random geometric conditions for the interpretation of diffraction data.
1b) Explanation of the static mechanical stability of the amorphous phase and of the nature of the free energy barrier to nucleation to the crystalline state.
1c) Explanation of the glass-forming ability and of the effect of composition on the stability of binary amorphous systems. We might generalize this condition and ask for an explanation of the interrelation between the constitution diagram and glass formation.
1d) Reasonable consistency with the gross properties, in particular with the density of the amorphous phase.

In a second group he mentions some additional properties which he considers as merely desirable.

2a) Explanation of the glass transition and its associated thermodynamics.
2b) Reproduction of the low-temperature vibrational spectra, if possible together with the low-temperature thermodynamic, thermal conductivity, and acoustic properties.
2c) Explanation of the major electronic (including electronic transport) and magnetic properties.
2d) Some support for theories of strength, plasticity, self-diffusion, annealing, etc.

Of course we might think of many other points to be added to this list. We claim that the microscopic theory of simple-metal glasses gives a satisfactory answer to points 1a) to 1d) and that some important steps have been made to respond to the second group of conditions. We hope that the results will contribute to a better understanding of amorphous alloys in general.

Acknowledgements. Many of the results presented here are due to the fruitful cooperation of Dr. L. von Heimendahl. Stimulating discussions with Prof. H. J. Güntherodt, Prof. H. Rudin, Dr. F. Sommer, Dr. M. Fischer, Prof. H. Beck, Prof. H. Ruppersberg, and Prof. K. Bennemann are gratefully acknowledged. The author thanks Prof. H. Rudin, Prof. H. Beck, Dr. F. Sommer, Dr. M. Grimsditch, and Prof. H. Ruppersberg for communicating unpublished material.

This work has been supported by the "Fonds zur Förderung der wissenschaftlichen Forschung in Österreich".

References

6.1 W.Buckel: Z. Phys. **138**, 136 (1954)
W.Buckel, R.Hilsch: Z. Phys. **138**, 109 (1954)

6.2 W.Klement, R.H.Willens, P.Duwez: Nature **187**, 809 (1960)

6.3 For recent reviews see, e.g., H.J.Güntherodt: In *Advances in Solid State Physics*, Vol. XVII, ed. by H.J.Treusch (Vieweg, Braunschweig, 1977) p. 25
See also *Proc. of the 3rd Int. Conf. on Rapidly Quenched Metals*, ed. by B.Cantor (The Metals Society, London 1978)

6.4 G.S.Cargill III: In *Solid State Physics, Advances in Research, and Applications*, Vol. 30, ed. by H.Ehrenreich, F.Seitz, D.Turnbull (Academic Press, New York 1975) p. 227

6.5 J.Hafner: Phys. Rev. B**15**, 617 (1977); B**19**, 5094 (1979)

6.6 A.Calka, M.Madhava, D.E.Polk, B.C.Giessen, H.Matya, J.Vander Sande: Scr. Metall. **11**, 65 (1977)

6.7 J.Hafner, L. von Heimendahl: Phys. Rev. Lett. **42**, 386 (1979)

6.8 J.Hafner: Phys. Rev. B**21**, 406 (1980)

6.9 B.C.Giessen, C.N.J.Wagner: In *Liquid Metals – Physics and Chemistry*, ed. by S.Z.Beer (Dekker, New York 1972) p. 660

6.10 W.B.Pearson: *The Crystal Chemistry and Physics of Metals and Alloys* (Wiley-Interscience, New York 1972)

6.11 D.E.Polk, B.C.Giessen: In *Metallic Glasses*, ed. by J.J.Gilman, H.J.Leamy (American Society for Metals, Metals Park, Ohio 1978) p. 1

6.12 W.Bauhofer, A.Simon: Phys. Rev. Lett. **40**, 1730 (1978)

6.13 A.Simon: Z. Anorg. Allg. Chem. **395**, 301 (1973)
A.Simon, H.J.Deiseroth, E.Westerbeck, B.Hillenköter: Z. Anorg. Allg. Chem. **423**, 203 (1976)

6.14 S.Rundqvist: Ark. Kemi. **20**, 67 (1972)

6.15 P.H.Gaskell: Nature **276**, 484 (1978); J. Phys. C**12**, 4337 (1979)

6.16 B.C.Giessen, J.Hong, L.Kabacoff, D.E.Polk, R.Raman, R.St.Armand: In *Proc. 3rd Int. Conf. on Rapidly Quenched Metals*, Vol. 1, ed. by B.Cantor (The Metals Society, London 1978) p. 249
R.St.Amand, B.C.Giessen: Scr. Metall. **12**, 1021 (1978)
F.Sommer, G.Duddek, B.Predel: Z. Metallkd. **69**, 587 (1978)
B.Predel, K.Hülse: J. Less-Common Met. **63**, 45 (1979)

6.17 L.E.Tanner, B.C.Giessen: Metall. Trans. **9**A, 67 (1978); Acta Metall. **27**, 1727 (1979)

6.18 W.H.Shull, D.G.Naugle, S.J.Poon, W.H.Johnson: Phys. Rev. B**18**, 3263 (1978)

6.19 G.S.Cargill III: In *Thin Film Phenomena – Interfaces and Interactions*, ed. by J.E.Baglin, J.M.Poate (The Electrochemical Society, Princeton 1978) p. 221
6.20 F.C.Frank, J.S.Kasper: Acta Crystallogr. **11**, 184 (1958); **12**, 483 (1959)
6.21 R.Ray, B.C.Giessen, N.J.Grant: Scr. Metall. **2**, 357 (1968)
J.Vitek, J.VanderSande, N.J.Grant: Acta Metall. **23**, 165 (1975)
B.C.Giessen, M.Madhava, D.E.Polk: Mater. Sci. Eng. **23**, 145 (1976)
6.22 R.Wang: Nature **278**, 700 (1979)
See also J.L.Finney: Nature **280**, 847 (1979)
6.23 H.Hansen, H.Anderko: *Constitution of Binary Alloys* (McGraw-Hill, New York 1958)
R.P.Elliott: *Constitution of Binary Alloys – 1st Supplement* (McGraw-Hill, New York 1965)
F.A.Shunk: *Constitution of Binary Alloys – 2nd Supplement* (McGraw-Hill, New York 1969)
W.G.Moffat: *Binary Phase Diagram Handbook* (General Electric, Schenectady 1976)
6.24 L. von Heimendahl: J. Phys. F**9**, 161 (1979)
6.25 J.Hafner: J. Phys. F**6**, 1243 (1976)
6.26 W.H.Young: In *Proc. of the 3rd Int. Conf. on Liquid Metals*, Conf. Series Vol. 30, ed. by R.Evans, D.A.Greenwood (The Institute of Physics, Bristol and London 1977) p. 7
6.27 J.Hafner: Phys. Rev. A**16**, 351 (1977)
6.28 J.Weeks, D.Chandler, H.C.Andersen: J. Chem. Phys. **54**, 5237 (1971)
6.29 C.H.Bennett, D.E.Polk, D.Turnbull: Acta Metall. **19**, 1259 (1971)
6.30 S.R.Nagel, J.Tauc: Phys. Rev. Lett. **35**, 380 (1975)
6.31 H.S.Chen: Acta Metall. **24**, 153 (1976); Mater. Sci. Eng. **23**, 151 (1976)
6.32 D.Turnbull: In *Solidification* (American Society for Metals, Metals Park, Ohio 1968) p. 1
6.33 W.H.Shull, D.G.Naugle: Phys. Rev. Lett. **39**, 1580 (1977)
J.E.Graebner, B.Golding, R.J.Schultz, F.S.L.Hsu, H.S.Chen: Phys. Rev. Lett. **39**, 1480 (1977)
6.34 B.Golding, J.E.Graebner, W.H.Haemmerle: In *Proc. Int. Conf. on Lattice Dynamics, Paris 1977*, ed. by M.Balkanski (Flammarion, Paris 1977) p. 348
6.35 H. von Löhneysen, F.Steglich: Phys. Rev. Lett. **39**, 1205 (1977)
6.36 L.Meisel, P.J.Cote: Phys. Rev. B**15**, 2970 (1977)
6.37 K.Froböse, J.Jäckle: J. Phys. F**7**, 2331 (1977)
6.38 S.R.Nagel: Phys. Rev. B**16**, 1694 (1977)
6.39 N.F.Mott: Philos. Mag. **26**, 1249 (1972)
J.E.Enderby, B.C.Dupree: Philos. Mag. **35**, 791 (1977)
6.40 R.Cochrane, R.Harris, J.Ström-Olsen, M.Zuckermann: Phys. Rev. Lett. **35**, 676 (1975)
R.Harris, M.Shalmon, M.Zuckermann: Phys. Rev. B**18**, 5906 (1978)
6.41 E.Esposito, H.Ehrenreich, C.D.Gelatt: Phys. Rev. B**18**, 3913 (1978)
6.42 W.A.Harrison: *Pseudopotentials in the Theory of Metals* (Benjamin, New York 1966)
6.43 V.Heine, D.Weaire: In *Solid State Physics, Advances in Research and Applications*, Vol. 24, ed. by H.Ehrenreich, F.Seitz, D.Turnbull (Academic Press, New York 1971) p. 247
6.44 J.C.Phillips, L.Kleinman: Phys. Rev. **116**, 287 (1959)
6.45 M.H.Cohen, V.Heine: Phys. Rev. **122**, 1821 (1961)
6.46 J.Hafner: Z. Phys. B**22**, 351 (1975); B**24**, 351 (1976)
J.Hafner, H.Eschrig: Phys. Status Solidi (b) **72**, 179 (1975)
6.47 P.Vashishta, K.S.Singwi: Phys. Rev. B**2**, 875 (1972)
6.48 J.M.Ziman: Adv. Phys. **13**, 89 (1964)
6.49 J.Hafner: Solid State Commun. **27**, 263 (1978)
6.50 J.Hafner: In *Proc. of the 3rd Int. Conf. on Liquid Metals*, Conf. Series Vol. 30, ed. by R.Evans, D.A.Greenwood (The Institute of Physics, Bristol and London 1977) p. 107
6.51 J.Hafner, F.Sommer: CALPHAD **1**, 351 (1977)
6.52 L.Pauling: *The Nature of the Chemical Bond*, 2nd ed. (Cornell University Press, Ithaca 1952) Sect. 12.5
6.53 H.Ruppersberg, W.Speicher: Z. Naturforsch. **31**a, 47 (1976)
W.Biltz, F.Weibke: Z. Anorg. Allg. Chem. **223**, 321 (1935)
6.54 N.W.Ashcroft, D.C.Langreth: Phys. Rev. **156**, 685 (1967)
6.55 K.Fuchs: Proc. R. Soc. A**151**, 585 (1935)

6.56 M. Ross, D. Seale: Phys. Rev. A**9**, 396 (1974)
I. H. Umar, A. Meyer, M. Watabe, W. H. Young: J. Phys. F**4**, 1691 (1974)
6.57 A. Simon: Proc. 3rd European Crystallographical Meeting, Zürich 1976, p. 335
6.58 A. Simon, W. Brämer, B. Hillenköter, H. J. Kullmann: Z. Anorg. Allg. Chem. **419**, 253 (1976)
6.59 R. Wang, M. D. Merz: Nature **260**, 35 (1976); Phys. Status Solidi (a) **39**, 697 (1977)
6.60 K. K. Mon, N. W. Ashcroft: Solid State Commun. **27**, 609 (1978)
6.61 F. Sommer: Private communication
6.62 J. A. Barker, D. Henderson: Rev. Mod. Phys. **48**, 587 (1976)
6.63 E. Thiele: J. Chem. Phys. **39**, 474 (1963)
6.64 M. S. Wertheim: Phys. Rev. Lett. **10**, 321 (1963); J. Math. Phys. **8**, 927 (1964)
6.65 N. F. Carnahan, K. E. Starling: J. Chem. Phys. **51**, 635 (1969)
G. A. Mansoori, N. F. Carnahan, K. E. Starling, T. W. Leland: J. Chem. Phys. **54**, 1523 (1971)
6.66 H. C. Andersen, D. Chandler, J. D. Weeks: J. Chem. Phys. **56**, 3812 (1972)
6.67 C. Regnaut, J. P. Badiali, M. Dupont: Phys. Lett. **74**A, 245 (1979)
R. Oberle, H. Beck: Solid State Commun. **32**, 959 (1979)
6.68 A. Isihara: J. Phys. A**1**, 539 (1968)
6.69 F. Sommer, B. Predel, D. Assmann: Z. Metallkd. **68**, 347 (1977)
F. Sommer: Z. Metallkd. **70**, 359 (1979)
6.70 R. Hultgren, P. D. Desai, D. T. Hawkins, M. Gleiser, K. K. Kelley, D. D. Wagmans: *Selected Values of the Thermodynamic Properties of Binary Alloys* (American Society for Metals, Metals Park, Ohio 1973)
6.71 J. D. Bernal: Proc. R. Soc. A**280**, 299 (1964)
6.72 J. F. Sadoc, J. Dixmier: Mater. Sci. Eng. **23**, 187 (1976)
6.73 T. Fukunaga, M. Misawa, K. Fukamichi, T. Masumoto, K. Suzuki: In *Proc. of the 3rd Int. Conf. on Rapidly Quenched Metals*, Vol. 2, ed. by B. Cantor (The Metals Society, London 1978) p. 325
6.74 M. Sakata, N. Cowlam, H. A. Davies: J. Phys. F**9**, L235 (1979)
6.75 C. N. J. Wagner: Workshop on Neutron Scattering from Amorphous Metals, ILL Grenoble, Oct. 1979 (unpublished)
6.76 T. M. Hayes, J. W. Allen, J. Tauc, B. C. Giessen, J. J. Hauser: Phys. Rev. Lett. **40**, 1282 (1978)
6.77 I. Vincze, D. S. Boudreaux, M. Tegze: Phys. Rev. B**19**, 4896 (1979)
6.78 J. L. Finney: Proc. R. Soc. A**319**, 479 (1970)
6.79 C. H. Bennett: J. Appl. Phys. **43**, 2727 (1972)
6.80 D. E. Polk: Scr. Metall. **4**, 117 (1970)
6.81 J. F. Sadoc, J. Dixmier, A. Guinier: J. Non-Cryst. Solids **12**, 46 (1973)
6.82 G. S. Cargill, III, S. Kirkpatrick: AIP Conf. Proc. **31**, 339 (1976)
6.83 L. von Heimendahl: J. Phys. F**5**, L141 (1975)
6.84 J. A. Barker, J. L. Finney, M. R. Hoare: Nature **257**, 120 (1975)
6.85 D. S. Boudreaux: Phys. Rev. B**18**, 4039 (1978)
6.86 Y. Waseda, H. Okezaki, T. Masumoto: In *Proc. Int. Conf. on the Structure of Non-Crystalline Materials, Cambridge 1976* (Taylor and Francis, London 1977) p. 202
6.87 D. S. Boudreaux, J. M. Gregor: J. Appl. Phys. **48**, 152, 5057 (1977)
6.88 R. Yamamoto, K. Haga, H. Shibuta, M. Doyama: J. Phys. F**8**, L179 (1978)
6.89 J. D. Weeks: Philos. Mag. **35**, 1345 (1977)
6.90 P. H. Gaskell: J. Non-Cryst. Solids **32**, 207 (1979)
6.91 J. L. Finney: Nature **266**, 309 (1977)
6.92 M. R. Hoare: Ann. N. Y. Acad. Sci. **279**, 186 (1976)
6.93 J. Farges, B. Raoult, G. Torchet: J. Chem. Phys. **59**, 3454 (1973)
6.94 C. L. Briant, J. J. Burton: Phys. Status Solidi (b) **85**, 393 (1978)
6.95 M. R. Hoare: J. Non-Cryst. Solids **31**, 157 (1978)
6.96 H. Rudin: Progress-Report AF-SSP-112, Institut für Reaktortechnik der ETH Zürich (1979) p. 36, and private communication
6.97 A. B. Bhatia, D. E. Thornton: Phys. Rev. B**2**, 3004 (1970)
6.98 C. N. J. Wagner: J. Non-Cryst. Solids **31**, 1 (1978)

6.98a P. Chieux: Zu *Neutron Diffraction*, ed. by H. Dachs, Topics in Current Physics, Vol. 6 (Springer, Berlin, Heidelberg, New York 1977)
6.99 H. Rudin: Private communication
6.100 H.S. Chen, K.A. Jackson: In *Metallic Glasses*, ed. by J.J. Gilman, H.J. Leamy (American Society for Metals, Metals Park, Ohio 1978) p. 74
6.101 D. Stroud, N.W. Ashcroft: Phys. Rev. B**5**, 371 (1971)
6.102 M. Hasegawa, W.H. Young: J. Phys. F**7**, 2271 (1977)
6.103 H. Beck, R. Oberle: In *Proc. 3rd Int. Conf. on Rapidly Quenched Metals*, Vol. 1, ed. by B. Cantor (The Metals Society, London 1978) p. 416
6.104 P. Oelhafen, M. Liard, H.J. Güntherodt, K. Berresheim, H.D. Polaschegg: Solid State Commun. **30**, 641 (1979)
6.105 J.D. Riley, L. Ley, J. Azoulay, K. Terakura: Phys. Rev. B**20**, 776 (1979)
6.106 U. Mizutani, K.T. Hartwig, T.B. Massalski, R.N. Hopper: Phys. Rev. Lett. **41**, 661 (1978)
6.107 J. Blétry: Z. Naturforsch. **33a**, 327 (1978)
6.108 O. Kubaschewski, E.Ll. Evans, C.B. Alcock: *Metallurgical Thermochemistry*, 4th ed. (Pergamon, London 1967)
6.109 H. Leitz, W. Buckel: Z. Phys. B**35**, 73 (1979)
6.110 P. Häußler, W.H.G. Müller, F. Baumann: Z. Phys. B**35**, 67 (1979)
6.111 M.H. Brodsky, M. Cardona: J. Non-Cryst. Solids **31**, 81 (1978)
6.112 R. Alben, L. von Heimendahl, P. Galison, M.L. Long: J. Phys. C**8**, L468 (1975)
6.113 L. von Heimendahl: Phys. Status Solidi (b) **86**, 549 (1978) and further references cited therein
6.114 J.J. Rehr, A. Alben: Phys. Rev. B**16**, 2400 (1977)
6.115 L. von Heimendahl, M.F. Thorpe: J. Phys. F**5**, 87 (1975)
6.116 H.J. Güntherodt, H. Beck (eds.): *Metallic Glasses II*, Topics in Applied Physics (Springer, Berlin, Heidelberg, New York in preparation)
6.117 A. Rahman, M.J. Mandell, J.P. McTague: J. Chem. Phys. **64**, 1564 (1976)
6.117a R. Mountain: In *Dynamics of Solids and Liquids by Neutron Scattering*, ed. by S. Lovesey and T. Springer, Topics in Current Physics, Vol. 3 (Springer Berlin, Heidelberg, New York 1977)
6.118 H. Mori: Prog. Theor. Phys. **33**, 423 (1965); **34**, 399 (1965)
6.119 H. Beck, D. Tomanek: Unpublished
D. Tomanek: Diplomarbeit, University of Basel (1979)
6.120 J.W. Tucker: Sol. State Commun. **18**, 43 (1976)
6.121 J. Hubbard, J.L. Beeby: J. Phys. C**2**, 556 (1969)
6.122 R. Haydock, V. Heine, M.J. Kelly: J. Phys. C**5**, 2845 (1972); C**8**, 2591 (1975)
6.123 P.E. Meek: Philos. Mag. **33**, 897 (1979)
6.124 J. Hafner: Unpublished
6.125 H. Eschrig, K. Feldmann, K. Hennig, L. Weiss: In *Neutron Inelastic Scattering, Grenoble* 1972 (IAEA, Vienna 1972) p. 157
6.126 M. Grimsditch, G. Güntherodt: Private communication
6.127 C.G. Windsor, H. Keyrandish, M.C. Narasimhan: Phys. Lett. **70**A, 485 (1979)
6.128 D.A. Smith: Phys. Rev. Lett. **42**, 729 (1979)
6.129 H.J. Güntherodt, H.U. Künzi: In *Metallic Glasses*, ed. by J.J. Gilman, H.J. Leamy (American Society for Metals, Metals Park, Ohio 1978) p. 247
6.130 R. Oberle, H.U. Künzi, H.J. Güntherodt, B.C. Giessen: Unpublished
6.131 J.J. Hauser, J. Tauc: Phys. Rev. B**17**, 3371 (1978)
6.132 A. Roll, H. Motz: Z. Metallkd. **48**, 272 (1957)
6.133 H.J. Güntherodt, H. Beck, P. Oelhafen, K.P. Ackermann, M. Liard, M. Müller, H.U. Künzi, H. Rudin, K. Agyeman: In *Electrons in Disordered Metals and Metallic Surfaces*, NATO-Advanced Study Institute, Gent 1979 (Plenum, New York 1980) p. 501
6.134 E. Hauser, R.J. Zirke, J. Tauc, J.J. Hauser, S.R. Nagel: Phys. Rev. Lett. **40**, 1733 (1978)
6.135 K. Sturm, E. Pajanne: J. Phys. F**3**, 199 (1973)
J. Hafner: Unpublished

Notes added in proofs for Chap. 6 see p. 261.

7. Electrical Transport in Glassy Metals

P. J. Cote and L. V. Meisel

With 12 Figures

Amorphous metals are convenient subjects for the study of electronic properties of disordered systems. They can be prepared in a single "phase" over wide ranges of composition, they have liquidlike structures and spherical Fermi surfaces, and they exist at low temperatures ($T<\Theta$) where nonlinearities in the temperature dependences of their resistivities are observed.

These features permit stringent tests for theories of transport in disordered metals. Attempts to test transport theories in liquid metals are hampered by the fact that competing theories [7.1] predict similar behavior at liquid metal temperatures ($T>\Theta$). Added complications in the analysis of electrical transport in liquid metals result from their large volume expansions and structural rearrangements on heating [7.2]; such effects are unimportant in most amorphous metals.

Current interest in amorphous alloys has generated a large body of electrical transport data, which exhibits interesting patterns of composition and temperature dependence. We present a review of these matters in Sect. 7.1.

The theoretical description of transport in amorphous systems is usually presented in terms of a diffraction model approach (i.e., Ziman-Faber theory [7.3, 4] and its extensions [7.5–12]) or a Green's function method [7.13–16]. These methods are characterized by a variety of approximation schemes and assumptions which permit computationally useful results. A discussion of some of the more popular schemes and a comparison of results with experiments will be given in Sect. 7.2.

7.1 Experimental Results

7.1.1 The Mooij Correlation

Mooij's analysis [7.17] of all available transport data in disordered transition metal alloys, encompassing a wide variety of crystalline and amorphous phases, established a correlation between the magnitude of the electrical resistance ϱ and the temperature coefficient of resistivity $\alpha=\varrho^{-1}(\partial\varrho/\partial T)$. This remarkable correlation is clearly illustrated in Fig. 7.1. His findings may be summarized as follows: i) At high resistivity ($\varrho>100\,\mu\Omega\text{cm}$), ϱ is insensitive to *details* of the *electronic* structure and the atomic arrangement. ii) Negative temperature

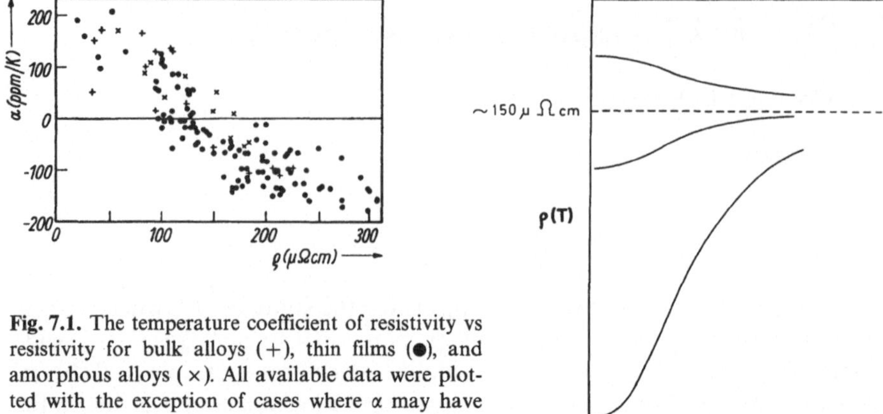

Fig. 7.1. The temperature coefficient of resistivity vs resistivity for bulk alloys (+), thin films (●), and amorphous alloys (×). All available data were plotted with the exception of cases where α may have been influenced by magnetic or structural transitions [7.17]

Fig. 7.2. Illustration of the general trends in the ϱ vs T curves found by *Mooij* for high resistivity amorphous and disordered crystalline alloys

coefficients of resistivity are the rule for alloys whose resistivities exceed 150 μΩcm. iii) All alloys (independent of residual resistivity or low temperature α) tend toward a temperature-independent resistivity of 150 to 200 μΩcm at high temperatures. This produces a sigmoidal shape in the ϱ vs T curves for positive and negative α cases as shown in Fig. 7.2. iv) The correlation obtains for alloys formed of elements from throughout the transition metal series, for all crystal structures, and amorphous phases as well.

These features are believed to be a result of the approach of the electron mean free path toward the interatomic spacing. For a monovalent free electron system with a mean free path equal to the interatomic spacing, one finds $\varrho^* \approx 200$ μΩcm, in agreement with Mooij's limiting resistivity. ϱ^* is commonly referred to as the "saturation" resistivity and Mooij's observations are called "saturation effects" [7.18].

Further manifestations of saturation effects are found in the surprisingly small relative changes in ϱ during an order-disorder transformation or upon melting in high-resistivity metals. The data of *Shacklette* and *Williams* [7.19] on the order-disorder transformation in vanadium carbide provide perhaps the clearest example. The order-disorder transformation in this system occurs at high temperatures where thermal disorder scattering (phonons) produces resistivities in the saturation region ($\varrho > 100$ μΩcm) in the ordered phase. The relative change in resistivity during the order-disorder transformation is small ($\Delta\varrho/\varrho \sim 0.1$) despite the high sensitivity to disorder indicated by the large phonon component of the ordered phase resistivity. This is in sharp contrast to the behavior observed in low-resistivity alloys such as CuAu [7.20] where order-disorder transformations produce $\Delta\varrho/\varrho \lesssim 1$.

Resistivity changes observed upon melting exhibit similar effects. *Mott* [7.1] compared the results for transition metals ($\Delta\varrho/\varrho \sim 0.2$) and noble metals ($\Delta\varrho/\varrho \gtrsim 1$) and concluded that the small values seen in the transition metals were a consequence of the short electron mean free paths (i.e., saturation effects) in these high-resistivity systems.

In the above systems, the possibility of strong *d* band effects, resulting from the overlap of the *d* band with the Fermi level, must be considered. It is often speculated that such *d* band effects are responsible for the anomalous behavior (e.g., negative α) seen in transition metal alloys [7.1, 16]. The apparent independence of the Mooij correlation to details of electronic structure suggests that this is not the case. Further evidence is available from the actinides and lanthanides.

The lanthanides and actinides generally exhibit high resistivities and strong magnetic scattering effects [7.20–24]. Yet, even at high temperatures, their resistivities seldom significantly exceed 200 μΩcm and, with relatively few exceptions, the ϱ vs T curves of rare earth elements and their alloys follow the trends shown in Fig. 7.2. Thus, if one allows for the resistivity anomalies associated with magnetic transitions, the rare earths and their alloys exhibit behavior consistent with the Mooij correlation. *Güntherodt* et al. [7.21] were led to essentially the same conclusion from their studies of the high-temperature electrical resistivity of the lanthanides.

More conclusive evidence that the Mooij correlation is not a unique feature of *d* band metals would be provided by observations of the same effects in noble or alkali metals. Since these elements are weak electron scatterers and generally form low-resistivity alloys, fewer examples of such effects are expected in these systems. Nevertheless, several good examples do exist.

i) Vapor-deposited amorphous noble metal alloys produced by *Korn* et al. [7.25, 26] provide examples of saturation effects in systems with no *d* bands at the Fermi energy. The highest resistivity samples ($\varrho > 100$ μΩcm) of these noble metal-polyvalent metal alloys exhibit negative α extending down to $T=0$; this is at variance with the prediction of standard transport theory which gives a positive value for α near $T=0$ for all amorphous systems. The negative α are thus in accord with the Mooij's correlation.

ii) Liquid Cs provides an example from the alkali metals. The electrical resistivity, thermopower, and Hall effect in liquid Cs have been measured [7.27] for densities ranging from 1.00 to 1.84 gm cm^{-3}. The principal conclusion of this study is that Cs exhibits nearly free electron behavior characteristic of a monovalent metal over the range of densities studied; the unusual feature is that the temperature coefficient of resistivity α assumes anomalous negative values when the electron mean free path approaches the interatomic spacing. In other words, liquid Cs also exhibits the Mooij correlation.

We therefore conclude that the Mooij correlation (with allowance made for variations in the precise value of ϱ^*) applies to all metallic systems. Furthermore, these effects are apparently related to the approach of the electron mean free path to the interatomic spacing. It should be emphasized

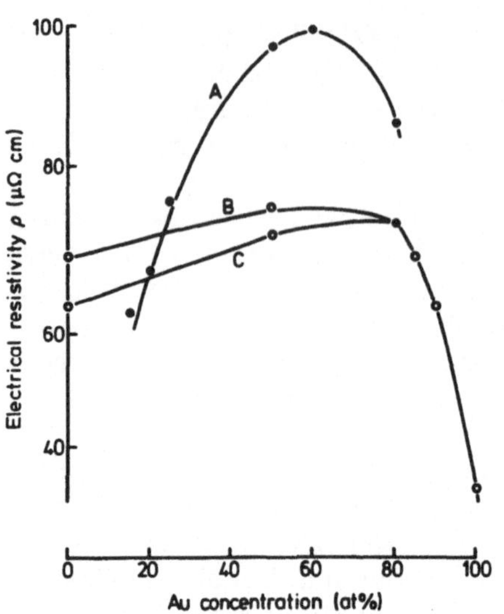

Fig. 7.3. Electrical resistivity of amorphous and liquid SnAu alloys as a function of composition. *A* amorphous (40 K); *B* liquid (1270 K); *C* liquid (920 K). Amorphous data taken from *Blasberg* et al. [7.26]; liquid data taken from *Busch* and *Güntherodt* [7.28]

that the indicated breakdown of standard transport theory at short electron mean free paths was not unexpected [7.1, 2]. What is somewhat surprising is the universality of the behavior and the relatively narrow range for ϱ^*. These results dictate that we consider the Mooij correlation in our discussions of the data and theories of transport for amorphous alloys.

7.1.2 Simple and Noble Metal Alloys

Amorphous simple and noble metal alloys have relatively low resistivities. Thus, saturation effects are less important and standard theory (i.e., Ziman-Faber theory) is expected to be applicable. Unfortunately, there is less information on these systems than on transition metal alloys because few can be prepared in the amorphous phase by the usual rapid solidification technique. Consequently, the most extensive studies deal with vapor-deposited and sputtered alloys.

The resistivity data of *Blasberg* et al. [7.26] for vapor-deposited AuSn alloys are typical of amorphous noble metal alloys. Their results and the liquid phase results obtained by *Busch* and *Güntherodt* [7.28] are compared in Fig. 7.3. Although the liquid and amorphous phase resistivities are not identical, many similarities do exist: the magnitudes of the resistivities are comparable, differing by no more than about 30%; both exhibit a maximum near 65 at.% Au, but the maximum in the amorphous phase is considerably sharper; negative temperature coefficients α (indicated by closed circles on the figure) occur in both alloys for concentrations near those corresponding to the maximum resistivity; however, the composition range for which negative α occur in the amorphous phase is much larger than that in the liquid phase.

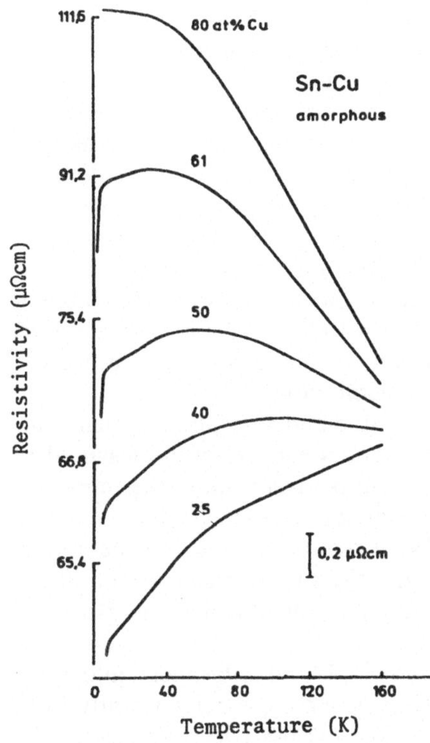

Fig. 7.4. Electrical resistivity of amorphous SnCu alloys as a function of temperature. All samples were annealed at 160 K prior to the measurements [7.25]

Considering that the amorphous phase resistivities were measured at 40 K and the liquid phase near 1000 K, and that the thermal expansion behavior of the liquid phase is complex [7.2], these similarities are quite striking.

The temperature dependences of the resistivity of vapor-deposited amorphous CuSn alloys for a number of concentrations were measured by *Korn* et al. [7.25] and are shown in Fig. 7.4 as another example. The liquid phase results [7.29] (not shown) exhibit a broad maximum at about 75 at. % Cu, resembling the liquid AuSn results. The amorphous phase data are sparse at the high Cu concentrations where the peak occurs in the liquid. However, if one recognizes that the resistivity for pure amorphous Cu should be close to that for liquid Cu (i.e., about 30 μΩcm), the amorphous results are consistent with a maximum resistivity near 70 to 80 at. % Cu as in the liquid. The composition dependence is stronger in the amorphous phase of CuSn as in AuSn. The liquid and amorphous phases of CuSn exhibit negative α for concentrations near those for maximum ϱ and, as with the AuSn alloys, the composition range for negative α is much larger in the amorphous phase.

Several observations concerning these results are in order since they will be considered again in the theory section. 1) Negative α occur for resistivities as low as 50 μΩcm in amorphous CuSn and AuSn. Since these negative α occur for small resistivities, they are not likely to be examples of the Mooij correlation.

Furthermore, the striking similarities between the amorphous and liquid phase results in these alloys support the application of established liquid metal theories to the amorphous phases. ii) Small maxima in the resistivity are seen between 30 and 100 K in CuSn (Fig. 7.4). Similar maxima are present in amorphous AuSn, and in several amorphous transition metal alloys as well (Sect. 7.2.3). We shall see that such maxima are consistent with the predictions of the standard diffraction model. However, in many of these alloys, superconducting fluctuations can obscure the interpretation of the data, particularly in negative α cases. It would therefore be interesting to examine the effects of magnetic fields on these curves. iii) The 80 at.% Cu curve in Fig. 7.4 exhibits a negative α for all temperatures which is contrary to the predictions of standard theory. As dicussed earlier (Sect. 7.2.1), this is an indication that saturation effects are present and occur even in noble metal alloys.

Similarities in the transport properties of liquid and amorphous systems can also be found in the thermopower data. Thermopower has been measured in liquid [7.30] and amorphous [7.31] CuSn. Both phases show thermopower varying from small negative values for Sn-rich compositions to large positive values for Cu-rich compositions. Ziman-Faber theory provides the theoretical basis for understanding the liquid results, and the similar results found in the amorphous phase indicate that standard theory will describe this aspect of transport in the amorphous phase.

Before we leave the subject of amorphous simple and noble metal alloys, we must point out that there are several systems which depart significantly from the trends seen in CuSn and AuSn alloys. We discuss two examples in the remainder of this section.

The concentration dependences for α and ϱ are similar in amorphous and liquid AuSi and are consistent with Ziman-Faber theory up to 30 at.% Si [7.32]. Beyond 30 at.%, the resistivity in the liquid decreases while that in the amorphous phase suddenly increases from about 100 μΩcm to more than 200 μΩcm. *Hauser* and *Tauc* [7.32] attributed this to an increase in tetrahedral coordination at the higher Si concentrations in the amorphous phase which must eventually lead to a breakdown of standard theory and the onset of semiconductorlike behavior since pure amorphous Si is semiconducting. Splitting of the main peak in the x-ray diffraction pattern observed at these Si concentrations supports this explanation. Thus, amorphous AuSi deviates from nearly free electron behavior at high Si concentrations and may be a suitable system for studying the transition from the metallic to the hopping regime for electron transport.

Amorphous MgZn alloys might be expected to exhibit properties predicted by the Ziman-Faber theory because it is free of complications generated by transition elements or semiconducting elements. However, sputtered MgZn films show unusual behavior in that both positive and negative α values occur and ϱ is strongly composition dependent, exhibiting a large peak ($\gtrsim$200 μΩcm) at $\sim$70 at.% Zn where the negative α values occur [7.32]. Simple liquid alloys generally show little composition dependence in either ϱ or α when the average electron-to-atom ratio is fixed [7.2] as in MgZn. These considerations led *Hauser* and *Tauc* to suggest that the anomalous peak in ϱ is related to the tendency to form the compound $MgZn_2$. This suggestion is similar to that made for liquid Mg alloys where strong deviations from nearly free electron behavior occur at compositions for which compounds form in the crystalline phase [7.2]. It appears then that besides reproducing the "normal" behavior of liquid alloys, amorphous alloys also exhibit the same types of anomalies that are observed in liquid alloys. Thus, the Ziman theory is probably not applicable and the correlation between negative α and the large resistivity ($\gtrsim 200\,\mu\Omega$cm) most likely represents another example of the Mooij correlation in a simple metal alloy system.

Summarizing the experimental results on this class of alloys: i) General trends in the electrical transport properties of the amorphous phase follow those of the liquid phase, and standard liquid metal transport theory appears to be valid in the majority of cases. ii) Significant differences in properties are observed in cases for which differences in the structures of the liquid and amorphous phases are seen (e.g., AuSi). iii) There are indications from the MgZn results that the type of anomalies which are produced by compound formation in the liquid (sharp peaks in ϱ and negative α) also occur in the amorphous phase.

7.1.3 Amorphous Transition Metal Alloys

Since amorphous alloys containing transition metals as major constituents are of technical importance and are relatively easy to prepare (by rapid quenching), there is a substantial quantity of resistivity data for these systems. The amorphous transition metal alloys fall into two classes, namely transition metals alloyed with metalloid atoms (e.g., PdSi or NiP) and transition metals alloyed with other transition metals or with noble metals (e.g., NbNi or ZrCu).

Metalloid-Containing Alloys

Most of the work on the transition metal-metalloid alloys deals with Fe, Ni, Pt, or Pd alloyed with B, Si, or P. Amorphous $Ni_{100-x}P_x$ alloys are typical and the temperature and composition dependences of their electrical resistivities measured by *Cote* [7.33] are shown in Figs. 7.5 and 7.6. Several features are seen. i) The temperature coefficient of resistivity α (measured at temperatures above the Debye temperature) changes from positive to negative values at $x \approx 24$. ii) Resistivitiy minima are seen near 15 K in most samples. iii) There are small resistivity maxima near 100 K for metalloid concentrations near those for which α changes sign. No maxima are observed at higher metalloid concentrations which exhibit higher resistivities. The thermopower of amorphous $Ni_{76}P_{24}$ has also been measured [7.34] and is linear over a large range of temperatures, with $S/T \cong 0.013\,\mu VK^{-2}$.

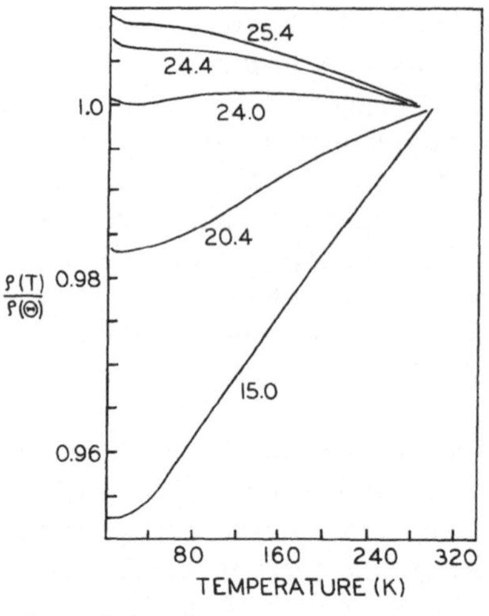

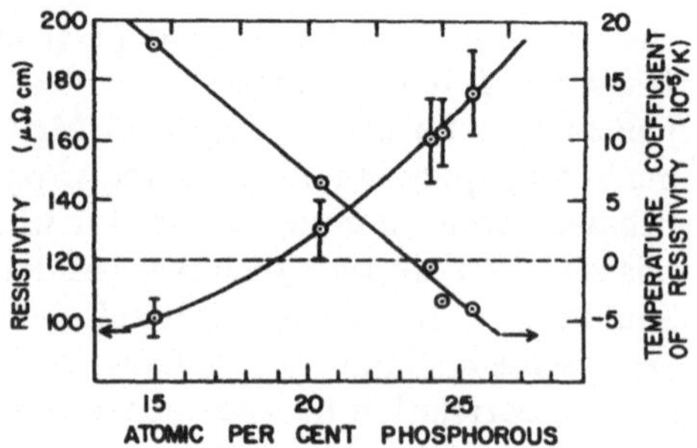

Fig. 7.6. Resistivity and temperature coefficient of resistivity of amorphous NiP at room temperature are shown as a function of phosphorus content [7.33]

Fig. 7.5. Relative resistivity (ϱ/ϱ_{293}) of amorphous NiP as a function of temperature for a range of compositions. Phosphorus content in atomic percent is labeled on each curve [7.33]

Similar results have been found in a number of related ternary alloys. The $(Ni_{50}Pd_{50})_{100-x}P_x$ system has been studied by *Boucher* [7.35] for $17<x<26$. α changes from positive to negative at $x\approx 24$ and ϱ increases rapidly with x as in $Ni_{100-x}P_x$. $(Ni_{100-x}Pt_x)_{75}P_{25}$ has been studied by *Sinha* [7.36] for $40<x<80$. ϱ varies only slightly with x as one would expect; however, α changes from small positive to negative values at $x\approx 50$. Neither $(Ni_{50}Pd_{50})_{100-x}P_x$ nor $(Ni_{100-x}Pt_x)_{75}P_{25}$ shows the small resistivity maxima seen in $Ni_{100-x}P_x$ for $x\approx 25$ and $T\approx 100$ K. $(Pd_{100-x}Cu_x)_{80}P_{20}$ has been studied by *Tangonan* [7.37] for $10\lesssim x\lesssim 50$ and exhibits a transition from positive to negative α at $x\approx 15$ which is accompanied by a small maximum near 100 K. Thermopower data are available [7.36] in $(Ni_{100-x}Pt_x)_{75}P_{25}$ for $40<x<80$. It is found to be only weakly dependent on x and $S/T\approx 0.007\,\mu VK^{-2}$ for $x=40$ which is comparable in magnitude to that found in amorphous NiP. In these four alloy systems, negative α occur whenever $\varrho\gtrsim 150\,\mu\Omega$cm, suggesting that saturation effects are present.

The Allied Chemical Corporation provides a wide variety of amorphous alloys, within this class of alloys, under the trademark "Metglas". These usually contain several transition metal and several metalloid constituents. Despite their greater complexity, these have electrical transport properties which are similar to those of the simpler alloys. (See examples given in [7.38].)

Metalloid-Free Alloys

Among transition metal-transition metal alloys, data are available for ZrNi [7.39], ZrCo [7.39], NbNi [7.40], and ZrPd [7.41]. The results of *Nagel* et al.

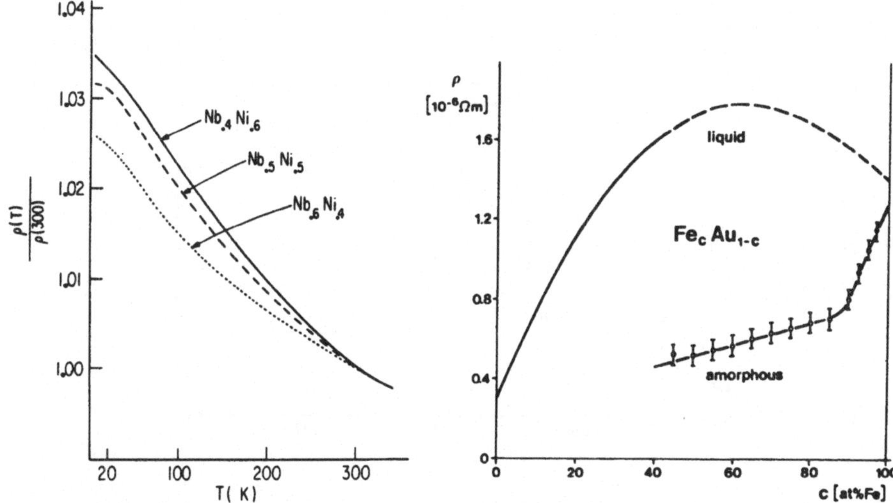

Fig. 7.7. Relative resistivity ϱ/ϱ_{300} for three NbNi samples [7.40]

Fig. 7.8. Resistivity of the amorphous Fe_cAu_{1-c} alloy series as a function of the concentration (lower curve [7.43]). The upper curve represents the resistivities of the corresponding liquid alloys [7.44]

[7.40] on amorphous NbNi are shown in Fig. 7.7. These results are typical for this class of alloys in that $\varrho \gtrsim 150\,\mu\Omega$cm, α is negative for all concentrations, and the ϱ vs T curves have a sigmoidal shape like that found by *Mooij* [7.17] to be characteristic of saturation effects (Fig. 7.2). The small maxima in the resistivity seen in NiP and in PdCuP at about 100 K are absent. (The maxima seen at ~10 K in many of these systems are probably generated by superconducting fluctuations.) The large resistivities and the similarities to the universal curves described by *Mooij* (Fig. 7.2) suggest that saturation effects are present and that standard transport theory is *not* applicable.

Transport has been studied in two binary transition metal-noble metal amorphous alloys, ZrCu [7.42] and FeAu [7.43]. ZrCu behaves exactly like the transition metal-transition metal alloys. However, transport data in amorphous FeAu alloys exhibit some new effects. The data of *Bergmann* and *Marquardt* [7.43] for vapor-deposited FeAu are shown along with those of *Güntherodt* and *Künzi* [7.44] for liquid FeAu in Fig. 7.8. The large difference in the resistivity of the two phases is attributed to the presence of ferromagnetism in the amorphous phase. The resistivity is relatively low so that standard transport theory might be expected to describe the composition and temperature dependences of the resistivity. The authors do, in fact, find satisfactory agreement between the data and the predictions of Ziman-Faber theory if effects of spin-up and spin-down *d* level scattering in the ferromagnetic phase are included.

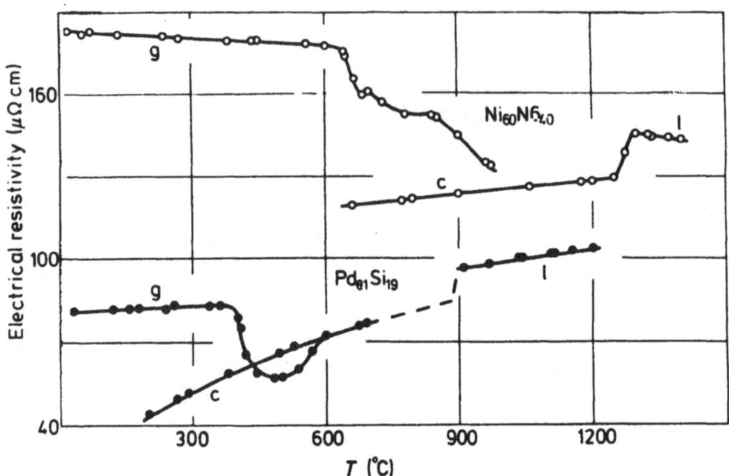

Fig. 7.9. Electrical resistivity of $Pd_{81}Si_{19}$ and $Ni_{60}Nb_{40}$ as a function of temperature for the glassy (g), crystalline (c), and liquid (l) phases [7.47]

It is appropriate to mention the extensive results of *Shull* et al. [7.45] on amorphous LaGa in this section. Negative α values occur when $\varrho > 150\,\mu\Omega\text{cm}$ and the negative α persist down to the lowest temperatures, which is in accord with the Mooij correlation. A close similarity between the amorphous and liquid phase transport properties is clearly indicated in this case because the anomalous positive Hall coefficient of liquid La is duplicated in the amorphous phase [7.46]. These positive Hall coefficients are unexplained at present.

Generally, strong similarities are found in the transport properties of amorphous and liquid alloys. The exceptional cases can usually be attributed to the existence of ordered phases in the amorphous metals (e.g., FeAu). The results of *Güntherodt* et al. [7.47] on amorphous and liquid NbNi and PdSi, shown in Fig. 7.9, are a convincing illustration of this point. The values for ϱ and α are approximately equal in both phases. Similar results are found with amorphous and liquid FeB [7.48]. This indicates that a proper theoretical description of one phase (liquid or amorphous) should be directly applicable to the other phase.

7.2 Theoretical Approaches to Electrical Transport in Amorphous Metals

At present, Ziman-Faber theory provides the only quantitative results for electrical transport in amorphous metals. The Boltzmann equation is the basis for this theory and is frequently questioned for strong scattering systems such as amorphous transition-metal-based alloys. More sophisticated approaches based upon the Kubo formula [7.49] have been successfully applied to transport in completely random liquids and crystalline alloys. In Sect. 7.3.1 we illustrate the type of approximations required to reduce the Kubo expression to manageable form according to *Chen*, *Weisz*, and *Sher* (CWS) [7.50] and relate these results to the Ziman-Faber theory. In Sect. 7.3.2 we present the Ziman-Faber theory of transport in amorphous metals and compare its predictions with experimental results. Another approach to transport in amorphous metals is based upon tunnelling in two-level systems. This approach is discussed in Chap. 8 of this volume and will not be treated here. In Sect. 7.3.3 we discuss theoretical approaches to the problem of transport in high-resistivity metals.

7.2.1 Reduction of the Kubo Formula and Its Relation to Ziman-Faber Theory

The Kubo formula [7.49] for the dc electrical conductivity is

$$\bar{\bar{\sigma}} = \frac{1}{\Omega} \lim_{\eta \to 0^+} \mathrm{Tr}\left\{ \int_0^\infty dt\, e^{-\eta t} \int_0^\beta d\lambda\, \varrho_e(H) \boldsymbol{j}(-i\hbar\lambda) \boldsymbol{j}(t) \right\}, \tag{7.1}$$

where the total current operator in the unperturbed Heisenberg picture

$$\boldsymbol{j}(t) = \exp(iHt/\hbar) \boldsymbol{j}(0) \exp(-iHt/\hbar) \tag{7.2}$$

the equilibrium density matrix

$$\varrho_e(H) = \exp(-\beta H)/\mathrm{Tr}\{\exp(-\beta H)\}, \tag{7.3}$$

H is the Hamiltonian before the electric field is turned on, $\beta = (k_B T)^{-1}$, k_B is Boltzmann's constant, T is the temperature, $\hbar = h/2\pi$ with h the Planck constant, and Ω is the system volume.

Equation (7.1) can be expressed in terms of Green's functions $G(E)$ and brought to the form

$$\bar{\bar{\sigma}} = \frac{e^2 h}{\pi m^2 \Omega} \mathrm{Tr}\left\{ \int dE \left(-\frac{df}{dE} \right) \varrho_e G^I(E) \boldsymbol{P} G^I(E) \boldsymbol{P} \right\}, \tag{7.4}$$

where

$$G^I(E) = \frac{1}{2i} [G(E) - G^\dagger(E)] = \mathrm{Im}\{G(E)\}, \tag{7.5}$$

and f is the Fermi-Dirac function. [Equation (7.4) is equivalent to (22) of CWS.] One then makes various approximations to compute the averaged product of Green's functions in (7.4).

The transport studies in liquids and disordered crystalline alloys which use the Kubo expression employ the single site approximation to determine the Green's functions. Discussions of these approximations are found, for example, in *Schwartz* and *Ehrenreich* [7.51] or *Elliott* et al. [7.52]. The coherent potential approximation (CPA) [7.53] is favored by many investigators, although the simpler form of the quasicrystalline approximation (QCA) [7.54] remains popular.

CWS treated disordered crystalline binary alloys within the CPA. In order to obtain manageable results they made the following assumptions (among others): i) Phonon scattering and electron-phonon interaction was treated as "elastic scattering". ii) The A and B atoms were distributed at random on a cubic lattice with local distribution functions given by $P_i(\eta)=\sqrt{2\pi\alpha_i}\exp(-\eta^2/2\alpha_i)$ with $i=A$, B and the parameters α_i varying in proportion to the temperature. iii) The Hubbard ellipse model [7.55] was taken for the pure crystal density of states and the velocity dispersion law was taken to be $v^2(E)=v_m^2(1-E^2)$ with E measured with respect to the center of the elliptical band in units of the bandwidth. Their result takes the form

$$\sigma=\frac{2e^2\hbar}{3\pi\Omega_c}\int d\eta\left(-\frac{df}{d\eta}\right)\int_{-1}^{1}dx\frac{N(x)v^2(x)\Delta^2(\eta)}{\{[\eta-\Lambda(\eta)-x]^2+\Delta^2(\eta)\}^2} \tag{7.6}$$

$$\cong\frac{\pi^2e^2\hbar v_m^2}{12\Omega_c}\frac{N^3(E_F)}{\Delta(E_F)}\left[1+\frac{6\Delta(E_F)}{\pi N(E_F)}\right], \tag{7.7}$$

where

$$v^2(x)=[N(E)N]^{-1}\sum_{\boldsymbol{k}}v^2(\boldsymbol{k})\delta[x-\varepsilon(\boldsymbol{k})], \tag{7.8}$$

$N(x)$ is the unperturbed density of states, $\boldsymbol{v}(\boldsymbol{k})=m^{-1}\langle\boldsymbol{k}|\boldsymbol{p}|\boldsymbol{k}\rangle$, $\Lambda(\eta)=\mathrm{Re}\{\Sigma(\eta+\mathrm{i}O\}$, $\Delta(\eta)=|\mathrm{Im}\{\Sigma(\eta+\mathrm{i}O\}|$, $\Sigma(\eta)$ is the electron self-energy, E_F is the Fermi energy, Ω_c is the volume per atom, N the number of atoms, e is the electron charge, and $\bar{\bar{\sigma}}=\sigma_{\bar{1}}$. The approximate equality appears in (7.7) because the term which varies like T^2 and higher order terms are neglected in the expansion of $df/d\eta$. If one is far from the band edges or if the imaginary part of the self-energy is small, (7.7) reduces to

$$\varrho\cong\frac{12\Omega_c}{\pi^2e^2\hbar v_m^2N^3(E_F)}\Delta(E_F), \tag{7.9}$$

which is the Boltzmann equation result for the same approximations. [CWS show that (7.9) is the free electron result even in the case of strong scattering.]

The agreement of (7.9) with the appropriate Boltzmann equation result suggests that (7.7) might provide a useful form for corrections to the Boltzmann equation results in the general case. In fact several authors [7.56, 57] have attempted to explain the Mooij correlation in terms of (7.7).

In order to complete the calculation of the resistivity one needs to compute the electron self-energy. CWS presented results of such calculations within CPA for a variety of scattering strengths and concentrations for a perfectly random binary crystalline alloy. (We know of no self-consistent calculations for realistic amorphous metallic structures.)

Before considering the Ziman-Faber theory, let us show explicitly how (7.9) goes into the appropriate form. For weak scattering, the self-energy can be expressed as a series in the matrix elements of the scattering potential $W(\boldsymbol{r}) = \Sigma w(\boldsymbol{r}-\boldsymbol{R}_i)$ as follows:

$$\begin{aligned}\Sigma(\boldsymbol{k},E) &= \langle \boldsymbol{k}|W|\boldsymbol{k}\rangle + \sum_{\boldsymbol{k}'(\neq \boldsymbol{k})} \langle \boldsymbol{k}|W|\boldsymbol{k}'\rangle G_0(\boldsymbol{k}')\langle \boldsymbol{k}'|W|\boldsymbol{k}\rangle \\ &\quad + \sum_{\boldsymbol{k}'(\neq \boldsymbol{k})}\sum_{\boldsymbol{k}'(\neq \boldsymbol{k}')} \langle \boldsymbol{k}|W|\boldsymbol{k}'\rangle G_0(\boldsymbol{k}')\langle \boldsymbol{k}'|W|\boldsymbol{k}''\rangle G_0(\boldsymbol{k}'')\langle \boldsymbol{k}''|W|\boldsymbol{k}\rangle + \dots \\ &\cong N\langle \boldsymbol{k}|w|\boldsymbol{k}\rangle + \sum_{\boldsymbol{q}\neq 0} \frac{|s(\boldsymbol{q})|^2 \langle \boldsymbol{k}|w|\boldsymbol{k}+\boldsymbol{q}\rangle\langle \boldsymbol{k}+\boldsymbol{q}|w|\boldsymbol{k}\rangle}{E-E_{k+q}-\mathrm{i}O^+},\end{aligned} \tag{7.10}$$

where $s(\boldsymbol{q}) = \Sigma \mathrm{e}^{\mathrm{i}\boldsymbol{q}\cdot\boldsymbol{R}_n}$ with $\boldsymbol{R}_n$ the position of the nth ion and $\langle \boldsymbol{k}|w|\boldsymbol{k}+\boldsymbol{q}\rangle$ is the pseudopotential matrix element (or the t matrix). Equation (7.9) with the second-order approximation of (7.10) gives a restricted form of Ziman-Faber theory in which the factor $(1-\cos\Theta)$ is taken as unity (i.e., the isotropic scattering form of Ziman-Faber theory is obtained). This is the appropriate form since the short-ranged random potential assumed by CWS, which makes the vertex corrections vanish, also leads to a vanishing "backscattering" term in the Boltzmann equation. One expects that self-consistent procedures will lead to Ziman-Faber theory in more general cases as well. In particular, for parabolic bands and weak scattering potentials of arbitrary range, *Edwards* [7.58] has shown that the vertex corrections indeed give rise to the required backscattering term. The higher order terms which are not included in (7.10) depend on combinations like $s(q_1)\ s(q_2)s(-q_1-q_2)$ and more complicated terms, and are usually neglected in approximations to the electron self-energy. The higher order terms and the correction factor $[1+6\Delta(E_F)/\pi N(E_F)]$ in (7.7) suggest generalizations of the Ziman-Faber theory.

7.2.2 The Diffraction Model: Ziman-Faber Theory

The diffraction model result for the resistivity ϱ of pure liquid metals [7.3, 5] is given as

$$\varrho = \frac{12\pi\Omega_c}{e^2\hbar V_F^2}\int_0^1 d\left(\frac{K}{2k_F}\right)\left(\frac{K}{2k_F}\right)^3 S^\varrho(\boldsymbol{K})|t(\boldsymbol{K})|^2, \tag{7.11}$$

where V_F is the Fermi velocity, $\boldsymbol{k}_F$ is the Fermi wave vector, $\boldsymbol{K}$ is the scattering vector, and the resistivity static structure factor $S^\varrho(\boldsymbol{K})$ is defined [7.59] in terms of the *Van Hove* [7.60] dynamical structure factor $S(\boldsymbol{K},\omega)$ as

$$S^{\varrho}(\boldsymbol{K})=\int_{-\infty}^{\infty} S(\boldsymbol{K},\omega)xn(x)d\omega\,, \tag{7.12}$$

with $x=\hbar\omega/k_{\mathrm{B}}T$ and $n(x)=(e^{x}-1)^{-1}$. The incorporation of the t matrix in (7.11) in the place of the pseudopotential matrix element is the *Evans* et al. [7.5] modification of the original form of the Ziman-Faber theory and may be expressed in terms of scattering phase shifts for nonoverlapping potentials (the "local approximation" [7.61]) as

$$t(K)=\frac{2\pi\hbar^{3}}{m(2mE_{\mathrm{F}})^{1/2}\Omega_{\mathrm{c}}}\sum_{l}(2l+1)\sin\eta_{l}(E_{\mathrm{F}})\mathrm{e}^{\mathrm{i}\eta_{l}(E_{\mathrm{F}})}P_{l}(\cos\Theta)\,, \tag{7.13}$$

where m is the electron mass and $\eta_{l}(E_{\mathrm{F}})$ is the phase shift for angular momentum quantum number l evaluated at E_{F}. Equation (7.11) was originally derived from the Boltzmann equation; however, *Evans* et al. [7.61] demonstrated that it could be derived directly from a force-force correlation expression for the resistivity (valid in the case of strong scattering) by making the local approximation.

In the usual formulation of Ziman-Faber theory which is given for liquids, the superscript ϱ is left off the static structure factor and no distinction is made with respect to the x-ray static structure factor $S^{x}(K)$. The distinction between $S^{\varrho}(K)$ and $S^{x}(K)$ becomes significant for temperatures less than the Debye temperature; for such temperatures S^{x} does *not* determine the temperature dependence of ϱ. With this distinction recognized, (7.11) applies to amorphous metals. Thus, the static structure factor plays a central role in determining the transport properties of amorphous metals.

Meisel and *Cote* [7.10] derived expressions for $S^{\varrho}(K)$ and $S^{x}(K)$ for amorphous metals having Debye phonon spectra. Their result for the resistivity static structure factor is

$$S^{\varrho}(K)=S^{\varrho}_{0}(K)+S^{\varrho}_{1}(K)+S^{\varrho}_{2}(K)+\dots, \tag{7.14}$$

where $S^{\varrho}_{n}(K)$ is an n phonon term. The elastic term is

$$S^{\varrho}_{0}(K)=a(K)e^{-2W(K)}, \tag{7.15}$$

where the geometrical structure factor

$$a(K)=\frac{1}{N}\sum_{\boldsymbol{m},\boldsymbol{n}}\exp[\mathrm{i}\boldsymbol{K}\cdot(\boldsymbol{m}-\boldsymbol{n})]\,; \tag{7.16}$$

$\boldsymbol{m}$ is the averaged position of the mth ion, $\exp[-2W(K)]$ is the Debye-Waller factor and

$$2W(K)=3[(\hbar K)^2/Mk_B\Theta](T/\Theta)^2\int_0^{\Theta/T} x[n(x)+\tfrac{1}{2}]\,dx\,, \tag{7.17}$$

where M is the ionic mass. The one-phonon term

$$S_1^\varrho(K)=\alpha(K)(\Theta/T)\int_0^1 (q/q_D)(q/q_D)^2 n(x)[n(x)+1)\int\frac{d\Omega}{4\pi}a(|\boldsymbol{K}+\boldsymbol{q}|)\,, \tag{7.18}$$

where $\alpha(K)=3(\hbar K)^2\exp[-2W(K)]/Mk_B\Theta$, and q_D is the Debye wave number, and $x=\hbar\omega/k_BT=(\hbar\omega_D/k_BT)q/q_D=(\Theta/T)q/q_D$.

At low temperatures ($T<\Theta$), $S_0^\varrho+S_1^\varrho$ gives an excellent approximation to S^ϱ; however, for $T>\Theta$ the multiphonon terms become significant. The simplest and most popular multiphonon approximation was suggested by *Sham* and *Ziman* [7.63] who accounted for the multiphonon series by replacing the Debye-Waller factor in S_1^ϱ by unity. Thus, in the Sham-Ziman approximation, $S^\varrho(K)=S_0^\varrho(K)+S_1^\varrho(K)$, with S_0^ϱ given in (7.15) and S_1^ϱ given by (7.18) except that $\alpha(K)=3(\hbar K)^2/Mk_B\Theta$. Another useful approximation to the multiphonon series was suggested by *Hernandez-Calderone* et al. [7.64]; they gave

$$\sum_{n=2}^{\infty} S_n^\varrho(K)\cong 1-[1+2W(K)]\mathrm{e}^{-2W(K)}\,. \tag{7.19}$$

For alloys, the product $S^\varrho(K)|t(K)|^2$ in (7.11) is replaced by a sum of concentration-dependent terms involving single-site t matrices of the individual constituents and partial structure factors. The partial structure factors S_{ij}^ϱ satisfy equations which are completely analogous to those for the pure case. For a binary alloys,

$$S^\varrho|t|^2\to\sum_{i=1}^{2} c_i|t_i|^2(1-c_i+c_iS_{ii}^\varrho)+c_1c_2(t_1^*t_2+t_1t_2^*)(S_{12}^\varrho-1)\,, \tag{7.20}$$

where c_i is the concentration of the ith component.

For transition-metal-based glassy metals, the largest contributions to the resistivity come from the transition metal d wave phase shift and the trends in the resistivity in these systems can be explained in terms of the transition metal partial structure factor evaluated in the backscattering region ($K\approx 2k_F$), i.e.,

$$\varrho\propto S_{TM}^\varrho(2k_F)\quad\text{(approximate)}. \tag{7.21}$$

Meisel and *Cote* also demonstrated [7.9, 10, 62] that at low temperatures ($T\approx 0$ K), $S^\varrho(K)\propto 1+BT^2$ where B is positive and independent of T in ranges of K where $a(K)$ is a continuous function and that at high temperatures $S^\varrho(K)\approx S^x(K)$. Although the low-temperature result was established for a

Debye spectrum, it is valid whenever the phonon density of states exhibits a quadratic dependence on frequency at low frequencies. This is the normal behavior of the phonon density of states and gives rise to the T^3 contribution to the low-temperature specific heat that is generally observed in crystalline and amorphous metals. The singularity in $a(K)$ at $K=0$ contributes a term to S_1^{ϱ} which gives the Bloch-Grüneisen result as in the crystalline case [7.8, 11] (i.e., a contribution which varies as T^5 at low temperatures, etc.). However, this term has not been observed in any amorphous metals analyzed thus far. It is expected to be small when backscattering dominates [7.62].

In discussions of the temperature dependence of the static structure factors at arbitrary T, it is useful to define [7.62] the averaged structure factors

$$A^{\varkappa}(K) \equiv \int_0^{q_D} dq\, q[n(x)+\tfrac{1}{2}] \int \frac{d\Omega}{4\pi} a(|\boldsymbol{K}+\boldsymbol{q}|) \Big/ \int_0^{q_D} dq\, q[n(x)+\tfrac{1}{2}] \tag{7.22}$$

and

$$A^{\varrho}(K) \equiv \int_0^{q_D} dq\, q^2 n(x)[n(x)+1] \int \frac{d\Omega}{4\pi} a(|\boldsymbol{K}+\boldsymbol{q}|) \Big/ \int_0^{q_D} dq\, q^2 n(x)[n(x)+1] \tag{7.23}$$

with x defined as in (7.18). Neglecting multiphonon terms, the static structure factors can thus be expressed as

$$\begin{aligned} S^{\varkappa}(K) &= a(K)\mathrm{e}^{-2W(K)} + 2W(K)A^{\varkappa}(K) \\ &\approx a(K)\mathrm{e}^{-2W(K)} + A^{\varkappa}(K)(1-\mathrm{e}^{-2W(K)}) \end{aligned} \tag{7.24}$$

and

$$S^{\varrho}(K) = a(K)\mathrm{e}^{-2W(K)} + \alpha(K)(T/\Theta)^2 A^{\varrho}(K) I_2(\Theta/T), \tag{7.25}$$

where

$$I_2(X) \equiv \int_0^x dx\, x^2 n(x)[n(x)+1]. \tag{7.26}$$

The forms of the Debye integrals $2W(K)$ and $I_2(X)$ are well known so that the temperature dependences of the static structure factors will be determined by that of the averaged structure factors for any given structural model. [Note that the multiphonon terms can be incorporated into (7.23), within the Sham-Ziman approximation, by simply replacing the Debye-Waller factor in $\alpha(K)$ by unity.] In particular, it is shown that $A^{\varrho}(K) \to a(K)$ as $T \to 0$ and that $A^{\varrho}(K) \approx A^{\varkappa}(K)$ for $T > \Theta$.

Many of the results in this extension of the diffraction model by *Meisel* and *Cote* have been obtained independently by others. *Markowitz* [7.65] pointed

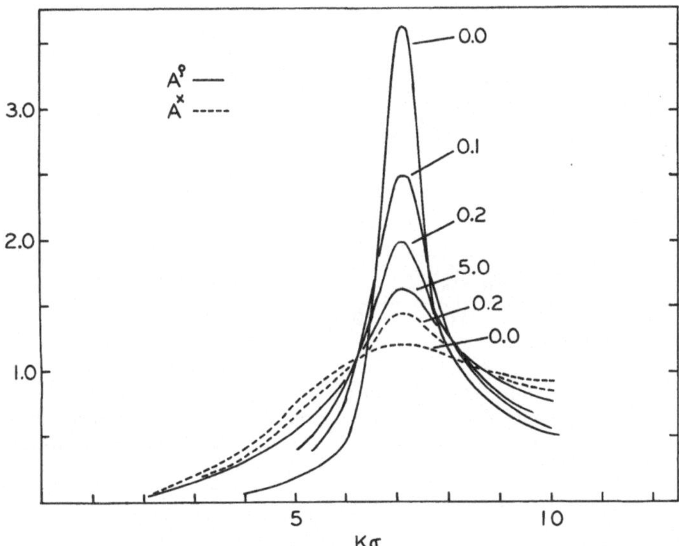

Fig. 7.10. The averaged structure factors for resistivity $A^{\varrho}(K\sigma)$ and for x-ray scattering $A^{x}(K\sigma)$ for $\eta=0.525$ and $\Theta=340$ K. The reduced temperature T/Θ is indicated for each curve. For $T/\Theta=5.0$, $A^{x}(K\sigma)$ is essentially identical to $A^{\varrho}(K\sigma)$. The curves for $T/\Theta=1.0$ (not shown) differ from the limiting curve ($T/\Theta=5.0$) by less than 2% [7.62]

out that the Debye-Waller factor is required in the elastic component of the resistivity. *Nagel*'s analysis [7.7] is similar to that presented here for high temperatures ($T>\Theta$). The formalism developed by *Froböse* and *Jäckle* [7.8] is essentially equivalent except that it was applied to a model with an Einstein phonon spectrum, the pseudopotential approximation, and step functions for the structure factors. *Ohkawa* [7.11] has also shown that the diffraction model predicts that ϱ increases as $+T^2$ at low temperatures in all cases when a Debye phonon spectrum is assumed.

The value of $2k_F$ in amorphous alloys generally lies in the vicinity of the first peak in the geometrical structure factor. Thus, it is convenient to introduce an analytic function which approximates the geometrical structure factors of amorphous metals in this region of K space. The Percus-Yevick hard sphere expressions [7.66] serve this purpose well. For example, the main peak in amorphous NiP can be fitted within a few percent by these expressions [7.67].

Figure 7.10 shows averaged structure factors computed [7.62] for a Percus-Yevick hard sphere model for a number of temperatures. The values selected for the packing fraction η and the hard sphere diameter σ are appropriate for NiP [7.67]. These results should be representative of all the metal-metalloid glassy alloys. The $T\to 0$ K limit for $A^{\varrho}(K)$ is $a(K)$ and for $T>\Theta$, $A^{\varrho}(K)\approx A^{x}(K)$ as stated above.

It is interesting to compare the predictions of this theory for the temperature dependence of $S^{x}(K)$ with experiment since this function can be measured

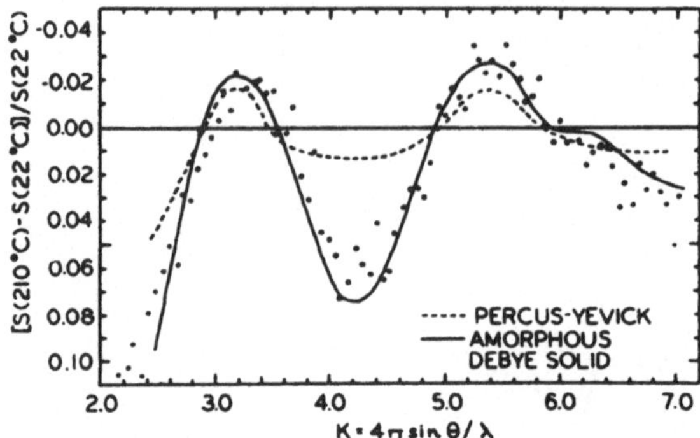

Fig. 7.11. Relative changes in the structure factor of amorphous $Ni_{75}P_{25}$ on heating from 22 to 210 °C. Data are compared with the model of an amorphous Debye solid ($\Theta = 340$ K). Also shown are the relative changes in the Percus-Yevick hard sphere alloy model due to thermal expansion [7.68]

directly. *Cote* et al. [7.68] have done this for amorphous $Ni_{75}P_{25}$ at two temperatures and their results are shown in Fig. 7.11. Near the peaks, $S^x(K)^{-1}\Delta S^x(K)/\Delta T$ is of the order of -10^{-4} K; it is positive and larger in the tails of the peaks. This behavior is analogous to that of Bragg peaks in crystalline materials where heating reduces peak intensities because of the Debye-Waller factor and increases the intensity between the Bragg peaks because of temperature diffuse scattering. (Measurements by *Waseda* and *Masumoto* [7.69] on amorphous PdSi and $Ni_{80}P_{20}$ generally corroborate these results.) The solid line shows the computed result. The good agreement with experiment indicates that the amorphous Debye solid is an appropriate model for amorphous metals.

Figure 7.12 shows $S^{\varrho}(T)/S^{\varrho}(\Theta)$ plotted against T/Θ for a series of K values. Equation (7.19) was used for the multiphonon series and the Percus-Yevick expression was used for $a(K)$ with parameters appropriate for NiP. (Note that the resistivity static structure factors are functions of K and T. We suppress either variable where convenient.) According to (7.11) a weighted average of these curves will give the temperature dependence of the resistivity; however, the major contributions to the scattering in these systems comes from the region near $2k_F$ [see (7.21)]. Thus, to define the approximate temperature and concentration dependences of the resistivity, we replace $S^{\varrho}(T)/S^{\varrho}(\Theta)$ by $\varrho(T)/\varrho(\Theta)$ and identify K with the appropriate $2k_F$. (One deduces the concentration dependence by defining a relationship between composition and $2k_F$. For example, one might define an effective charge for each ion and compute $2k_F$ from the free electron model. In general, the connection between concentration and $2k_F$ is difficult to establish a priori.)

The following characteristic features are exhibited: i) all curves increase as $+T^2$ from their $T=0$ values; ii) negative slopes occur for $T \gtrsim \Theta/2$ when

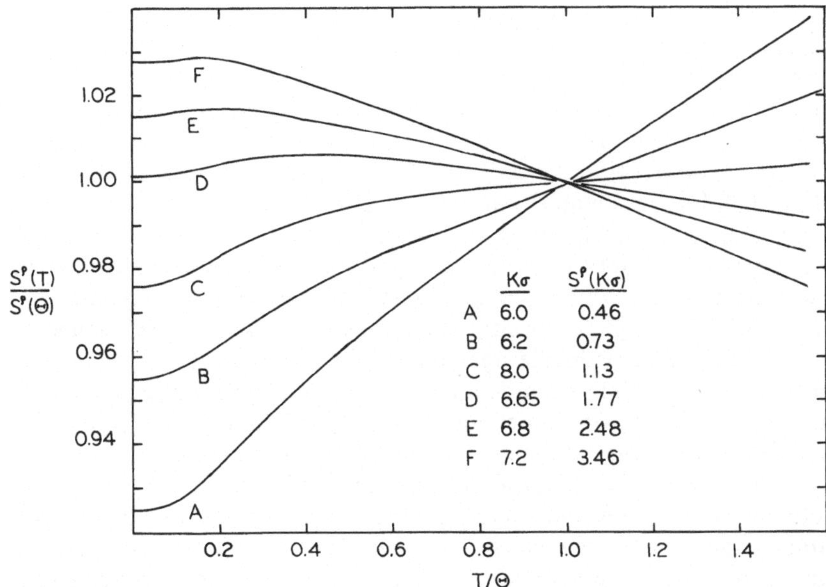

Fig. 7.12. Computed $S^{\varrho}(T)/S^{\varrho}(\Theta)$ for a variety of $K\sigma$ values, where $\eta=0.525$ and $\Theta=340$ K [7.62]. In this example, $K_p\sigma \cong 7.2$

$K \approx K_p$, the position of the main peak in $a(K)$, and positive slopes occur at all T for K outside the vicinity of K_p; iii) small maxima (of the order of tenths of a percent) are exhibited when $K \approx K_p$ and the largest maximum occurs for the nearly flat curve; iv) the amplitude of the variations of $S^{\varrho}(T)$, and the size of the maxima are sensitive to Θ and to the sharpness of the main peak in $a(K)$; and v) the curves which display only positive slopes have a sigmoidal shape.

Comparisons between some of the diffraction model results and the observed temperature dependence of ϱ have already been drawn in Sect. 7.1. The amorphous noble metal alloys show most of the features illustrated in Fig. 7.12 including resistivity maxima of the predicted magnitude and at the predicted temperature, and appropriate sign changes in the temperature coefficient of resistivity α with variation of concentration. The explanation of the latter effect in liquids is one of the major achievements of the original Ziman-Faber theory. Some indications of resistivity maxima are also found among the high-resistivity transition metal alloys despite the presence of strong saturation effects (to be discussed in Sect. 7.2.3).

Several additional observations on the temperature dependences are in order. The predicted sigmoidal shape (Fig. 7.12) in the positive α cases appears to be characteristic of amorphous metals (e.g., Fig. 7.5). The magnitude of α is typically of the order of 10^{-4}/K in agreement with the computed values. And, as first shown by *Hasegawa* [7.70], ϱ generally increases as $+T^2$ at low temperatures in the positive α cases.

Next let us consider the magnitude of the resistivity. [Implicit in these arguments is the assumption that the phase shifts (or the t matrices) are weak functions of T and K in comparison to S^{ϱ}.] There are a number of examples which indicate that Ziman-Faber theory yields the observed magnitude of ϱ as well. *Meisel* and *Cote* [7.67] found that Ziman-Faber theory gives the correct magnitude and composition dependence of ϱ (and the thermopower [7.34]) in amorphous NiP for "reasonable" values of k_F. *Froböse* and *Jäckle* [7.8] found good agreement with experiment for ϱ in amorphous CuSn using a pseudopotential approach; however, an Einstein phonon spectrum was required to obtain agreement with the observed α. *Waseda* and *Chen* computed ϱ and α for amorphous $Cu_{60}Zr_{40}$ [7.71] from (7.11) using measured partial structure factors and obtained reasonable agreement with experiment. These results, combined with the results on the temperature and composition dependences, indicate that the diffraction model contains much of the essential physics of transport in the lower resistivity amorphous alloys. Several questions remain, however. For example, calculations by *Dunleavy* and *Jones* [7.12] suggested that multiple scattering effects are significant in determining the t matrices, even in relatively low resistivity metals. There is also the inherent problem of determining the appropriate value of $2k_F$ and its concentration dependence, especially for transition metals. Moreover, high resistivity alloys show saturation effects, which are clearly beyond the scope of Ziman-Faber theory. In the last section we discuss some of the approaches which address these questions.

7.2.3 Transport in High Resistivity Metals

Transport in high resistivity metals ($\varrho \gtrsim 100\,\mu\Omega$cm) provides a difficult theoretical challenge. The Boltzmann equation is suspect in this limit. Localization and a metal-nonmetal transition are expected to occur [7.72]. Moreover, *Dunleavy* and *Jones* [7.12] have shown that multiple scattering effects play a significant part in determining the t matrix of (7.11) for liquid metals with resistivities as small as $40\,\mu\Omega$cm. There is, as yet, no generally accepted treatment of transport which addresses all these problems.

One approach to the treatment of high resistivity metals, which includes multiple scattering effects, was discussed in Sect. 7.2.1. Beginning with the Kubo formula, one makes a number of simplifying assumptions, computes the Green's functions, and then the conductivity. The problem with this prescription is, of course, to retain the essential physics and obtain manageable equations to solve. Most of the efforts along these lines begin with (7.7) which was originally derived by *Chen*, *Weisz*, and *Sher* (CWS) [7.13]. We list some of the important results of this approach for high resistivity metals.

i) CWS showed that as the strength of the effective scattering potential in a completely random alloy increases, an energy gap in the electron density of states begins to form and that, if the Fermi energy falls near this gap, one can obtain negative temperature coefficients of resistivity α. Since the resistivity generally increases with increasing scattering potential, this observation is consistent with the Mooij correlation. The opening of a gap in the density of states is a very interesting aspect of the CWS results with its obvious implications in regard to metal-nonmetal transitions [7.72].

ii) *Brouers* and *Brauwers* [7.16] applied the CWS approach to a two-band model. They showed that negative α occur when E_F is near a peak in the d band density of states. Since the resistivity is greatest for E_F in the peaks in the density of states, this result is also consistent with the Mooij correlation.

iii) *Harris* et al. [7.14] extended the discussions of CWS and made a separation of the scattering matrix into a pure phonon scattering part, a potential scattering part, and a "mixing term". The mixing term, produced by interaction between potential and phonon scattering, was shown to be negative. They suggested that this might be responsible for the Mooij correlation.

iv) *Richter* and *Schiller* [7.15] addressed the Mooij correlation directly. Following *Harris* et al. [7.14] they separated the self-energy into three terms and computed α and ϱ for a variety of conditions in a random binary alloy. Their results show a Mooij-like correlation. However, examination of their density of states and resistivity versus temperature curves indicates that the negative α values occurred for effective scattering strong enough to initiate gap formation and for E_F near the energy gap, just as in the analysis of CWS.

These theoretical studies, which completely ignore the structure of disordered materials, show many features of saturation and the Mooij correlation. They are not able to explain some details of the ϱ vs T curves, such as their sigmoidal shapes or their $\varrho \propto 1 - AT^2$ behavior in the $T \to 0$ limit. They also require special values of E_F with respect to gaps in the electronic density of states or with respect to d band maxima in the density of states. These special conditions on E_F obtain in few high resistivity metals. Thus, it is difficult to see how these results can explain the universal features of the Mooij correlation. Despite these shortcomings, the above attempts at treating transport directly from the Kubo formula, which does not fail in strong scattering systems, seem promising.

Another approach to the theory of high resistivity metals has been made by *Cote* and *Meisel* [7.73]. They noticed that, although Ziman-Faber theory could not explain the Mooij correlation, it continued to give a good approximation to the concentration dependence of the resistivity even in high resistivity amorphous and liquid alloys. The largest contributions to the resistivity in these alloys result from the elastic component of the scattering. This suggested that saturation effects might occur in the phonon scattering contributions to the resistivity before they become important in the elastic term. Thus, the question of electron mean free path effects on the electron-phonon interaction was considered as a basis for the Mooij correlation.

These effects had been considered earlier in *Ziman*'s discussion [7.74] of the "failure of the adiabatic approximation". The theory was originally introduced by *Pippard* [7.75] in the context of ultrasonic attenuation. *Ziman* stated: "*Phonons whose wavelengths $2\pi/q$ exceed the electron mean free path Λ are* *ineffective electron scatterers.*" We refer to this statement as the Pippard-Ziman condition. Pippard's work contains an analytic expression (the Pippard function) for the reduction of the electron-phonon interaction which occurs at short electron mean free paths (i.e., for $q\Lambda \lesssim 2\pi$). This idea forms the basis for the standard analysis of ultrasonic attenuation [7.76] and thermal conductivity data [7.77] in high resistivity metals. Although these ideas are well known, their application to electrical transport is recent [7.73, 78]. A similar approach in the context of the $s-d$ scattering model was also given in [7.78].

The effect of the Pippard-Ziman condition is to produce a low-frequency cutoff into the integrals over the phonon spectrum which occur in the expressions for the resistivity static structure factor. The form of the Pippard function or the Pippard-Ziman condition implies that the cutoff wave number q_c should be given by $q_c = 2\pi/\Lambda$. If one were to carry this approximation to the extreme, the electron-phonon interaction would be completely turned off for $\Lambda = 2\pi/q_D$, i.e., for electron mean free path approximately equal to the nearest neighbor spacing, in accord with the *Ioffe* and *Regel* [7.79] condition and corresponding to a resistivity of about 200 μΩcm in a monovalent free electron metal.

In the context of the model discussed in Sect. 7.3.2, *Cote* and *Meisel* [7.73] obtained a modified expression for the inelastic part of the static resistivity structure factor which depends explicitly on the electron mean free path,

$$S_1^{\varrho}(K,\Lambda) = \alpha(K)\frac{\Theta}{T}\int_0^1 d\left(\frac{q}{q_D}\right)\left(\frac{q}{q_D}\right)^2 F(q\Lambda)n(x)[n(x)+1]\int\frac{d\Omega}{4\pi}a(|\boldsymbol{K}+\boldsymbol{q}|) \tag{7.27a}$$

$$\cong \alpha(K)\frac{\Theta}{T}\int_\gamma^1 d\left(\frac{q}{q_D}\right)\left(\frac{q}{q_D}\right)^2 n(x)[n(x)+1]\int\frac{d\Omega}{4\pi}a(|\boldsymbol{K}+\boldsymbol{q}|) \tag{7.27b}$$

$$\equiv \alpha(K)\frac{\Theta}{T}A^{\varrho}(K,\Lambda)\, I_2\left(\frac{\Theta}{T}\right), \tag{7.27c}$$

where x is defined as in (7.18), the Pippard function [7.75]

$$F(a) \equiv \frac{6}{\pi a}\left[\frac{a^2 \tan^{-1} a}{3(a-\tan^{-1} a)} - 1\right],$$

the cutoff in (7.27b) is given by $\gamma \equiv 2\pi/\Lambda q_D$. and (7.27c) defines the mean free path dependent generalization of the averaged resistivity structure factor $A^{\varrho}(K,\Lambda)$. For high temperatures, $T > \Theta/2$, a good approximation is

$$S_1^{\varrho}(K,\Lambda) = \alpha(K)\frac{T}{\Theta}\int_\gamma^1 d\left(\frac{q}{q_D}\right)\int\frac{d\Omega}{4\pi}a(|\boldsymbol{K}+\boldsymbol{q}|) \tag{7.28a}$$

$$= \alpha(K)\frac{T}{\Theta}A^{\varrho}(K,\Lambda). \tag{7.28b}$$

For $K > k_F$ (the important region in amorphous metals) and $T > \Theta/2$,

$$A^{\varrho}(K,\Lambda) = \int_{\gamma}^{1} d\left(\frac{q}{q_D}\right) \int \frac{d\Omega}{4\pi} a(|\boldsymbol{K}+\boldsymbol{q}|) \approx (1-\gamma) A^{\varrho}(K), \tag{7.29}$$

where $A^{\varrho}(K) \equiv A^{\varrho}(K,\infty)$ is the temperature-dependent averaged structure factor defined in (7.23). [Equation (7.29) will also apply to crystalline alloys.]

Neglecting multiphonon terms, which are expected to be reduced even more than the one-phonon terms,

$$\varrho \cong (1-\gamma)\varrho_{ip} + \varrho_0 e^{-2W^*}, \tag{7.30}$$

where ϱ_{ip} is the ideal phonon resistivity (for $\Lambda \to \infty$) which is proportional to T at high temperatures, ϱ_0 is the residual resistivity, and $2W^*$ is an averaged Debye-Waller exponent.

Now

$$\begin{aligned} \varrho &= 3/N(E_F)e^2 v_F \Lambda \\ &= \gamma[3q_D/2\pi N(E_F)e^2 v_F] \equiv \gamma \varrho^*, \end{aligned} \tag{7.31}$$

where the second equality follows from the definition of γ and (7.31) defines the saturation resistivity ϱ^*. (As previously stated, $\varrho^* \approx 200\,\mu\Omega\text{cm}$ for a monovalent free electron metal.)

Combining (7.30) and (7.31) we find

$$\varrho = (\varrho_{ip} + \varrho_0 e^{-2W^*})/(\varrho_{ip}/\varrho^* + 1). \tag{7.32}$$

Equation (7.32) has the following features: i) Negative temperature coefficients of resistivity α are predicted for all temperatures when ϱ_0 is large. This follows from the fact that the averaged Debye-Waller factor determines α when $\gamma \approx 1$, which leads to $\varrho \propto 1 - BT^2$ at low temperatures in contrast to the predictions of Ziman-Faber theory (Sect. 7.3.2). Equation (7.32) will also not display the small maxima predicted in negative α cases by Ziman-Faber theory if $\gamma \gtrsim 0.5$. ii) There is only a very weak temperature dependence of ϱ if $\gamma \approx 1$ whether scattering is primarily elastic (from $\varrho_0 e^{-2W^*}$) or primarily inelastic (from ϱ_{ip}). iii) Negative α occurs for $\gamma \lesssim 0.5$ no matter what the electron per atom ratio may be. iv) Small $\Delta\varrho/\varrho$ on melting or on disordering are predicted when ϱ is large. Thus, the primary anomalies of high resistivity metals, namely saturation and the Mooij correlation, can be explained within the framework of Ziman-Faber theory by incorporating the Pippard-Ziman condition.

For high enough resistivity the elastic component is no longer expected to be approximated by Ziman-Faber theory either. The approach to electron localization in these cases might be described in the context of (7.7), which was taken from CWS and (7.10). Another approach might follow the lines described above for inelastic scattering. The analog of the Pippard-Ziman condition was suggested [7.73] to be: Periodicities in the radial distribution function of an amorphous metal corresponding to interatomic spacings larger than Λ are ineffective scatterers. This condition would lead to a cutoff in $S^e(K)$ at small momentum transfers, a region of little importance in determining the resistivity of amorphous and liquid high resistivity metals.

The Pippard-Ziman form of the Ziman-Faber theory fits the experimental data with values of ϱ^* equal about 200 μΩm, the value for a free electron monovalent metal. A major question is why this should be so in complicated systems such as transition metal alloys. There are also questions related to the importance of multiple scattering. Some authors [7.80] suggest that a discription including higher order correlation functions must be devised to treat high resistivity metals. However, the work of *Dunleavy* and *Jones* indicates that the Ziman-Faber form of the theory (7.11) may be retained with the effects of multiple scattering included in the t matrix. This is important because the major successes of the diffraction model comprise the explanations for the concentration and temperature dependences of the resistivity and thermopower directly in terms of the structure factor.

7.3 Concluding Comments

The focus of this chapter has been on the temperature and composition dependence of the electrical resistivity of amorphous metals. Magnetic effects were not considered here since they are discussed in the chapter on magnetism in this volume.

The data presented were selected as representative of the various classes of amorphous metals. The two theoretical approaches discussed in detail are believed to be the most successful or promising with respect to electrical transport in amorphous metals. Both may be viewed as derived from the exact Kubo formula. The Ziman-Faber theory does not treat multiple scattering but is correct to second order in the t matrix if the scattering is not too strong, and accounts for structural effects in terms of the pair correlation functions. The *Chen, Weisz,* and *Sher* approach is self-consistent and treats multiple scattering correctly but has only been applied to random binary alloys. Ziman-Faber theory appears to be valid for low resistivity alloys and the CWS theory results suggest that self-consistent Green's function analyses may be appropriate for high resistivity alloys. However, a variety of questions has surfaced and it is possible that a full understanding of transport in amorphous alloys may require one of the many other models which have been proposed in the literature.

Acknowledgements. The authors are grateful to Dr. H. Beck for his helpful suggestions on several points in this review. The authors also acknowledge the able assistance of Ellen Fogarty in the preparation of this manuscript.

References

7.1 N.F.Mott: Philos. Mag. **26**, 1249 (1972)
7.2 T.E.Faber: *Liquid Metals* (Cambridge University Press, London 1972)
7.3 J.M.Ziman: Philos. Mag. **6**, 1013 (1961)
7.4 T.E.Faber, J.M.Ziman: Philos. Mag. **11**, 153 (1965)
See also C.C.Bradley, T.E.Faber, E.G.Wilson, J.M.Ziman: Philos. Mag. **7**, 865 (1962)
7.5 R.Evans, D.A.Greenwood, P.Lloyd: Phys. Lett. A **35**, 57 (1971)
7.6 O.Dreirach, R.Evans, H.-J.Güntherodt, H.V.Künzi: J. Phys. F **2**, 709 (1972)
7.7 S.R.Nagel: Phys. Rev. B **16**, 1694 (1977)
7.8 K.Froböse, J.Jäckle: J. Phys. F **7**, 2331 (1977)
7.9 P.J.Cote, L.V.Meisel: Phys. Rev. Lett. **39**, 102 (1977)
7.10 L.V.Meisel, P.J.Cote: Phys. Rev. B **16**, 2978 (1977)
7.11 F.J.Ohkawa: Tech. Rpt. Inst. Solid State Phys. A **842** (1977)
7.12 H.N.Dunleavy, W.Jones: J. Phys. F **8**, 1477 (1978)
7.13 A.Chen, G.Weisz, A.Sher: Phys. Rev. B **5**, 2897 (1972)
7.14 R.Harris, M.Shalmon, M.Zuckerman: Phys. Rev. B **18**, 5906 (1978)
7.15 J.Richter, W.Shiller: Phys. Status Solidi (b) **92**, 511 (1979)
7.16 F.Brouers, M.Brauwers: J. Phys. **36**, L17 (1975)
7.17 J.H.Mooij: Phys. Status Solidi A **17**, 521 (1973)
7.18 Z.Fisk, G.W.Webb: Phys. Rev. Lett. **36**, 1084 (1976)
7.19 L.W.Shacklette, W.S.Williams: Phys. Rev. B **7**, 5041 (1973)
7.20 G.T.Meaden: *Electrical Resistance of Metals* (Plenum Press, New York 1965)
7.21 H.-J.Güntherodt, E.Hauser, H.V.Künzi, R.Evans, J.Evers, E.Kaldis: J. Phys. F **6**, 1513 (1976)
7.22 B.Delley, H.Beck, H.V.Künzi, H.-J.Güntherodt: Phys. Rev. Lett. **40**, 193 (1978)
7.23 K.H.J.Buschow, H.J.van Daal: AIP Conf. Proc. **5**, 1464 (1972)
7.24 B.R.Coles: Adv. Phys. **7**, 40 (1958)
7.25 D.Korn, W.Murer, G.Zibold: Phys. Lett. **47**A, 117 (1972)
7.26 E.Blasberg, D.Korn, H.Pfeifle: J. Phys. F **9**, 1821 (1979)
7.27 V.Even, W.Freyland: J. Phys. F **5**, L104 (1975)
7.28 G.Busch, H.-J.Güntherodt: Phys. Kondens. Mater. **6**, 325 (1967)
7.29 A.Roll, H.Motz: Z. Metallkd. **48**, 435 (1957)
7.30 J.E.Enderby, R.A.Howe: Philos. Mag. **18**, 923 (1968)
7.31 D.Korn, W.Murer: Z. Phys. B **27**, 309 (1977)
7.32 J.J.Hauser, J.Tauc: Phys. Rev. B **17**, 3371 (1978)
7.33 P.J.Cote: Solid State Commun. **18**, 1311 (1976)
7.34 P.J.Cote, L.V.Meisel: Phys. Rev. B**20**, 3030 (1979)
7.35 B.Boucher: J. Non-Cryst. Solids **7**, 277 (1972)
7.36 A.K.Sinha: Phys. Rev. B **1**, 4541 (1970)
7.37 G.L.Tangonan: Phys. Lett. A **54**, 307 (1975)
7.38 J.A.Rayne, R.A.Levy: In *Amorphous Magnetism II*, ed. by R.A.Levy, R.Hasegawa (Plenum Press, New York 1972) p. 319
W.Teoh, N.Teoh, S.Arajs: In *Amorphous Magnetism II*, ed. by R.A.Levy, R.Hasegawa (Plenum Press, New York 1972) p. 327
7.39 K.H.Buschow, N.M.Beekmans: Phys. Rev. B **19**, 3843 (1979)
7.40 S.R.Nagel, J.Vassiliou, P.M.Horn, B.C.Giessen: Phys. Rev. B **17**, 462 (1978)
7.41 G.R.Gruzalski, J.A.Gerber, D.J.Sellmyer: Phys. Rev. B **19**, 1469 (1979)
7.42 T.Mizoguchi, S.von Molnar, G.S.Cargill III, T.Kudo, N.Shiotani, H.Sekizawa: In *Amorphous Magnetism II*, ed. by R.A.Levy, R.Hasegawa (Plenum Press, New York 1972) p. 513
7.43 G.Bergmann, P.Marquardt: Phys. Rev. B **18**, 326 (1978)
7.44 H.-J.Güntherodt, H.U.Künzi: Phys. Kondens. Mater. **16**, 117 (1973)
7.45 W.H.Shull, D.G.Naugle, S.J.Poon, W.L.Johnson: Phys. Rev. B **18**, 3263 (1978)
7.46 P.C.Colter, T.W.Adair III, D.G.Naugle, W.L.Johnson: J. Phys. **39**, C6–955 (1978)
7.47 H.-J.Güntherodt, M.Müller, R.Oberle, C.Hauser, H.U.Künzi, M.Liard, R.Müller: Inst. Phys. Conf. Ser. **39**, 436 (1978)

7.48 H.-J.Güntherodt, H.U.Künzi, M.Liard, R.Müller, R.Oberle, H.Rudin: Inst. Phys. Conf. Ser. **30**, 342 (1977)
7.49 R.Kubo: J. Phys. Soc. Jpn. **12**, 570 (1957)
7.50 A.Chen, G.Weisz, A.Sher: Phys. Rev. B **5**, 2897 (1972)
7.51 L.Schwartz, H.Ehrenreich: Ann. Phys. **64**, 100 (1971)
7.52 R.J.Elliott, J.A.Krumhansl, P.L.Leath: Rev. Mod. Phys. **46**, 465 (1974)
7.53 P.Soven: Phys. Rev. **156**, 809 (1967)
7.54 M.Lax: Rev. Mod. Phys. **23**, 287 (1951); Phys. Rev. **85**, 621 (1952)
7.55 J.Hubbard: Proc. R. Soc. (London) A **281**, 401 (1964)
7.56 R.Harris, M.Shalmon, M.Zukermann: J. Phys. F **7**, L259 (1977)
7.57 J.Richter, W.Schiller: Phys. Status Solidi (b) **92**, 511 (1979)
7.58 S.F.Edwards: Philos. Mag. **3**, 1020 (1958); **4**, 1171 (1959)
See also B.Velicky: Phys. Rev. **184**, 614 (1969)
7.59 G.Baym: Phys. Rev. **135**, A1691 (1964)
7.60 L.van Hove: Phys. Rev. **95**, 249 (1954)
7.61 R.Evans, B.L.Gyorffy, N.Szabo, J.M.Ziman: In *The Properties of Liquid Metals*, ed. by S.Takeuchi (Wiley, New York 1973)
7.62 L.V.Meisel, P.J.Cote: Phys. Rev. B **17**, 4625 (1978)
7.63 L.J.Sham, J.M.Ziman: *Solid State Physics*, Vol. 15, ed. by F.Seitz, D.Turnbull (Academic Press, New York 1963)
7.64 I.Hernandez-Calderone, J.S.Helman, H.Vucetich: Phys. Rev. B **14**, 2310 (1976)
7.65 D.Markowitz: Phys. Rev. B **15**, 3617 (1977)
7.66 J.K.Percus, G.J.Yevick: Phys. Rev. **110**, 1 (1958)
J.K.Percus: Phys. Rev. Lett. **8**, 462 (1962)
7.67 L.V.Meisel, P.J.Cote: Phys. Rev. B **15**, 2970 (1977)
7.68 P.J.Cote, G.P.Capsimalis, L.V.Meisel: Phys. Rev. B **16**, 4651 (1977)
7.69 Y.Waseda, T.Masumoto: Sci. Rep. Res. Inst. Tohoku Univ. **27**A (1978)
7.70 R.Hasegawa: Phys. Lett. **36**A, 425 (1971)
7.71 Y.Waseda, H.S.Chen: Phys. Status Solidi (b) **87**, 777 (1978)
7.72 See, for example, N.F.Mott, E.A.Davis: *Electronic Processes in Non-Crystalline Materials* (Clarendon Press, Oxford 1971)
7.73 P.J.Cote, L.V.Meisel: Phys. Rev. Lett. **40**, 1586 (1978)
7.74 J.M.Ziman: *Electrons and Phonons* (Clarendon Press, Oxford 1960)
7.75 A.B.Pippard: Philos. Mag. **46**, 1104 (1955)
7.76 See, for example, C.Kittel: *Quantum Theory of Solids* (Wiley, New York 1964)
7.77 P.G.Klemens: *Solid State Physics*, Vol. 7, ed. by F.Seitz, D.Turnbull (Academic Press, New York 1958)
J.E.Zimmerman: J. Phys. Chem. Solids **11**, 299 (1959)
7.78 N.Morton, B.W.James, G.H.Wostenholm: Cryogenics **18**, 131 (1978)
7.79 A.F.Ioffe, A.R.Regel: Prog. Semicond. **4**, 237 (1960)
7.80 See, for example, E.Esposito, H.Ehrenreich, C.D.Gelatt, Jr.: Phys. Rev. B **18**, 3913 (1978)

8. Low-Energy Excitations in Metallic Glasses

J. L. Black

With 7 Figures

The purpose of this chapter is to review our present understanding of metallic glasses at very low temperatures ($T \leqq 1$ K). In this regime there are notable similarities between metallic glasses and other amorphous solids such as the insulating glasses. There are also, however, some significant and interesting differences. The theoretical model which encompasses both the similarities and the differences is the tunnelling model of *Anderson* et al. [8.1] and *Phillips* [8.2]. This model postulates the existence of a new class of characteristically "glassy" excitations called tunnelling levels (TLS) and furnishes a scheme for calculating their effects upon experimentally accessible quantities. In the insulating glasses, it is the coupling between TLS and phonons which leads to most of the interesting consequences. In metallic glasses, as we shall see, there is an additional coupling between the TLS and the conduction electrons, and this accounts for some interesting new physics.

If we momentarily disregard the novel consequences of this TLS electron interaction, then the most noteworthy development in metallic glasses at low temperatures is the confirmation of the generality of the tunnelling model itself. Originally formulated with melt-cooled insulating glasses in mind, this model has now proved to be applicable to a much wider class of amorphous solids. On the other hand, there can be no doubt that the structure of metallic glasses on an atomic scale differs considerably from that of insulating glasses [8.3, 4]. Thus, the mere existence of TLS in metallic glasses adds considerable impetus to the search for a common origin of these ubiquitous excitations and provides important clues about the role of such factors as bonding and atomic coordination.

In what follows, however, there will be little attention paid to microscopic derivations of the tunnelling model. This reflects the fact that there is no microscopic theory. Instead we shall rely upon the wealth of experimental information available for both metallic and insulating glasses in constructing the appropriate phenomenological theory. The first step will be to review this procedure for the insulating glasses. Then the experimental evidence in metallic glasses will be critically surveyed with emphasis upon those features which do not fit into the scheme developed for insulators. These features will then lead to the introduction of TLS-electron coupling, whose consequences will be extensively discussed. The focus throughout will be to critically examine the extent to which low-energy excitations in metallic glasses fit into the TLS scheme as it has been developed for insulating glasses.

8.1 The Tunnelling Model

8.1.1 General Considerations

The tunnelling model [8.1, 2] is based on a very plausible picture of the glassy state, which has been most clearly discussed in a recent review by *Anderson* [8.5]. The basic idea is that a glass finds itself trapped for kinetic reasons in a region of configuration space far removed from the true ground state (which is crystalline). Since this region of configuration space is characterized by a relatively high energy, there must be numerous other energetically equivalent regions into which the glass could have been trapped. Most of these states are very far away and, hence, inaccessible. The existence of numerous roughly equivalent but inaccessible "ground states" accounts for the observation of finite zero-point entropy ($S \neq 0$ at $T=0$) in glasses [8.1]. On the other hand, it is certainly plausible that not all of these states are so distant as to be *totally* inaccessible. Any roughly equivalent states into which the system can make transitions within an experimental time scale will contribute to the specific heat rather than to the zero-point entropy. It is these configurations which lead to "tunnelling states" (or TLS for "two-level systems").

Within this general picture, it is apparent that a TLS is undoubtedly some small group of atoms undergoing a local rearrangement. The number of atoms involved must be reasonably small in order to minimize the distance between the states in configuration space. On the other hand, the requirement of roughly equivalent energies will surely become easier to satisfy as the number of participating atoms increases. It is this competition between accessibility and degeneracy which determines the size of the TLS.

But what specifically are these states? There is no easy answer, but the search for specific rearrangement modes must take into account the "universality" of TLS. As shown by *Zeller* and *Pohl* [8.6] and *Stephens* [8.7], these excitations appear to be intrinsic (i.e., not identifiable with any specific impurities) features of a wide variety of glasses. Subsequent work by *Jones* et al. [8.8] has shown that there is probably at least one counterexample, amorphous As. On the other hand, there is a growing body of evidence for the existence of TLS in systems which might not normally be considered glasses. These include superionic conductors [8.9–11] neutron-irradiated quartz [8.12], and certain crystals which contain small "disordered regions" [8.13, 14]. All of this suggests that "glassiness', as we presently understand it, is neither totally sufficient nor absolutely necessary for the occurrence of TLS. As has been emphasized by *Phillips* [8.15], there appears to be a need for at least some short-range disorder in conjunction with bonding constraints which are not "too restrictive".

8.1.2 Two-Level Model

Putting such microscopic matters aside, consider a point of local minimum on the potential energy surface of a glass. We are now interested in finding nearby minima with roughly the same energy. The magnitude [8.7] of the linear specific heat in glasses tells us that the probability per atom of finding even one such nearby state is exceedingly small (on the order of 10^{-5}). Thus it is an excellent approximation to neglect the possibility of finding a third minimum in the vicinity. If we imagine slicing a "minimum difficulty" (in terms of distance and potential barrier) path between the two chosen minima, we obtain Fig. 8.1. Now the important concern is to determine the excitations of this double minimum in isolation from all other degrees of freedom. Each well alone has a series of vibrational states beginning with a ground state (called ψ_L in the "left" well, ψ_R in the "right" well) and followed by excited states at intervals of $\hbar\omega_0$, the vibrational energy. Since $\hbar\omega_0$ is likely to be comparable to the Debye energy, excitations among these single-well states are not important in the temperature regime of interest ($T<1$ K). The remaining possibility is to take advantage of the near degeneracy of the two ground states ψ_L and ψ_R, but it is still necessary that we be able to cause *transitions* between these states; otherwise they will be unobservable. Barring the unlikely possibility of classical transitions over the barrier (which at 1 K requires a barrier so low that we are back to a single well with $\hbar\omega_0$ energy spacings), it is quantum-mechanical tunnelling which allows these states to be seen.

To see how tunnelling accomplishes this, consider the Hamiltonian [8.1, 2, 16, 17] for our two-state system written in terms of the basis states ψ_L and ψ_R,

$$H=\frac{1}{2}\begin{pmatrix} \Delta & -\Delta_0 \\ -\Delta_0 & -\Delta \end{pmatrix}+\begin{pmatrix} \gamma & 0 \\ 0 & -\gamma \end{pmatrix}e, \tag{8.1}$$

where $\Delta_0=\hbar\omega_0\exp(-\lambda)$, $\lambda=d\hbar^{-1}\sqrt{2mV}$, e is the strain field at the site of the TLS, and $2\gamma=d\Delta/de$. Here m is the effective mass of the rearranging atoms and all other quantities are defined in Fig. 8.1. The quantity Δ_0 is the tunnelling energy[1], which results from overlap of ψ_L into the right well and vice versa. The "deformation potential" γ establishes contact between the TLS and the outside world. It originates in the dependence of the TLS part of the potential energy surface (see Fig. 8.1) upon the relative positions of nearby atoms. We are neglecting a much smaller coupling term proportional to $d\Delta_0/de$.

The next step is to diagonalize the strain-independent part of (8.1) in order to obtain the eigenstates of the unperturbed TLS. In doing this it is convenient to rewrite (8.1) in terms of spin-1/2 operators ($S_\alpha=(1/2)\sigma_\alpha$, where σ_α are the Pauli matrices)

$$H=\Delta S_z-\Delta_0 S_x+2\gamma e S_z\,. \tag{8.2}$$

Carrying out the diagonalization, (8.2) becomes

$$H=ES_z+(2MS_x+DS_z)e\,, \tag{8.3}$$

1 Most authors [8.1, 2, 16, 17] have unfortunately written (8.1) with $+\Delta_0$ instead of the correct $-\Delta_0$. This sign error does not, however, affect any observable quantities.

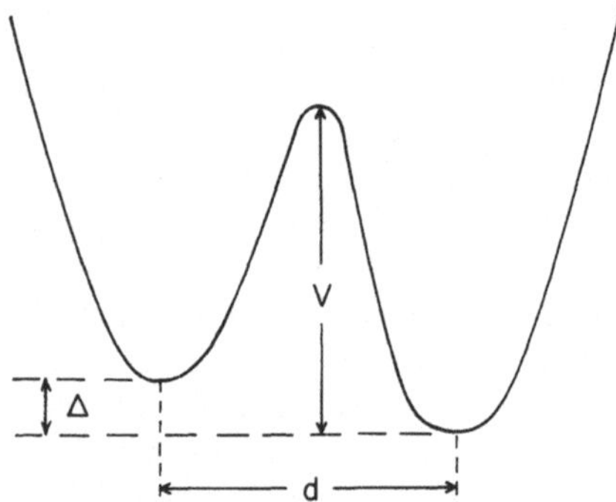

Fig. 8.1. Part of potential energy surface producing a tunnelling state characterized by a barrier height V, asymmetry energy Δ, and generalized distance d

where $E^2=\Delta^2+\Delta_0^2$, $M=(\Delta_0/E)\gamma$, and $D=2(\Delta/E)\gamma$. In going from (8.2) to (8.3), the Hamiltonian has been reexpressed in terms of the ground state $\psi_g=\psi_R\cos\theta+\psi_L\sin\theta$ and the excited state $\psi_e=\psi_L\cos\theta-\psi_R\sin\theta$ with $\tan 2\theta=\Delta_0/\Delta$. This simple derivation illustrates two important points. First of all, the energy splitting E can never be less than Δ_0. This fact, together with the experimental requirement $E<1$ K, means that the barrier V cannot become arbitrarily small. It is necessary to have some sort of barrier in order to obtain sufficiently closely spaced energy levels. Secondly, consider the symmetric case $\Delta=0$ (which implies $E=|\Delta_0|$). In this case, ψ_g and ψ_e are even and odd linear combinations of ψ_L and ψ_R and are not localized in either well. Such states are the easiest to observe because "observation" implies the ability to cause a transition from ψ_g to ψ_e by means of the off-diagonal coupling MeS_x. For a fixed value of E, this quantity M ranges from its maximum value γ, in this symmetric case to its minimum value of zero in the high-barrier case ($\Delta_0=0$, $E=|\Delta|$). Thus it is clear that the experimentally important states have a barrier distribution whose minimum value is determined by the requirement of small energy spacings and whose maximum value is determined by observability.

To complete the construction of the model, it is only necessary to make explicit assumptions about the distribution of the parameters Δ, Δ_0, and γ over the total potential energy surface. The usual assumptions [8.16–18] are that the distribution function is a) independent of Δ, b) dependent on Δ_0 only through the overlap parameter λ, which is uniformly distributed, and c) strongly peaked at a dominant value of γ. These assumptions are at least qualitatively consistent with most experiments below 1 K. The result is a useful expression for the partial density of states for TLS with energy splitting E and off-diagonal coupling $r=\Delta_0^2/E^2$,

$$P(E,r)=\tfrac{1}{2}\bar{P}r^{-1}(1-r)^{-1/2}, \tag{8.4}$$

where $0\leqq r\leqq 1$ and $\bar{P}$ is an experimentally determinable density of states on the order of $10^{19}-10^{20}\,\mathrm{eV^{-1}\,cm^{-3}}$.

8.1.3 Experimental Consequences

In this section we review the implications of the above ideas for low-temperature experiments in insulating glasses as a prerequisite for understanding the recent results in metallic glasses. For example, the TLS contribution to the specific heat is obtained by summing up the Schottky type contribution of each pair of levels to give

$$\begin{aligned} C_v &= k_B \int_0^\infty dE \int_{r_0}^1 dr\, P(E,r)(\beta E/2)^2 \operatorname{sech}^2(\beta E/2) \\ &= \frac{\pi^2}{6} k_B^2 T \bar{P} \left[\int_{r_0}^1 \frac{dr}{r} (1-r)^{-1/2} \right], \end{aligned} \tag{8.5}$$

where r_0 is a maximum barrier height cutoff which can be a function of the time scale of the experiment [8.18] and $\beta = (k_B T)^{-1}$. A roughly linear term in the specific heat below 1 K has been observed in a wide variety of insulating glasses, as reported by *Stephens* [8.7]. Furthermore the deduced magnitude of the density of states [$\bar{P}$ times the integral in brackets in (8.5)] varies by no more than a factor of 5 from one glass to another. This density of states implies a concentration of TLS with $E < 1$ K which ranges between 1 and 10 TLS per million atoms. The question of whether this density of states should be an increasing function of the experimental time scale is still controversial [8.18–20].

Aside from the specific heat, most of the principal consequences of the tunnelling model involve the interaction of TLS with phonons. A phonon with frequency $\omega = \hbar^{-1} E$ can be resonantly absorbed by a TLS of splitting E in its ground state. Likewise this same phonon can stimulate the emission of another phonon of the same frequency and wave vector if it happens to encounter the TLS in its excited state. Both of those processes are governed by the term proportional to M in (8.3) and their difference leads to a phonon mean free path given by [8.1, 2]

$$l^{-1} = \frac{\pi\omega}{\varrho v^3} \bar{P} \gamma^2 \tanh(\beta\hbar\omega/2), \tag{8.6}$$

where ϱ is the mass density and v is the sound velocity. The hyperbolic tangent expresses the TLS population difference and leads to a negative temperature coefficient of ultrasonic attenuation. Measurements [8.21, 22] on the insulating glass SiO_2 have shown good agreement with (8.6). Furthermore this expression for the mean free path permits a calculation of the phonon contribution to the thermal conductivity. (The localized TLS carry no heat current themselves.) Using the Debye spectrum and summing over all acoustic phonon modes, the

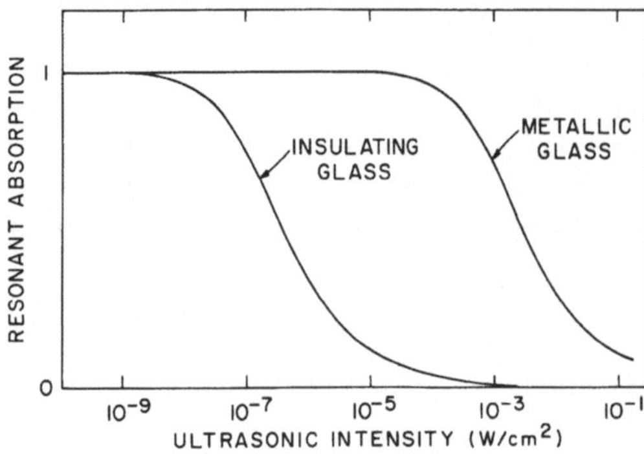

Fig. 8.2. Typical metallic vs insulating behavior in the saturation of ultrasonic absorption (normalized to its low-power value) at temperatures near 10 mK and frequencies near 1 GHz [8.25, 26, 51]

elementary transport theory result $\kappa=(1/3)Cvl$ becomes [8.1, 2]

$$\kappa=\frac{\varrho k_B^3 v}{6\pi\hbar^2\bar{P}\gamma^2}T^2. \tag{8.7}$$

The T^2 temperature dependence of (8.7) is in excellent agreement with experiment for a wide variety of glasses [8.7].

A related feature of phonon propagation in glasses is the sound velocity. The resonant interaction responsible for (8.6) leads to a measurable temperature-dependent shift in this quantity. A Kramers-Kronig analysis applied to the scattering law of (8.6) yields [8.22–24] a relative velocity shift $\Delta v/v$ of the form

$$\frac{\Delta v}{v}=\frac{\bar{P}\gamma^2}{\varrho v^2}\ln(T/T_0), \tag{8.8}$$

where T_0 is some arbitrary reference temperature. Velocity shifts of this form have been widely observed in insulating glasses [8.22–24]. Note that both the attenuation and the velocity allow a determination of $\bar{P}\gamma^2$. Values obtained by these two different methods tend to agree within 50%.

One of the most successful features of the tunnelling model is its ability to explain the saturation of the resonant attenuation in (8.6). As shown in Fig. 8.2, "saturation" is the decrease in l^{-1} as the flux of phonons increases. This effect originates in the tendency of the beam of phonons to equalize the populations in the TLS states ψ_g and ψ_e. If these populations were to become equal, the resonant phonon absorption and emission processes would completely cancel each other, and there would be no net absorption. In practice, however, there are always other processes which attempt to restore the TLS populations to their thermal equilibrium values, thereby counteracting this tendency toward equalization. The competition between these two tendencies determines the saturation threshold I_c, which is the value of phonon energy flux [W/cm^2] at which l^{-1} falls to $1/\sqrt{2}$ of its unsaturated value. In the simplest steady-state theory [8.22, 25, 26], this critical intensity is given by

$$I_c = \frac{\hbar^2 \varrho v^3}{2\gamma^2} T_1^{-1} T_2^{-1}, \tag{8.9}$$

where T_1 and T_2 are the longitudinal and transverse relaxation times of the TLS [8.27].

The longitudinal relaxation rate, which governs the direct return of the TLS population to equilibrium by means of interactions with thermal phonons, is given by the one-phonon process [8.16]

$$T_1^{-1} = \frac{M^2 E^3}{2\pi \varrho v^5 \hbar^4} \coth(\beta E/2). \tag{8.10}$$

The transverse rate, which determines the "homogeneous linewidth" of the TLS, is dominated by strain-mediated "spin-spin" interactions among the TLS. This leads to a width T_2^{-1} which is roughly linear in T, independent of E, and much greater than T_1^{-1} [8.25, 28, 29].

Ultrasonic experiments in insulating glasses are in good agreement with these consequences of the tunnelling model. The observation by *Hunklinger* et al. [8.30] and *Golding* et al. [8.26] of intensity-dependent resonant absorption of the type shown in Fig. 8.2 has allowed determination of the saturation threshold in SiO_2 and borosilicate glasses. The magnitude of this quantity is in reasonable agreement[2] with estimates based upon (8.9) using experimentally determined values of the quantities γ, T_1, and T_2 [8.25]. Furthermore *Hunklinger* and *Arnold* [8.25, 32] have used a weak probing pulse to measure the depression of the TLS population difference caused by a second pulse of high intensity. By changing the frequency of one of the pulse, these authors were able to map out the "burned hole" caused by the intense pulse, thereby allowing a direct measurement of the homogeneous width T_2^{-1}. Such two-pulse techniques have also yielded information about T_1 by allowing observation (via the weak pulse) of the time-dependent recovery of the population difference to its equilibrium value, as has been reviewed by *Golding* and *Graebner* [8.27]. This gives values of T_1 near 1 µs for $T \simeq 1$ K and $\omega/2\pi \simeq 1$ GHz.

The resonant interaction between TLS and phonons leads to another important effect which is usually called "phonon echoes". A phonon echo is the spontaneous generation of a coherent phonon pulse by tunnelling levels which have been excited previously by two or more ultrasonic pulses. This emitted pulse is the result of a momentary "phase-coherent tunnelling" among the various excited TLS. This process is completely analogous to the generation of spin echoes by the momentary phase coherence of precessing magnetic moments [8.33]. *Golding* and *Graebner* [8.34] first observed phonon echoes in SiO_2 glass at temperatures below 100 mK. These experiments have yielded a

2 There can be, however, some important corrections [8.26, 29] to (8.9) when the duration τ_p of the ultrasonic pulse is shorter than T_1. These effects increase I_c by roughly a factor of T_1/τ_p [8.31].

wealth of detailed information about the strength of the phonon-TLS coupling γ and about the dynamical processes which are summarized under the names of T_1 and T_2 [8.35]. In particular, the decay of the echoes with increasing pulse separation is readily explicable in terms of a "spectral diffusion" process arising from the "spin-spin interactions" among the TLS [8.29, 35].

Finally there is a nonresonant TLS-phonon process which deserves some emphasis. This is the relaxational attenuation of phonons due to tunnelling systems [8.36]. The origin of this effect is the fact that a phonon-induced elastic stress provides a periodic (at the phonon frequency ω) modulation of the TLS energy splitting, as is seen in the term proportional to D in (8.3). As the TLS population difference attempts to adjust to this changing splitting, there is a corresponding adjustment in the local strain field. Thus each TLS contributes a frequency-dependent term to the elastic modulus of the solid, and this frequency dependence is determined by T_1^{-1}, the relaxation rate of the population difference. The result is a contribution to the attenuation of the form [8.36]

$$l^{-1}=\int_0^{\infty} dE \int_{r_0}^{1} dr\, P(E,r)\, \mathrm{sech}^2(\beta E/2)\, \frac{\beta D^2}{4\varrho v^3}\, \frac{\omega^2 T_1}{1+\omega^2 T_1^2}, \tag{8.11}$$

where the density of states $P(E,r)$ is given by (8.4). Likewise there is a contribution to the sound velocity which is given by [8.36]

$$\frac{\Delta v}{v}=-\int_0^{\infty} dE \int_{r_0}^{1} dr\, P(E,r)\, \mathrm{sech}^2(\beta E/2)\, \frac{\beta D^2}{8\varrho v^2}\, \frac{1}{1+\omega^2 T_1^2}. \tag{8.12}$$

These relaxational contributions have been observed in insulating glasses, but primarily at temperatures above 1 K [8.25, 36]. This restriction is easy to understand in terms of the temperature dependence of the relaxation time for those TLS which contribute to (8.11) and (8.12), which must satisfy $E \leqq k_B T$. Reliable estimates based on (8.10) yield [8.29] the result that T_1 varies as T^{-3} and exceeds 1 µs at $T=1$ K. This implies $\omega T_1 \gg 1$ for frequencies near 1 GHz whenever $T \leqq 1$ K. Thus relaxational effects, which become comparable to resonant effects only when $\omega T_1 \simeq 1$, are large only above 1 K in insulating glasses.

This concludes our discussion of the tunnelling model and its experimental consequences for insulating glasses. We have emphasized those experimental probes which rely upon the coupling between TLS and long-wavelength acoustic phonons and have neglected the analogous effects resulting from the coupling between TLS and electric fields [8.22, 37–39]. This has been done in anticipation of the discussion of metallic glasses, where electric field probes have not been used. We have seen that the overall agreement between theory and experiment is good enough to allow a meaningful discussion of the dynamics of the tunnelling levels, as summarized by the relaxation times T_1 and T_2. In the next section, we shall ask whether this picture of low-energy excitations in amorphous solids applies equally well to metallic glasses.

8.2 Low-Temperature Properties of Metallic Glasses

8.2.1 Specific Heat

The first place to look for evidence of tunnelling levels in metallic glasses is the low-temperature specific heat. The primary obstacle here is to distinguish TLS effects from the large electronic contribution, which is also linear in the temperature. For example, the coefficient of T in the TLS linear specific heat [8.7] of SiO_2 glass is roughly 0.1 mJ/K^2/mole, whereas the electronic coefficient for crystalline Pd is 9 mJ/K^2/mole [8.40]. This difficulty can, however, be overcome if the measurements are made at temperatures sufficiently far below T_c in superconducting metallic glasses. An early attempt by *Comberg* et al. [8.41] failed to detect a linear term in amorphous indium films, although some "excess" lattice specific heat was found.[3] The first direct observation of a linear term was accomplished by *Graebner* et al. [8.42] in measurements on ZrPd glass well below its T_c of 2.53 K. The magnitude of the reported linear term is only a factor of 3 smaller than the corresponding term in SiO_2 glass [8.7].

8.2.2 Thermal Conductivity

More widespread evidence for the existence of TLS in metallic glasses comes from measurements of the low-temperature thermal conductivity. Here again, the crucial question is whether the TLS effects can be distinguished from electronic effects. Two questions must be addressed: a) Which excitations carry the heat? b) Which scattering processes limit the mean free paths of the heat carriers? Electrons can contribute to both a) and b), thereby complicating the simple result of (8.7), which applies only to the scattering of phonon carriers by TLS. As in the case of the specific heat, measurements in superconducting metallic glasses have played an important role in separating these contributions.

Matey and *Anderson* [8.43] reported the first evidence of a "glassy" T^2 contribution to κ in measurements on PdSi (nonsuperconducting). These authors used the Wiedemann-Franz relation [8.40] between thermal and electrical conductivity [$\kappa \sim T\sigma$] to subtract off the electronic contribution as a heat carrier. In fact, the electrical resistance is typically large enough in metallic glasses to estimate this contribution as less than 10% of the phonon part of the heat current. The authors also discounted the importance of electron-phonon contributions to the phonon mean free path on the basis of magnitude

3 Similar negative results for amorphous Bi had been found by G. Krauss, W. Buckel: Z. Physik B20, 147 (1975).

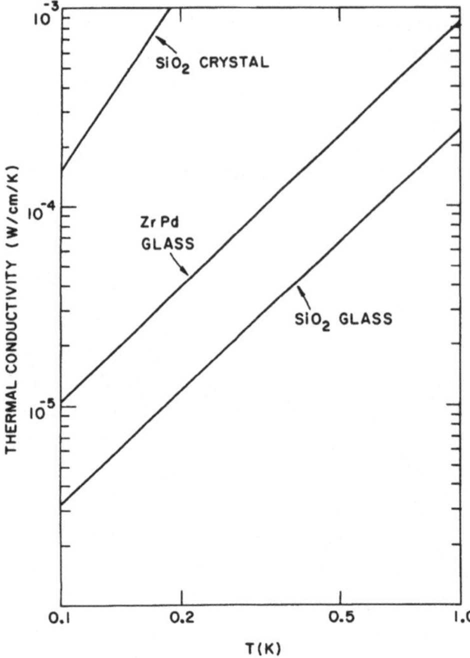

Fig. 8.3. Thermal conductivity of the metallic glass ZrPd [8.42], the insulating glass SiO_2 [8.6, 7], and crystalline SiO_2 [8.6]

estimates. They supported this position with the observation that the phonon mean free path was observed to show no sharp temperature dependence near the superconducting transition in the metallic glass TiBe.

Von Löhneysen and *Steglich* [8.44] made extensive use of the superconducting transition in amorphous and crystalline PbCu films in order to separate the various contributions to the thermal conductivity. Well below T_c, where electronic contributions as carriers and scatterers should be completely absent, they observed a phonon conductivity which is suppressed relative to crystalline PbCu superconductors below T_c. The temperature dependence of κ in this regime is consistent with (8.7). At temperatures near but below T_c, the authors observed a rapid increase in κ which they attributed to electrons as carriers. Above T_c, the roughly $aT+bT^2$ behavior was explained as an impurity-limited electron current (aT) plus an electron-limited phonon current (bT^2).

Graebner et al. [8.42] finally confirmed all the indications of TLS-limited phonon conduction by observing an almost perfect T^2 law well below T_c in the metallic glass ZrPd. Their results are compared with insulating glasses and crystals in Fig. 8.3. Furthermore, at temperatures just below T_c, their results showed a peak in the phonon conductivity, indicating the possible importance of electron-phonon scattering. *Raychaudhuri* and *Hasegawa* [8.45] have subsequently verified both the TLS-limited phonon transport ($T \ll T_c$) and the electron-limited phonon transport ($T < T_c$) in measurements on ZrBe glass.

To summarize, there is now little doubt that a number of metallic glasses exhibit TLS-dominated phonon thermal conductivity at low temperatures. The magnitude of the T^2 term [8.42–45] indicates that the TLS-phonon coupling strength $\bar{P}\gamma^2$ in metallic glasses is comparable to the values measured in insulating glasses [8.18, 21–26]. For superconducting metallic glasses near T_c and for normal metallic glasses, the question of electrons serving as a) heat carriers and b) phonon scatterers has aroused considerable interest and some controversy. It is safe to say that electronic effects of both types have been observed in amorphous metals [8.42–45]. This fact serves as a warning against attempting to extract quantitative information about the phonon-TLS interaction unless the thermal conductivity measurements are made well below T_c.

8.2.3 Ultrasonic Attenuation and Sound Velocity

The advantage of ultrasonic measurements over thermal conductivity is that one is able to study the influence of various scattering mechanisms upon monochromatic phonons. The limitation until very recently has been the restriction to a small number of nonsuperconducting metallic glasses (such as PdSi and NiP) on the basis of available sample sizes. Recent experiments by *Weiss* et al. [8.46] have begun to overcome this obstacle, but most available experimental information is for systems in which conduction electrons abound. It is therefore necessary to address the question of the magnitude of the direct electronic contribution to ultrasonic properties at frequencies less than 1 GHz. The expression given by *Pippard* [8.47] for the ultrasonic attenuation due to electron-phonon interactions in the dirty limit (electronic mean free path $\ll$ phonon wavelength) is

$$l^{-1} = \tfrac{1}{5} nmv_F^2 (\varrho v^3)^{-1} \omega^2 \tau\,, \tag{8.13a}$$

where n, m, v_F, and τ are the electronic density, mass, Fermi velocity, and relaxation time. Equation (8.12) can be expressed in terms of the electrical conductivity ($\sigma = ne^2\tau/m$) as

$$l^{-1} = \tfrac{1}{5} \sigma (mv_F/e)^2 (\varrho v^3)^{-1} \omega^2\,. \tag{8.13b}$$

Using values appropriate[4] for PdSi, the estimated l^{-1} is $3 \times 10^{-2}\,(\omega/2\pi)^2\,\mathrm{cm}^{-1}$, where the ultrasonic frequency $\omega/2\pi$ is measured in GHz. For frequencies near 1 GHz, this electronic attenuation is a negligibly small fraction of the typically observed value of $1\,\mathrm{cm}^{-1}$ [8.48–51] in metallic glasses at low temperatures. Note, however, that this contribution becomes nonnegligible at ultrasonic frequencies high enough (>100 GHz) to correspond to the dominant phonons in a thermal conductivity experiment at $T = 2$ K, which is consistent with the discussion above. As far as the ultrasonic velocity is concerned, the electronic

4 These values are $\varrho = 10\,\mathrm{g/cm^3}$, $v = 2 \times 10^5$ cm/s, $\sigma^{-1} = 100\,\mu\Omega$-cm, and $v_F = 10^8$ cm/s.

contribution [8.47] analogous to (8.12) is temperature and frequency independent and thus not interesting.

Bellessa and co-workers [8.52] reported the first ultrasonic evidence for TLS in metallic glasses with the observation of logarithmic temperature dependence in the sound velocity of nonsuperconducting PdSi, CoP, and NiP. The TLS scattering strength $\bar{P}\gamma^2$, deduced from (8.8), was found to be about one order of magnitude smaller than the values typical of insulating glasses. *Matey* and *Anderson* [8.53] have pointed out that this scattering strength is not large enough to account for the observed TLS-limited phonon thermal conductivities in PdSi and CoP. It must be borne in mind, however, that this argument relies upon the ability to accurately subtract off all electronic effects in these normal metals.

At somewhat higher temperatures, *Bellessa* et al. [8.52] observed that the velocity goes through a maximum (near 2 K) followed by a "high-temperature" region in which the velocity decreases linearly in T. Such behavior is usually attributed to the relaxational process in (8.12), which begins to make an appreciable negative contribution when $T_1^{-1} > \omega$. Similar behavior above 1 K is, in fact, seen in insulating glasses [8.22–25], where T_1^{-1} is due to the one-phonon process described by (8.10). On the other hand, *Golding* et al. [8.48] reported significant negative deviations from the lnT law of (8.8) at temperatures as low as 0.3 K in PdSi, suggesting that the relaxation times in metallic glasses may be considerably shorter than in insulators.

The first low-temperature ultrasonic attenuation measurements in a metallic glass were done by *Golding* et al. [8.48] for longitudinal phonons in glassy PdSi. They noted that $dl^{-1}/dT<0$ at their lowest temperatures, strongly suggesting a resonant scattering term of the form (8.6). They were unable, however, to verify this hypothesis by saturating this part of the absorption. These authors also reported an additional nonsaturable absorption term which increased very slowly with increasing T and fitted fairly well to a $\ln T$ law. *Doussineau* et al. [8.49] substantiated the existence of this additional "$\ln T$" absorption not only in PdSi but also in glassy NiP. They were, however, unable to detect any evidence of resonant attenuation. At this point, then, the attenuation evidence to support the interpretation of thermal conductivity and sound velocity results in terms of resonant absorption by TLS was very suggestive but incomplete. Furthermore the $\ln T$ attenuation defied explanation in terms of any of the processes familiar from attenuation in insulating glasses.

This situation was clarified with the report by *Doussineau* et al. [8.50] of a successful saturation experiment using transverse phonons in PdSi glass at temperatures as low as 60 mK. This was accomplished by means of an acoustic flux which exceeded the typical critical intensity in insulators by several orders of magnitude (see Fig. 8.2). *Golding* et al. [8.51] subsequently reported successful saturation for the transverse phonon attenuation in PdSi down to a temperature of 10 mK. In the latter experiment, the negative temperature dependence ($dl^{-1}/dT<0$) of the resonant attenuation was clearly visible at low powers and disappeared at high powers. Here, too, the observed saturation

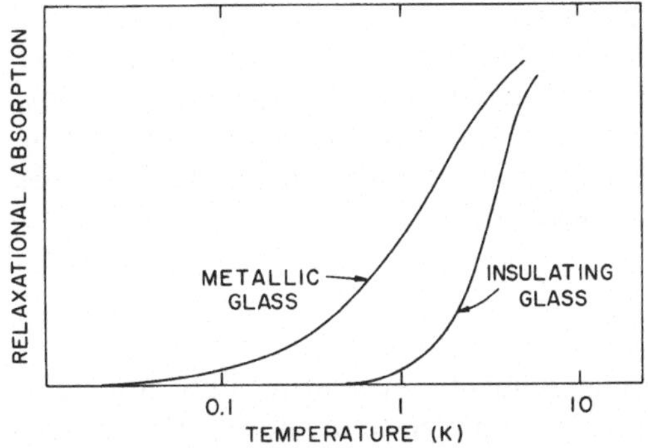

Fig. 8.4. Temperature dependence of the nonresonant "ln T" ultrasonic absorption in metallic glasses [8.49–51] compared with the corresponding absorption in insulating glasses [8.25, 36]

threshold was very high compared to insulating glasses. In both of these experiments, a residual (i.e., nonsaturable) $\ln T$ absorption of the type shown in Fig. 8.4 was observed. As we shall see shortly, this absorption is probably attributable to relaxational absorption (8.11) with a T_1 determined by the conduction electrons.

The successful observation of saturation in PdSi immediately suggests two other nonlinear effects which would be present in metallic glasses if the analogy with insulators were strictly correct. The first of these is the ability of a strong ultrasonic pulse to "burn a hole" (cf. Sect. 8.1) in the TLS population difference, which is then detected by a weak probing pulse [8.22, 25]. Such effects have been ruled out in PdSi for pulses as close together as 100 ns at $T = 10$ mK [8.51]. The second phenomenon is the phonon echo (cf. Sect. 8.2), which is not seen in PdSi [8.54] even though the temperatures studied in [8.51] certainly are below the temperatures at which such echoes appear in SiO_2 [8.27, 34, 35]. The absence of both of these effects strongly suggests a short relaxation time for the TLS in metallic glasses.

There are also several noteworthy recent ultrasonic attenuation experiments in metallic glasses. *Doussineau* and *Robin* [8.55] reported saturation of the resonant attenuation in glassy NiP and observed a relaxational term in general accord with the behavior in PdSi. *Araki* et al. [8.56] have observed an unexpectedly large saturable absorption mechanism in PdSi at very low frequencies ($10\,\mathrm{MHz} < \omega/2\pi < 90\,\mathrm{MHz}$) and relatively high temperatures

Table 8.1. Classification of metallic glass experiments

Insulatorlike behavior	Unusual behavior
Linear specific heat [8.42]	Size of saturation threshold [8.50, 51, 55]
T^2 thermal conductivity [8.42–45]	"lnT" attenuation [8.48–51]
lnT sound velocity [8.52, 53]	Deviations from lnT sound velocity [8.48, 51]
Saturation of resonant attenuation [8.50, 51, 55]	Absence of echoes and two-pulse saturation [8.51]
	Quantitative discrepancies in resonant attenuation law [8.53, 56]

($T > 0.3$ K). This term seems to have much less frequency and temperature dependence than required by (8.6). Finally *Weiss* et al. [8.46] and *Thomas* et al. [8.57] reported the first measurements of low-temperature ultrasonic attenuation in superconducting amorphous metals. We shall have more to say about the significance of these experiments when we discuss the coupling between conduction electrons and TLS in the next section.

8.2.4 Summary

A summary of the low-temperature experiments relevant to TLS in metallic glasses is presented in Table 8.1. We have attempted to divide experiments into a) those aspects which are readily explicable in terms of the "insulating glass" ideas presented in Sect. 8.2 and b) those aspects which are new features of metallic glasses. One qualitative feature of many of the "puzzles" is that they could be the result of an unusually short lifetime (T_1) of the TLS in metallic glasses. This point will be pursued further in Sect. 8.4.

8.3 Interaction Between Tunnelling States and Conduction Electrons

The foregoing discussion has revealed a number of unexpected ultrasonic features associated with tunnelling levels in metallic glasses. The question now arises: "How many of these puzzles are attributable to effects of metallic electrons upon the TLS?" In answering this question, it is first necessary to find a reasonable generalization of (8.3) which includes TLS-electron interactions. Once this is done, it is fairly straightforward to understand most of the puzzles in terms of a rapid "Korringa" relaxation rate for the TLS. This leads to some interesting predictions for the ultrasonic properties of superconducting metallic glasses. Finally, armed with good evidence for the existence of a TLS-electron coupling, it is interesting to consider further consequences, such as electrical resistivity and many-body effects.

8.3.1 Origin of the TLS-Electron Coupling

In Sect. 8.1, we justified the TLS-phonon interaction as a consequence of the dependence of the asymmetry energy Δ upon local strains. It is obvious that Δ must also depend upon the local occupation of valence electron states in a metallic glass, thus yielding a TLS-electron coupling. A derivation is perhaps easier to visualize in analogy with the electron-phonon interaction [8.58]. The key idea is that the two states ψ_L and ψ_R corresponding to Fig. 8.1 are distinguished from each other by atomic (more correctly "ionic") displacements. The electron-ion interaction causes the electrons to experience a

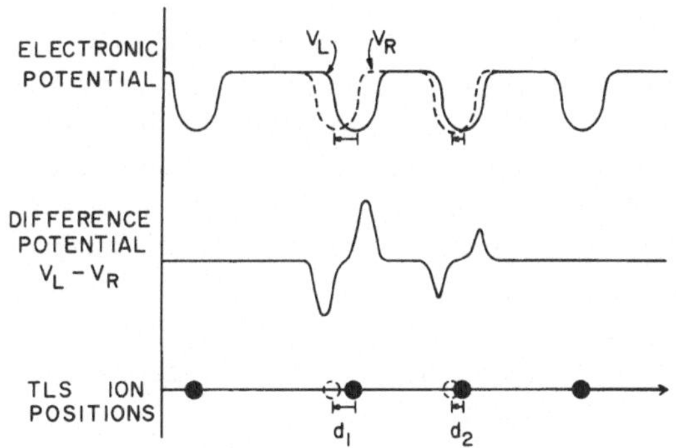

Fig. 8.5. Origin of the electron-TLS coupling. Ion displacements (d_1, d_2) change the electronic potential from V_L to V_R, producing a difference potential

different potential for the TLS state ψ_L than for the state ψ_R. This situation is shown schematically in Fig. 8.5 for a TLS consisting of two ions.

Expressing this electron-ion interaction in terms of electronic wave functions labelled by α, the interaction involving a single TLS is

$$\mathscr{H}' = \sum_{\alpha,\alpha'} [\langle\alpha, L|V_{\mathrm{el}}|\alpha', L\rangle(\tfrac{1}{2}+S_z) + \langle\alpha, R|V_{\mathrm{el}}|\alpha', R\rangle(\tfrac{1}{2}-S_z)]c_\alpha^+ c_{\alpha'}, \tag{8.14}$$

where S_z is defined in (8.2), c_α^+ creates an electron in the state α, and $\langle\alpha, L|V_{\mathrm{el}}|\alpha', L\rangle$ is the matrix element[5] of the electron-ion interaction when the TLS state is ψ_L. As a tractable example, we now suppose that the electronic states are plane waves, $V^{-1/2}\mathrm{e}^{ikr}$, where V is the total volume. This yields [8.59, 60]

$$\mathscr{H}' = V^{-1}\sum_{k,q}\left[\sum_{i=1}^{N'} \mathrm{e}^{\mathrm{i}qR_i^0} 2\mathrm{i}\sin(qd_i/2)\right]\mu(q)S_z c_k^+ c_{k+q}, \tag{8.15}$$

where the N' atoms of the TLS each have two positions: $R_i = R_i^0 \pm d_i/2$. In (8.15), $\mu(q)$ is the spatial Fourier transform of the ionic potential, and we have neglected an unimportant term which is independent of S_z. Finally, we perform the diagonalization which leads from (8.2) to (8.3), giving

$$\mathscr{H}' = \frac{1}{N}\sum_{k,q}[V_{\|}(q)S_z + V_{\perp}(q)S_x]c_k^+ c_{k+q}, \tag{8.16}$$

with

$$\left.\begin{matrix} V_{\|}(q) \\ V_{\perp}(q) \end{matrix}\right\} = \frac{2\mathrm{i}\mu(q)}{\Omega}\sum_{i=1}^{N'} \mathrm{e}^{\mathrm{i}qR_i^0}\sin(qd_i/2)\left\{\begin{matrix} \Delta/E \\ \Delta_0/E \end{matrix}\right., \tag{8.17}$$

where $\Omega = V/N$ is the atomic volume and N is the total number of atoms.

5 We have neglected matrix elements of the form $\langle\alpha, L|V|\alpha', R\rangle$, which have been considered in [8.59], but which are smaller by a factor of $\langle R|L\rangle = \exp(-\lambda)$ than the matrix elements we have retained.

Thus conduction electrons can couple both diagonally ($V_{\|}$) and off-diagonally ($V_{\perp}$) to the TLS, just as in the case of phonons (8.3). Using the estimates $\mu(q_F)/\Omega = 1$ eV, $q = k_F = 1$ Å^{-1}, and $d = 0.1$ Å, we find that the matrix elements $V_{\|}$ and $V_{\perp}$ should have magnitudes up to about 0.1 eV. Of course, the actual electronic states in the metallic glass will not be the plane waves on which (8.15–17) are based. Nonetheless, the general form of (8.16) will survive, with the label α instead of the wave vector k. The magnitudes of the coupling constants will depend upon the amplitude of the true electronic wave functions at the TLS site. For simplicity, however, we shall use the approximate form (8.16) throughout the remainder of this section.

8.3.2 Korringa Relaxation Process

Perhaps the most important consequence of the TLS-electron coupling is a huge enhancement of the TLS relaxation rate T_1^{-1} relative to the one-phonon value (8.10). The dominant relaxation process at low temperatures is a TLS transition between ψ_g and ψ_e accompanied by an electronic transition from $k+q$ to k. This is the analog of the Korringa relaxation of nuclear magnetic moments in contact with conduction electrons [8.61] and is governed by the off-diagonal matrix element $V_{\perp}$. For example, the rate of transitions $\psi_g \to \psi_e$ is given in lowest-order perturbation theory by

$$R_{ge}(E,T) = \frac{2\pi}{\hbar} N^{-2} \sum_{k,q} \frac{1}{4} V_{\perp}^2(q) f_{k+q}(1-f_k)\, \delta(\varepsilon_{k+q} - \varepsilon_k - E), \tag{8.18}$$

where ε_k is the electronic energy measured from the Fermi energy ε_F and f_k is $[\exp(\beta\varepsilon_k)+1]^{-1}$. Performing the sums, replacing $V_{\perp}(q)$ by its average over the Fermi surface, and using $R_{eg}(E) = R_{ge}(-E)$, the expression for T_1^{-1} becomes

$$T_1^{-1}(E,T) = R_{eg} + R_{ge} = \frac{\pi}{4\hbar}(\tilde{\varrho}V_{\perp})^2 E \coth\frac{\beta E}{2}, \tag{8.19}$$

where $\tilde{\varrho} = N(\varepsilon_F)\Omega$ is the electronic density of states per atom at the Fermi level.

A few comments about the validity of (8.19) in metallic glasses are in order. The use of lowest-order perturbation theory is permitted only when higher-order terms become successively smaller. It turns out that these terms represent an expansion in the dimensionless coupling constants $\tilde{\varrho}V_{\perp}$ and $\tilde{\varrho}V_{\|}$, which are found from experiment to be less than 1. In the absence of serious divergences from many-body effects (see below), this is sufficient justification. A second question is whether (8.19) remains valid when the electronic lifetime τ becomes much shorter than $E/\hbar$, as must surely be the case in metallic glasses. An analogous question arises for the electronic contribution to the phonon lifetime when $\omega\tau \ll 1$. For the latter case, the analog of (8.19) remains valid even though

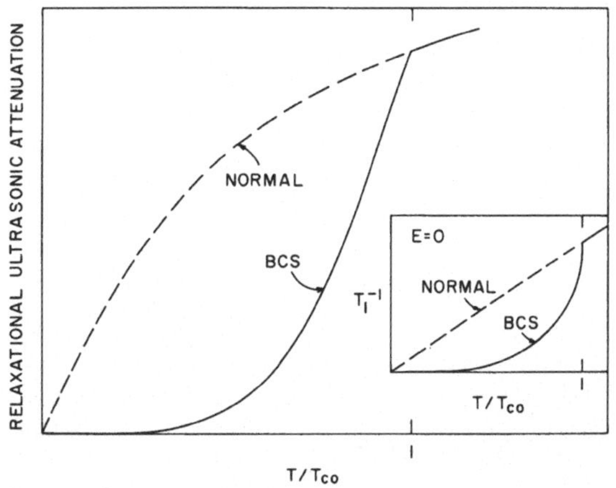

Fig. 8.6. Predicted relaxational attenuation due to Korringa relaxation in the normal and superconducting states. Inset shows the Korringa rate T_1^{-1} for the special case $E=0$. T_{c0} is the superconducting transition temperature in zero magnetic field

$\omega\tau \ll 1$ as long as $kl_e \gtrsim 1$, where k is the phonon wave vector and l_e is the electronic mean free path [8.47, 62]. This indicates that (8.19) itself remains valid even when $E\tau\hbar^{-1} \ll 1$ as long as $k_0 l_e \gtrsim 1$, where k_0 is roughly the inverse of the "radius" of the TLS. Thus (8.19) should not be seriously modified unless the size of the TLS greatly exceeds the electronic mean free path, which is unlikely for a TLS containing at most a few atoms.

It is instructive to compare the magnitude of (8.19) with that of (8.10). Taking $\varrho = 10\,\mathrm{g/cm^3}$, $v = 4\times10^5\,\mathrm{cm/s}$, and $M = 1\,\mathrm{eV}$, the one-phonon process (8.10) yields a T_1 of 26 µs at $T=1\,\mathrm{K}$ and $E/h = 1\,\mathrm{GHz}$. By comparison, (8.19) yields a value of 0.5 ns for $V_\perp = 0.1\,\mathrm{eV}$ and $\tilde{\varrho} = 1\,\mathrm{eV^{-1}}$ at the same values of T and E. This difference of over 4 orders of magnitude is largely the result of the high electronic density of states at ε_F, which must be compared with the relatively small phonon density of states at energies corresponding to 1 GHz.

Ultrasonic experiments are strongly affected by this greatly shortened T_1 in metallic glasses. *Golding* et al. [8.51] have argued, in fact, that most of the puzzles in Table 8.1 are explicable in terms of (8.19) with $\tilde{\varrho}V_\perp \leqq 0.2$ for PdSi glass. For example, (8.19) and (8.9) (with $T_2 = T_1$) yield a saturation threshold of $2\times10^{-3}\,\mathrm{W/cm^2}$ at $T=10\,\mathrm{mK}$ and $\omega/2\pi = 1\,\mathrm{GHz}$ which is in good agreement with the observed value of $1\times10^{-3}\,\mathrm{W/cm^2}$. Furthermore the predicted 5 ns value of T_1 at this temperature and frequency certainly explains the absence of two-pulse "hole-burning" effects or phonon echoes. In the former case, the burned hole recovers too quickly to be detected by the weak probing pulse. In the latter case, any phase coherence of the excited TLS is lost before an echo can be produced.

Similarly (8.19) strongly influences the behavior of the relaxational attenuation and velocity contributions of (8.11, 12). As discussed earlier, these contributions become comparable to the resonant terms (8.6, 8) only when T_1 is short enough to satisfy $T_1^{-1} \simeq \omega$, where ω is the ultrasonic frequency. With $\tilde{\varrho}V_\perp = 0.2$, this occurs at temperatures as low as 0.1 K in PdSi, thus explaining [8.51] both the nonsaturable $\ln T$ attenuation of Fig. 8.4 and the negative deviations of the

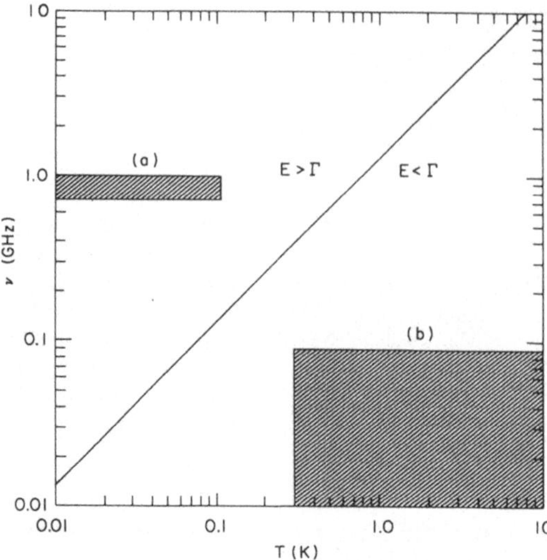

Fig. 8.7. Comparison of the TLS width ($\Gamma = \hbar T_1^{-1}$) with the TLS level splitting ($E = h\nu$) for various temperatures and resonant frequencies ν, where T_1^{-1} is computed from (8.19) with $\tilde{\varrho} V_\perp = 0.2$. Experimental evidence for the usual kind of resonant attenuation has been obtained in region (a) [8.50, 51], whereas quite unusual T and ν dependence has been seen in region (b) [8.56]

sound velocity [8.48, 51] away from (8.8). The temperature dependence of the attenuation resulting from (8.4, 11, 19) is very important. At low temperatures, the relationship $T_1^{-1} \sim E$ leads to absorption which increases linearly with T, as opposed to the T^3 variation in insulating glasses [8.25, 36]. This linear dependence in a metallic glass has been observed by *Doussineau* and *Robin* [8.55]. As T increases to the point where $T_1^{-1} \simeq \omega$, the temperature dependence becomes much weaker than linear, as is illustrated by the "normal" curve in Fig. 8.6. This weak T dependence comes from the broad distribution of relaxation times [cf. (8.4)] inherent in the tunnelling model[6] and is in reasonable quantitative agreement [8.51] with the observed "ln T" dependence [8.48–51] in metallic glasses.

The remaining ultrasonic puzzles of Table 8.1 are a) the discrepancy between the magnitude of $\bar{P}\gamma^2$ as deduced from $\Delta v/v$ and κ [8.53] and b) the very low-frequency "resonant" attenuation observed in PdSi [8.56]. It is conceivable that both of these mysteries are related to (8.19). Concerning a) for example, the relaxational contribution to $\Delta v/v$ deduced from (8.4, 12) is negative and varies as $\ln T$ over a large temperature range for sufficiently small T_1. This could contribute to the small deduced values of $\bar{P}\gamma^2$, which are usually attributed to the resonant interaction (8.8) alone. Concerning b), it is useful to refer to Fig. 8.7, which shows the ranges of resonant frequency ($\nu = \omega/2\pi = E/h$) and temperature for which $E > T_1^{-1}\hbar$ and $E < T_1^{-1}\hbar$, based on (8.19). It is interesting to note that the puzzling low-frequency results [8.56] fall into a region where the TLS width $\hbar T_1^{-1}$ greatly exceeds the splitting E. It is no surprise that the usual resonant absorption law (8.6) no longer seems to apply.

6 Here particular attention must be paid to the r integration in (8.11) and to the fact that $T_1^{-1} \sim V_\perp^2 \sim \Delta_0^2/E^2 = r$, which follows from (8.17).

8.3.3 Ultrasonics in Superconducting Metallic Glasses

The foregoing discussion indicates that many ultrasonic properties of metallic glasses are influenced by a short TLS lifetime of electronic origin. This hypothesis suggests a number of interesting phenomena at or below T_c in superconducting glasses. *Black* and *Fulde* [8.63] have calculated the consequences of the BCS (Bardeen, Cooper, Schriefer) superconducting state upon the Korringa relaxation process of (8.19) and have found that the result depends on the relative sizes of E and Δ_s, the BCS energy gap. When $E \ll 2\Delta_s$, Korringa relaxation can occur only through scattering of thermally excited quasiparticles, and this leads to

$$T_1^{-1} = \frac{\pi}{\hbar}(\tilde{\varrho} V_\perp)^2 k_B T (e^{\beta \Delta_s(T)} + 1)^{-1} \tag{8.20}$$

whose temperature dependence is shown in the inset of Fig. 8.6. When $E > 2\Delta_s$, an additional relaxation channel opens, in which the excited TLS creates two quasiparticles from a Cooper pair. In this case, the total rate T_1^{-1} is actually enhanced over (8.19) as a result of coherence factors [8.64] and the singular quasiparticle density of states.

Resonant ultrasonic experiments at frequencies less than 1 GHz generally satisfy $\omega \ll \Delta_s/\hbar$ and are thus governed by (8.20). Equations (8.9) (with $T_2 = T_1$) and (8.20) imply that the saturation threshold should be a rapidly increasing function of temperature below T_c. This steep temperature dependence should also be evident in hole-burning [8.25] and phonon echo [8.35] experiments. Furthermore all of these experiments should become magnetic-field sensitive below T_c, as has been suggested in [8.63]. Finally resonant ultrasonic experiments at frequences high enough to satisfy $\hbar\omega = 2\Delta_s$ should exhibit discontinuous behavior at the threshold for creation of two quasiparticles from a Cooper pair. To date, however, there has been no report of resonant ultrasonic attenuation in any superconducting metallic glass.

The experiments have been more successful in observing relaxational ultrasonic effects below T_c [8.46, 57]. The theoretical predictions based on (8.4, 11) and the generalization of (8.19) below T_c are shown in Fig. 8.6. The relaxational attenuation due to TLS is predicted to sharply decrease as T decreases below T_c and to increase again with the application of a magnetic field [8.63]. The experiment of *Weiss* et al. [8.46] on ZrPd is in qualitative agreement with some, but not all, of these predictions. They found that the attenuation begins to sharply decrease with decreasing T at some temperature near, $T_c/2$. Furthermore, the magnetic field dependence in this low-temperature regime agrees with the predictions. The surprising result, however, is that the steep temperature dependence begins well below T_c rather than at T_c, as predicted. Recent work by *Thomas* et al. [8.57] on partially amorphous NbZr seems to confirm these experimental findings. One possible explanation for this apparent discrepancy is the neglect of nonelectronic contributions to T_1^{-1} in generating Fig. 8.6. For example, the one-phonon process (8.10) can be comparable to

(8.19) at T_c when T_c is above a few degrees K, thereby shifting the electronic effects to lower temperatures.

To summarize, the electronic influence on the low-temperature TLS behavior is well established through ultrasonic experiments which are sensitive to a short TLS lifetime. Efforts to alter this lifetime by means of a superconducting transition are just beginning, but they seem likely to produce a number of interesting new effects in metallic glasses.

8.3.4 Electrical Resistance and Many-Body Effects

Having now seen that conduction electrons can strongly affect the tunnelling states, it is natural to ask whether the TLS can influence electronic properties, transport in particular. This question was, in fact, raised by *Cochrane* et al. [8.65] in connection with anomalies in the electrical resistance long before there was any direct evidence for the existence of TLS in metallic glasses. They suggested that the widespread occurrence of a resistance minimum (i.e., $dR/dT<0$ below some temperature) in metallic glasses is caused by a divergence in the rate of electronic scattering from TLS as the temperature is lowered, just as in the Kondo effect [8.66]. Their analysis, however, is based upon the questionable assumption that the electrons possess some internal "spin" label which couples to the TLS in complete analogy with the magnetic exchange interaction of the Kondo effect. The primary interest here is to decide whether the more readily justifiable electron-TLS interaction of the form (8.14) or (8.16) leads to divergent many-body effects of the type suggested. Numerous alternative explanations (see Chap. 7) for $dR/dT<0$, which have nothing to do with TLS, will not be considered here.

The interaction (8.16) does, in fact, lead to singular many-body effects, which originate in the divergent response of an electron gas to a localized perturbation with internal dynamics ($E \neq 0$ and $[S_x, S_z] \neq 0$). These effects are, however, definitely *not equivalent* to the Kondo effect. This distinction was first pointed out by *Kondo* [8.59, 67] himself, who analyzed the lowest-order singular terms of perturbation theory for (8.16). He found that the lowest-order divergent term in the electronic elastic scattering rate is

$$\tau^{-1} = A(\tilde{\varrho}V_{\|})^2(\tilde{\varrho}V_{\perp})^2\left(\ln\frac{T}{W}\right)^2, \tag{8.21}$$

where A is a positive constant and W is the conduction band width. The distinction between this result and the Kondo effect is already apparent in that (8.21) is a fourth-order term ($V_{\|}^2 V_{\perp}^2$), whereas the lowest-order Kondo effect divergence occurs in third order [$\tau^{-1} \sim -J^3 \ln(T/W)$] [8.66]. Despite such differences, the scattering rate (8.21) with $A>0$ is qualitatively consistent with a resistance minimum. It should be noted, however, that (8.21) is obtained only when $V_{\perp}$ is taken to be proportional to the very small matrix elements neglected in (8.17) and mentioned in footnote 5, p. 181.

The usefulness of a nearly-free electron calculation for the electrical resistivity in these highly disordered metals is highly questionable. It is far more likely, in fact, that some effects of incipient electron localization are responsible for the observed temperature dependences. Thouless [8.68] has shown that under certain circumstances the electrical resistance of weakly localized electrons can be proportional to a positive power of the *inelastic* electron scattering time. For long, thin wires the resistance goes as the square root of the inelastic time [8.68]. Qualitatively similar predictions have been obtained for bulk disordered metals by *Imry* [8.69] and by *Jonson* and *Girvin* [8.70]. Recent experiments [8.7] have given results which are consistent with Thouless' predictions, having a weak temperature dependence ($R \sim T^{-1/2}$) at low temperatures. Since most other mechanisms of inelastic scattering (e.g., phonons) have a much stronger T-dependence, it has been suggested [8.68] that inelastic scattering from the TLS may be the dominant mechanism.

It is thus important to know the overall magnitude of the inelastic part of the TLS contribution to the electronic scattering rate τ^{-1}. A reasonable estimate of this quantity is obtained by calculating the lowest-order inelastic scattering process and integrating over the TLS distribution using (8.4). This yields [8.60]

$$\tau^{-1} = \frac{\pi}{2\hbar} \frac{\bar{P}}{N(\varepsilon_F)} (\tilde{\varrho} V_\perp)^2 k_B T \int\limits_0^\infty dx (1 + \tfrac{2}{3} \sinh^2 x)^{-1} , \qquad (8.22)$$

where $\bar{P}$ and $N(\varepsilon_F)$ are the TLS and electronic densities of states, respectively. Taking the value $\tilde{\varrho} V_\perp = 0.2$ from ultrasonic results [8.51], this inelastic contribution to τ^{-1} turns out to be $4 \times 10^8 T\,\mathrm{s}^{-1}$ [8.60]. Unfortunately this calculation leads to a value of τ which is at least one order of magnitude too large to be consistent with the experiments [8.71]. There remains the interesting possibility, however, that many-body enhancements of (8.22) may occur. In fact perturbation theory calculations by *Black* and *Gyorffy* [8.72] and by *Leiter* and *Morandi* [8.73] suggest that logarithmic enhancements in the inelastic scattering channels may occur. Whether these effects are large enough and whether there are compensating effects in the higher order corrections to (8.11) and (8.19) have not yet been definitely answered. There remains much work to be done in this area.

8.3.5 Other Effects

The interaction between TLS and conduction electrons has additional consequences which may become important for future experiments in metallic glasses. For example, both Kondo's perturbation theory [8.59] and the scaling theory [8.72] agree that the atomic tunnelling is accompanied by a "polaron-like" cloud of conduction electrons, which leads to a temperature-dependent reduction of the tunnelling frequency

$$\Delta_0 \rightarrow \Delta_0 (T/W)^\alpha , \qquad (8.23)$$

where $\alpha=(1/2)(\tilde{\varrho}V_{||})^2$. The conduction electrons may further influence the TLS by inducing an "*RKKY*-like" [8.74] interaction between the ith and jth TLS, of the form $JS_z^iS_z^j$ with

$$J \sim V_{||}^2 \varepsilon_F^{-1}(k_F r)^{-3}\,.$$

Such an interaction could probably compete in magnitude with the strain-mediated interaction which influences many low-temperature properties of insulating glasses [8.29]. Whether such an influence could be detected in the presence of the very short Korringa T_1 is, however, less certain. Finally, there is the intriguing possibility that TLS may contribute an attractive part to the effective electron-electron interaction, thereby enhancing the tendency toward superconductivity. A recent calculation by *Vujicic* et al. [8.75] indicates that the TLS contribution to the coupling parameter $\lambda[T_c \sim \exp(-\lambda^{-1})]$ can be an appreciable fraction of the phonon contribution when $E \ll \hbar\omega_D$, where ω_D is the Debye frequency. If the magnitude of this effect is large enough, it may be possible to detect the influence of TLS by means of superconducting tunnelling spectroscopy [8.76].

8.4 Conclusions

One central theme of this discussion has been the similarity between low-energy atomic excitations in metallic and insulating glasses. This kinship has been most clearly demonstrated in the low-temperature thermal conductivity, sound velocity, and saturable ultrasonic attenuation. The behavior of these quantities in metallic glasses conforms nicely to the tunnelling model predictions, which were originally developed to explain similar features in insulating glasses. On the other hand, the magnitude of the saturation threshold, the "$\ln T$" relaxational ultrasonic attenuation, and the conspicuous absence of phonon echoes and two-pulse saturation effects suggest a real distinction between metallic and insulating glasses. Our point, however, has been that these distinguishing features are very likely to be consequences of the coupling between tunnelling states and conduction electrons. This thesis implies that the principal difference between low-energy excitations in metallic and insulating glasses is the environment to which they couple, leaving open the exciting possibility of a universal microscopic explanation for the excitations themselves.

A second major theme has been the exploration of new physical effects derived from the interaction between tunnelling states and conduction electrons. We have indicated the origin of a short Korringa relaxation time and have discussed some of its consequences for ultrasonics in glasses. We have also discussed the question of tunnelling state influence upon the electrical resistance and the related question of many-body effects in metallic glasses, many aspects of which remain to be explored. Finally we have seen that the interplay between superconductivity and tunnelling states opens up exciting new possibilities, which have only begun to be explored.

Acknowledgments. The author wishes to thank W. Arnold, V. J. Emery, P. Fulde, B. Golding, J. E. Graebner, B. L. Gyorffy, B. I. Halperin, and J. Jäckle for helpful discussions. This work was supported by the Division of Basic Energy Sciences, Department of Energy, under Contract No. EY-76-C-02-0016.

References

8.1 P.W.Anderson, B.I.Halperin, C.M.Varma: Philos. Mag. **25**, 1 (1972)
8.2 W.A.Phillips: J. Low Temp. Phys. **7**, 351 (1972)
8.3 D.Weaire: Contemp. Phys. **17**, 173 (1976)
8.4 G.S.Cargill: "Structure of Metallic Alloy Glasses", in *Solid State Physics*, Vol. 30, ed. by H.Ehrenreich, F.Seitz, D.Turnbull (Academic Press, New York 1975) p. 227
8.5 P.W.Anderson: "Lectures on Amorphous Systems", in *Physics of Ill-Condensed Matter*, ed. by R.Balian, R.Maynard, G.Toulouse (North-Holland, Amsterdam 1979)
8.6 R.C.Zeller, R.O.Pohl: Phys. Rev. B**4**, 2029 (1971)
8.7 R.B.Stephens: Phys. Rev. B**8**, 2896 (1973); B**13**, 852 (1976)
8.8 D.P.Jones, N.Thomas, W.A.Phillips: Philos. Mag. B**38**, 271 (1978)
8.9 P.J.Anthony, A.C.Anderson: Phys. Rev. B**14**, 5198 (1976)
8.10 P.Doussineau, R.G.Leisure, A.Levelut, J.-Y.Prieur: In *Phonon Scattering in Condensed Matter*, ed. by H.Maris (Plenum Press, New York 1980) p. 57
8.11 U.Strom, M. von Schickfus, S.Hunklinger: Phys. Rev. Lett. **41**, 910 (1978)
8.12 C.Laermans: Phys. Rev. Lett. **42**, 250 (1979)
8.13 L.F.Lou: Solid State Commun. **19**, 335 (1976)
8.14 J.M.Peech, R.O.Pohl, A.K.Raychaudhuri: In *Phonon Scattering in Condensed Matter*, ed. by H.Maris (Plenum Press, New York 1980) p. 45
8.15 W.A.Phillips: J. Non-Cryst. Solids **31**, 267 (1978)
8.16 J.Jäckle: Z. Phys. **257**, 212 (1972)
8.17 B.I.Halperin: Ann. N. Y. Acad. Sci. **279**, 173 (1976)
8.18 J.L.Black: Phys. Rev. B**17**, 2740 (1978)
8.19 W.M.Goubau, R.A.Tait: Phys. Rev. Lett. **34**, 1220 (1974)
8.20 R.B.Kummer, R.C.Dynes, V.Narayanamurti: Phys. Rev. Lett. **40**, 1187 (1978)
8.21 B.Golding, J.E.Graebner, R.J.Schutz: Phys. Rev. B**14**, 1660 (1976)
8.22 S.Hunklinger, M. von Schickfus: In *Amorphous Solids: Low-Temperature Properties*, Topics in Current Physics, Vol. 24, ed. by W.A.Phillips (Springer, Berlin, Heidelberg, New York 1980)
8.23 B.Golding, J.E.Graebner, A.B.Kane: Phys. Rev. Lett. **37**, 1248 (1976)
8.24 L.Piche, R.Maynard, S.Hunklinger, J.Jäckle: Phys. Rev. Lett. **32**, 1426 (1974)
8.25 S.Hunklinger, W.Arnold: In *Physical Acoustics*, Vol. 12, ed. by R.N.Thurston, W.P.Mason (Academic Press, New York 1976) p. 155
8.26 B.Golding, J.E.Graebner, R.J.Schutz, B.I.Halperin: Phys. Rev. Lett. **30**, 223 (1973)
8.27 B.Golding, J.E.Graebner: In *Amorphous Solids: Low-Temperature Properties*, Topics in Current Physics, Vol. 24, ed. by W.A.Phillips (Springer, Berlin, Heidelberg, New York 1980)
8.28 J.Joffrin, A.Levelut: J. Phys. (Paris) **36**, 811 (1976)
8.29 J.L.Black, B.I.Halperin: Phys. Rev. B**16**, 2879 (1977)
8.30 S.Hunklinger, W.Arnold, S.Stein, R.Nava, K.Dransfeld: Phys. Lett. **42**A, 253 (1972)
8.31 W.Arnold, J.L.Black, G.Weiss: In *Scattering in Condensed Matter*, ed. by H.Maris (Plenum Press, New York 1980) p. 77
8.32 W.Arnold, S.Hunklinger: Solid State Commun. **17**, 883 (1975)
8.33 A.Abragam: *The Principles of Nuclear Magnetism* (Clarendon, Oxford 1961) Chaps. 2, 3
8.34 B.Golding, J.E.Graebner: Phys. Rev. Lett. **37**, 852 (1976)

8.35 J.E.Graebner, B.Golding: Phys. Rev. B**19**, 964 (1979)
8.36 J.Jäckle, L.Piche, W.Arnold, S.Hunklinger: J. Non-Cryst. Solids **20**, 365 (1976)
8.37 M. von Schickfus, S.Hunklinger, L.Piché: Phys. Rev. Lett. **35**, 876 (1975)
8.38 L.Bernard, L.Piché, G.Schumacher, J.Joffrin, J.Graebner: J. Phys. Lett. (Paris) **39**, L-126 (1978)
8.39 C.Laermans, W.Arnold, S.Hunklinger: J. Phys. C**10**, L161 (1977)
8.40 C.Kittel: *Introduction to Solid State Physics*, 4th ed. (Wiley, New York 1971)
8.41 A.Comberg, S.Ewert, H.Wühl: Z. Phys. B**25**, 173 (1976)
8.42 J.E.Graebner, B.Golding, R.J.Schutz, F.S.L.Hsu, H.S.Chen: Phys. Rev. Lett. **39**, 1480 (1977)
8.43 J.R.Matey, A.C.Anderson: J. Non-Cryst. Solids **23**, 129 (1977); Phys. Rev. B**16**, 3406 (1977)
8.44 H. v. Löhneysen, F.Steglich: Phys. Rev. Lett. **39**, 1205 (1977); Z. Phys. B**29**, 89 (1978)
8.45 A.K.Raychaudhuri, R.Hasegawa: Phys. Rev. B**2**, 479 (1980)
8.46 G.Weiss, W.Arnold, K.Dransfeld, H.J.Güntherodt: Solid State Commun. **33**, 111 (1980)
8.47 A.B.Pippard: *The Dynamics of Conduction Electrons* (Gordon and Breach, New York 1965) Chap. 11
8.48 B.Golding, J.E.Graebner, W.H.Haemmerle: In *Proc. of the Int. Conf. on Lattice Dynamics*, ed. by M.Balkanski (Flammarion, Paris 1977) p. 348
8.49 P.Doussineau, A.Levelut, G.Bellessa, O.Bethoux: J. Phys. Lett. (Paris) **38**, L-483 (1977)
8.50 P.Doussineau, P.Legros, A.Levelut, A.Robin: J. Phys. Lett. (Paris) **39**, L-265 (1978)
8.51 B.Golding, J.E.Graebner, A.B.Kane, J.L.Black: Phys. Rev. Lett. **41**, 1487 (1978)
8.52 G.Bellessa: J. Phys. C**10**, L285 (1977)
G.Bellessa, O.Bethoux: Phys. Lett. **62**A, 125 (1977)
G.Bellessa, P.Doussineau, A.Levelut: J. Phys. Lett. (Paris) **38**, L-65 (1977)
8.53 J.R.Matey, A.C.Anderson: Phys. Rev. B**17**, 5029 (1978)
8.54 B.Golding, J.E.Graebner: Private communication
8.55 P.Doussineau, A.Robin: In *Phonon Scattering in Condensed Matter*, ed. by H.Maris (Plenum Press, New York 1980) p. 65
8.56 H.Araki, G.Park, A.Hikata, C.Elbaum: In *Phonon Scattering in Condensed Matter*, ed. by H.Maris (Plenum Press, New York 1980) p. 69; Solid State Commun. **32**, 625 (1979)
8.57 N.Thomas, W.Arnold, G.Weiss, H. v. Löhneysen: Solid State Commun. (to be published)
8.58 L.J.Sham, J.M.Ziman: "The Electron-Phonon Interaction", in *Solid State Physics*, Vol. 15, ed by F.Seitz, D.Turnbull (Academic Press, New York 1964) p. 223
8.59 J.Kondo: Physica (Utrecht) **84**B, 40 (1976)
8.60 J.L.Black, B.L.Gyorffy, J.Jäckle: Philos. Mag. B**40**, 331 (1979)
8.61 J.Korringa: Physica (Utrecht) **16**, 601 (1950)
8.62 R.E.B.Makinson: Proc. Cambridge Philos. Soc. **34**, 474 (1938)
8.63 J.L.Black, P.Fulde: Phys. Rev. Lett. **43**, 453 (1979)
8.64 P.G. de Gennes: *Superconductivity of Metals and Alloys* (Benjamin, New York 1966) Chap. 4
8.65 R.Cochrane, R.Harris, J.Ström-Olson, M.Zuckerman: Phys. Rev. Lett. **35**, 676 (1975)
8.66 G.Gruner, A.Zawadowski: Rep. Prog. Phys. **37**, 1497 (1974)
8.67 J.Kondo: Physica (Utrecht) **84**B, 207 (1976)
8.68 D.J.Thouless: Phys. Rev. Lett. **39**, 1167 (1977); Solid State Commun. **34**, 683 (1980)
8.69 Y.Imry: Phys. Rev. Lett. **44**, 469 (1980)
8.70 M.Jonson, S.M.Girvin: Phys. Rev. Lett. **43**, 1447 (1979)
8.71 N.Giordano, W.Gilson, D.E.Prober: Phys. Rev. Lett. **43**, 725 (1979);
P.Chaudhuri, H.-U.Habermeier: Solid State Commun. **34**, 687 (1980)
8.72 J.L.Black, B.L.Gyorffy: Phys. Rev. Lett. **41**, 1595 (1978)
8.73 H.Keiter, G.Morandi: Phys. Rev. B**22**, 5004 (1980)
8.74 M.A.Ruderman, C.Kittel: Phys. Rev. **96**, 99 (1954)
8.75 G.M.Vujicic, V.L.Aksenov, N.M.Plakida, S.Stamenkovic: Phys. Lett. **73**A, 439 (1979)
8.76 W.L.McMillan, J.M.Rowell: In *Superconductivity*, ed. by R.D.Parks (Dekker, New York 1969) Chap. 11

9. Superconductivity in Metallic Glasses

W. L. Johnson

With 12 Figures

9.1 Preview

9.1.1 Historical Background

The earliest systematic studies of amorphous metals were carried out by *Buckel* and *Hilsch* [9.1, 2] beginning in 1954. The experiments were aimed at investigating the influence of extreme atomic disorder on the superconducting properties of simple metals. In quenching certain metallic vapors onto a cryogenically cooled substrate, it was found that the resulting thin films had a metastable noncrystalline structure. The superconducting transition temperatures of these noncrystalline films were measured and found in all cases to be higher than those of the corresponding bulk crystalline metals. Particularly fascinating was the observation of superconductivity ($T_c \approx 6$ K) in amorphous Bi [9.2]. Crystalline Bi is a nonsuperconducting semimetal. These pioneering experiments served to illustrate that the electronic structure of metals can be significantly altered by the absence of long-range atomic order. Although these studies were carried out 25 years ago, only recently has a fundamental understanding of the electronic structure in noncrystalline metals begun to be developed. In the intervening period, a microscopic theory of superconductivity has also emerged and, with the advent of sophisticated computer methods, it is now possible to calculate, with reasonable accuracy, the superconducting properties of simple metals from first principles. A detailed comparison of superconductive tunnelling data with predictions of the strong coupling theory has established beyond question the essential correctness of the Eliashberg equations [9.3].

In applying current knowledge to the systematics of noncrystalline superconductors, one finds that a few remarkably simple concepts can account for a large body of existing data. It is even reasonable, as will be seen, to state that the phenomenology of amorphous superconductors is far simpler than that of crystalline materials. The latter claim follows from the lack of long-range order in amorphous materials.

In the remainder of this chapter, the terms "amorphous", "noncrystalline", and "glassy" will be used loosely and interchangeably to describe materials which lack long-range order. There is one convention which will nevertheless be observed. The term "metallic glass" will be used to refer only to materials prepared by cooling liquid metals or alloys. This usage is consistent with the traditionally accepted definition of a glass.

9.1.2 Amorphous Materials

Since the earliest work on cryoquenched thin films, a variety of amorphous materials have been shown to exhibit superconductivity. The thin film studies have been extended to cover a wide variety of simple metals. *Bergmann* has recently reviewed this work [9.4]. In 1972, *Collver* and *Hammond* [9.5] systematically surveyed the transition metals of the 4*d* and 5*d* series using the cryoquenching technique. Whereas the noncrystalline phases of simple metals crystallize on heating to room temperature [9.1], *Collver* and *Hammond* found that certain amorphous transition metal alloys could be heated to room temperature and above while remaining amorphous.

In 1960, *Duwez* and collaborates showed that certain alloys could be quenched from the melt to form a glass [9.6, 7]. Superconductivity in metallic glasses was first reported by this author and colleagues in 1975, in alloys of lanthanum and gold [9.8]. At present, a large number of metallic glasses are known to exhibit superconductivity. The author has surveyed the literature on these materials in a recent review [9.9]. All of the superconducting metallic glasses studied are stable at room temperature; many do not crystallize at temperatures ranging as high as 1000 K. The bulk nature of these samples together with stability against crystallization have permitted experiments not possible on thin films.

In the remainder of this chapter, the properties of amorphous superconducting materials will be outlined and discussed. The microscopic origin of superconductivity is examined in Sect. 9.2. The microscopic theory is outlined and compared to experimental data. Simple metals and transition metals are treated in separate subsections. In Sect. 9.3, the influence of disorder on the thermodynamic characteristics of the superconducting state is described within the framework of the Ginzburg-Landau theory. Finally, Sect. 9.4 surveys those properties of amorphous superconductors which offer potential advantages in technical applications.

9.2 Origins of Superconductivity

9.2.1 Microscopic Theory

a) BCS Superconductors

The BCS theory [9.10] approximates the electrons in a superconductor as a nearly free electron gas in which individual electrons weakly interact pairwise through an attractive phonon mediated potential. The only material-dependent parameters contained in the theory are [9.10]

$D(\varepsilon_F)$ is the electronic density of states at the Fermi level,

$\langle\omega\rangle$ is the Debye frequency (or a suitably averaged phonon frequency),

V is the effective matrix element for phonon-mediated electron-electron interaction.

The parameters characterizing the superconducting state are related to these material-dependent parameters through explicit analytical expressions. For example, the superconducting transition temperature is given by

$$T_c = 1.14\langle\omega\rangle e^{-[1/D(\varepsilon_F)V]}. \tag{9.1}$$

The dimensionless parameter $\lambda = D(\varepsilon_F)V$ is frequently referred to as the electron-phonon coupling constant. The BCS theory assumes that $\lambda \ll 1$ in deriving the above expression for T_c. When this *weak-coupling* approximation breaks down, (9.1) must be replaced by a strong-coupling expression.

Within the weak-coupling approximation, the central problem in explaining the behavior of T_c is that of determining the microscopic parameters $\langle\omega\rangle$, $D(\varepsilon_F)$, and V for real metals. This problem is complicated by the fact that these parameters are interdependent. For example, variations in $D(\varepsilon_F)$ influence both $\langle\omega\rangle$ and V through the screening effects of electrons. A comprehensive analysis of T_c must include these effects. A simple model applicable to amorphous nearly free electron metals will be presented in Sect. 9.2.3.

b) Strong-Coupling Superconductors

The BCS theory treats electrons and phonons in a simple average way. In a real metal, the phonon states are distributed in energy according to a spectral function $F(\omega)$. Furthermore, the inelastic scattering of electrons by phonons depends both on the phonon momentum $\hbar\boldsymbol{q}$, and on the phonon frequency ω. A typical process is illustrated in Fig. 9.1. The process is subject to energy conservation and, in the case of a crystalline metal, to momentum conservation. An average of electron-phonon scattering processes over the possible initial and final electron states $\boldsymbol{K}_i$ and $\boldsymbol{K}_f$ can be carried out which leaves only the ω dependence of the matrix element. The ω-dependent average electron-phonon matrix element is called $\alpha(\omega)$ [9.11], and the function $\alpha^2(\omega)F(\omega)$ is called the Eliashberg function [9.12]. This function serves as a basis for defining the coupling constant for a strong-coupling superconductor which is given by

$$\lambda \equiv 2\int \frac{d\omega\,\alpha^2(\omega)F(\omega)}{\omega}. \tag{9.2}$$

McMillan [9.11] has shown that λ can be rewritten as

$$\lambda_M = \frac{D(\varepsilon_F)\langle I^2\rangle}{M\langle\omega^2\rangle_M}, \tag{9.3}$$

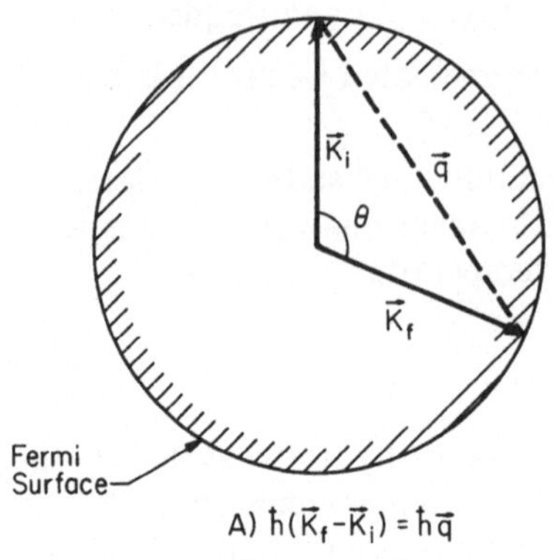

Fig. 9.1. Schematic illustration of a first-order electron-phonon scattering process. For a crystalline material, both energy and momentum are conserved. Only energy conservation is required for an amorphous material

where $\langle I^2 \rangle$ is an average squared electronic matrix element (see Sect. 9.2.3a for a detailed definition) with the average taken over all possible electron-phonon scattering processes, M is the ionic mass, and $\langle \omega^2 \rangle_M$ is given by

$$\langle \omega^2 \rangle_M = \frac{\int d\omega \alpha^2(\omega) \omega F(\omega)}{\int \frac{d\omega \alpha^2(\omega) F(\omega)}{\omega}} . \tag{9.4}$$

The average $\langle \omega^2 \rangle_M$ is often called the McMillan mean square phonon frequency. Finally, *McMillan* showed that T_c is to a good approximation given by the expression

$$T_c = \left(\frac{\langle \omega^2 \rangle^{1/2}}{1.45} \right) \exp \left[- \frac{1.04(1+\lambda_M)}{\lambda_M - \mu^*(1+0.62\lambda_M)} \right] , \tag{9.5}$$

where μ^* is an effective coupling constant for the repulsive Coulomb interaction between electrons. In most metals, one can take $\mu^* \approx 0.13$ [9.11]. The above expression for T_c should be reasonable even for the case where $\lambda \approx 1$ (i.e., the strong-coupling limit).

c) The Case of Amorphous Materials

The major modification of the BCS and strong-coupling theories required for describing amorphous metals is a generalization of the description of electron and phonon states. The lack of long-range order in amorphous metals leads to a breakdown in the concept of crystal momentum. The wave functions of electrons can no longer be described as Bloch states. If we choose to represent the wave functions in terms of linear combinations of plane waves, then the nth state can be written

$$\psi_n(\boldsymbol{r}) = \sum_{\boldsymbol{K}} c_{\boldsymbol{K}} e^{i(\boldsymbol{K}_n + \boldsymbol{K}) \cdot \boldsymbol{r}} \quad (\text{all } \boldsymbol{K}), \tag{9.6}$$

where $\boldsymbol{K}$ is no longer restricted to be a reciprocal lattice vector. The important point for the theory of superconductivity is the fact that we can still define a time-reversed state corresponding to $\psi_n(\boldsymbol{r})$. This state is simply

$$\psi_n^{\mathrm{tr}}(\boldsymbol{r}) = \sum_{\boldsymbol{K}} c_{\boldsymbol{K}}^* e^{-i(\boldsymbol{K}_n + \boldsymbol{K}) \cdot \boldsymbol{r}}. \tag{9.7}$$

The BCS theory does not depend in any essential way on the exact nature of the electronic states. *Anderson* [9.13] has shown that only the existence of time-reversed states is essential for the theory. One can thus proceed to reconstruct the theory of superconductivity with the proviso that one use n, not $\boldsymbol{K}$, to label electron wave functions. A similar convention must be applied in labelling the normal vibrational modes, since these too do not have well-defined momentum. One can still define $F(\omega)$ which represents the distribution of vibrational modes as a function of frequency. Furthermore, the electron-phonon matrix elements retain their significance in terms of the generalized $\psi_n(\boldsymbol{r})$ functions and vibrational modes. The matrix element $\alpha(\omega)$ represents an average over all scattering processes where an electron is scattered from state n to n' with the accompanying emission or absorption of a phonon with frequency ω.

Several arguments have been put forth regarding changes in the form of the function $\alpha^2(\omega)F(\omega)$ which result from disorder [9.14, 15]. In particular, *Bergmann* and *Rainer* [9.15] have argued that the low-frequency contributions to $\alpha^2(\omega)F(\omega)$ are enhanced by disorder. As can be seen from (9.2), these should in turn result in an increased λ, i.e., amorphous superconductors should be strong-coupling materials. This conclusion was shown to be consistent with a large number of data collected in tunnelling experiments on thin films of simple metals [9.16–18]. On the other hand, results of tunnelling experiments [9.9] on transition metals are in good agreement with the BCS weak-coupling theory. Although λ is apparently enhanced by disorder, the enhancement does not always give $\lambda > 1$. Amorphous superconductors may be of the weak- or strong-coupling type depending on the material of interest.

In what follows, it will be assumed that the microscopic parameters in the BCS and McMillan expressions for T_c can be meaningfully defined as outlined above. An attempt will be made in Sect. 9.2.3 to analyze the expected variation of these parameters throughout the periodic table. It will be shown that a simple analysis is perhaps more realistic for amorphous superconductors than for crystalline superconductors.

9.2.2 Phenomenology

a) Free Electron Metals

Most amorphous simple metals and alloys studied have been produced by the cryoquench technique in thin film form. A list of such simple metals and alloys is given in Table 9.1. The metals and alloys are listed in order of increasing average valence. A few entries in the table describe results obtained on liquid quenched amorphous alloys or metallic glasses. In contrast to thin films, these metallic glasses are generally stable at ambient temperature. The superconducting transition temperature T_c and other properties characteristic of each material are listed in the table. Several trends in T_c are observed. First, there is a tendency for T_c to increase with valency. This has been pointed out by the author previously [9.9]. Second, for fixed valency, T_c tends to increase with atomic number. These two observations will be discussed in more detail in Sect. 9.2.3a.

Data from tunnelling experiments are also summarized in Table 9.1. These data include the energy gap coefficient $R=[2\Delta(0)/k_B T]$ where $\Delta(0)$ is the measured energy gap in the $T=0$ limit. The BCS theory predicts a ratio of $R=3.52$ whereas $R>3.52$ is predicted by the strong-coupling theory. Notice that several entries in Table 9.1 give $R\sim 4$–5 suggesting strong-coupling effects [9.4]. On the other hand, tunnelling data are not available on the lower valence

Table 9.1. Summary of properties of amorphous superconductors based on simple metals. The gap coefficient $2\Delta(0)/k_B T_c$ and coupling constant λ are taken from tunnelling data

Alloy	T_c[K]	Preparation method	$\frac{2\Delta(0)}{k_B T_c}$	λ	Valence Z	Ref.
$Au_{84}Si_{16}$	<1.2	LQ	—	—	1.48	[9.19, 20]
$Mg_{70}Zn_{30}$	~0.7[a]	LQ	—	—	2.0	[9.20, 21]
Be	9.95[b]	VQ	3.5	0.5	2.0	[9.22 a]
$Be_{90}Al_{10}$	7.2	VQ	—	—	2.1	[9.22 c]
$Be_{70}Al_{30}$	6.1	VQ	3.6	—	2.3	[9.22 c]
$Cd_{90}Ge_{10}$	1.6	VQ	2.7–3.0	—	2.3	[9.22 b]
$Ca_{60}Al_{40}$	<1.2	LQ	—	—	2.4	[9.23, 24]
Ga	8.4	VQ	4.6	1.94-2.25	3.0	[9.1, 4]
$Tl_{90}Te_{10}$	4.2	VQ	—	1.7	3.3	[9.4, 25]
$Pb_{50}Sb_{25}Au_{25}$	5.0	LQ	—	—	3.5	[9.24]
$Pb_{90}Cu_{10}$	6.5	VQ	4.75	2.0	3.7	[9.2, 4]
$Sn_{90}Cu_{10}$	6.76	VQ	4.46	1.84	3.7	[9.2, 4, 16, 26]
Bi	5	VQ	4.6	2.2–2.5	5	[9.1, 4]

[a] Estimated from superconducting fluctuation data for $T>1.2$.

[b] Extrapolated to infinite thickness. There is some controversy as to whether pure Be films are amorphous. Alumin additions appear to stabilize the amorphous phase.

LQ Prepared by liquid quenching of alloy to metallic glass.

VQ Prepared by vapor quenching onto cryogenically cooled substrate.

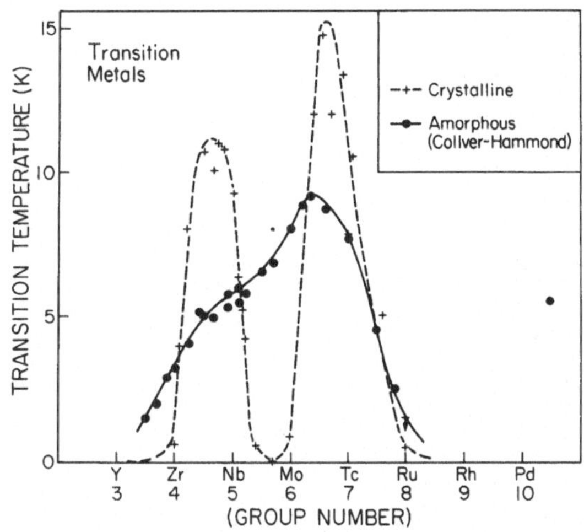

Fig. 9.2. Superconducting transition temperature as a function of average group number for crystalline transition metals [9.27] together with corresponding results for amorphous transition metal films [9.5]

materials which generally have lower T_c and should thus have correspondingly smaller λ. It is thus not possible to generalize concerning strong-coupling effects. In a few cases, a complete determination of the Eliashberg function $\alpha^2(\omega)F(\omega)$ was carried out from tunnelling data. In those cases, λ could be determined explicitly. The values listed in the table are in most cases rather large ($\lambda \gtrsim 1$).

b) Transition Metals

The earliest systematic study of amorphous transition metals was carried out by *Collver* and *Hammond* [9.5] again using the cryoquench technique. In this study, transition metals of the 4*d* and 5*d* series were cryoquenched together with alloys of metals of neighboring groups. In this manner, a continuous series of amorphous metals and alloys were obtained as a function of *d* band occupation. *Matthias* had earlier described the systematics of superconductivity in crystalline transition metals in terms of valency rules [9.27]. Plotting T_c as a function of average group number (e/a) as shown in Fig. 9.2, *Matthias* found most superconducting alloys to occur for (e/a) values near ~ 4.5 and ~ 6.5. For the corresponding amorphous alloys, *Collver* and *Hammond* observed a T_c vs (e/a) plot exhibiting a single broad maximum. This is also shown for comparison in Fig. 9.2. A maximum T_c value is obtained near $(e/a) \sim 6.8$ which roughly corresponds to a half-filled *d* band if the transition metals are taken to have a single *s*-like free electron. *Collver* and *Hammond* claimed that the cryoquenched films were amorphous on the basis of a precipitous drop in electrical resistivity which occurred on heating to room temperature. This was interpreted as evidence for crystallization. Only in a few cases did the amorphous alloy films remain stable at room temperature. In these cases, structural analysis carried out using electron diffraction confirmed the amor-

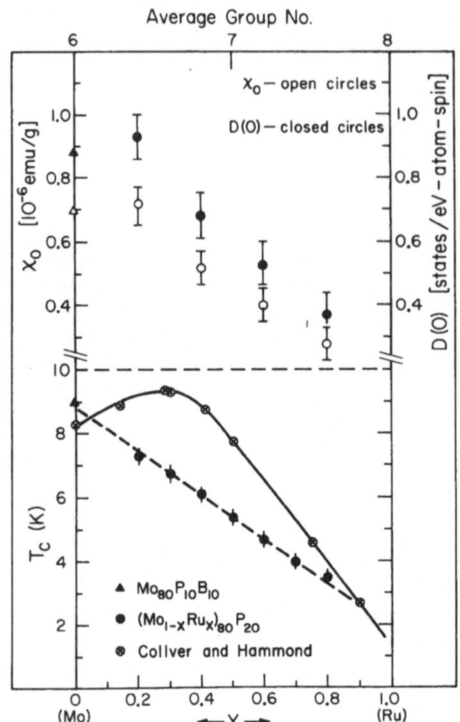

Fig. 9.3. (Above) Variation of the temperature-independent part of the magnetic susceptibility, χ_0, and the density of states $D(\varepsilon_F)$ estimated from χ_0 for the amourphous $(Mo_{1-x}Ru_x)_{80}P_{20}$ alloy series. (Below) Variation of T_c for the same alloy series compared with that obtained in [9.5] for amorphous $Mo_{1-x}Ru_x$ thin films

phous nature of the films. In later publications [9.28], these data were interpreted in terms of an "atomiclike" parameter which varies smoothly with *d* band occupation.

Superconducting metallic glasses based on transition metals and obtained by melt quenching were reported by the author and colleagues in 1975 [9.8, 9]. The earliest alloys studied contained 60–80 at.% of an early transition metal (ETM) with the balance of the alloy being a late transition metal (LTM) (e.g., $La_{80}Au_{20}$, $Zr_{70}Pd_{30}$, $Nb_{60}Rh_{40}$). It was later found [9.29, 30] that alloys based on closely neighboring transition metals could form metallic glasses with the addition of metalloid or glass-forming elements (e.g., B, P, Si). For example, alloys such as $(Mo_{1-x}Ru_x)_{80}P_{20}$ form glasses with $0.2 \leqq x \leqq 0.8$. This series of metallic glasses can be compared with $Mo_{1-x}Ru_x$ cryoquenched films studied by *Collver* and *Hammond* as shown in Fig. 9.3. A similar variation of T_c is observed with (e/a) although the curve appears shifted with respect to that of *Collver* and *Hammond.* This shift can in part be attributed to the effects of electron transfer [9.30] between the transition metal and the metalloid component. A summary of properties of several representative superconducting metallic glasses is given in Table 9.2.

The bulk nature of metallic glasses makes possible a variety of measurements to determine the microscopic parameters which enter the theory of

Table 9.2. Composition and properties of several metallic glasses and amorphous thin films based on transition metals. In the table, γ is the coefficient of the linear contribution to the low-temperature specific heat above T_c

Alloy	T_c[K]	θ_D[K]	$\gamma\left[\frac{mJ}{mole\text{-}K^2}\right]$	$D(0)\left[\frac{states}{eV\text{-}atom\text{-}spin}\right]$	$\frac{2\Delta(0)}{K_B T}$	$\left(\frac{dH_{c2}}{dT}\right)_{T_c}\left[\frac{kOe}{K}\right]$	Ref.
Metallic glasses							
$La_{80}Au_{20}$	3.5	96	—	0.80[a]	—	23±0.5	[9.8, 33]
$La_{80}Ga_{20}$	3.8	109.6	6.10	0.70[a]	—	22.5±0.5	[9.34, 40]
$Zr_{75}Rh_{25}$	4.55	—	—	—	—	26.3±0.5	[9.35, 102]
$Zr_{70}Pd_{30}$	2.4	180	4.7	0.99[a]	2.94[a]	26.5±0.5	[9.37, 38]
$Zr_{70}Be_{30}$	2.8	—	—	—	—	23.8	[9.36]
$Zr_{70}Co_{30}$	3.3	—	—	—	—		[9.39]
$Nb_{60}Rh_{40}$	4.8	—	—	—	—	31	[9.37]
$(Mo_{1-x}Ru_x)_{80}P_{20}$							
$x=0.2$	7.31	—	—	0.93[b]	—	24.5	[9.9, 29]
0.4	6.18	265	4.1	0.68[b], 0.51[a]	—	25.5	[9.9, 29, 32]
0.5	5.40	—	—	—	3.5±0.1[c]	—	[9.9, 29, 31]
0.6	4.68	267	3.6	0.53[b], 0.46[a]	—	26.8	[9.9, 29, 32]
0.8	3.43	—	—	0.37[b]	—	27.6	[9.9, 29]
$(Mo_{0.6}Ru_{0.4})_{82}B_{18}$	6.05	286	4.0	0.50[a]	—	23.6	[9.30, 37, 116]
$(W_{0.5}Ru_{0.5})_{80}P_{20}$	4.57	—	—	—	—	22.2	[9.30]
$(Mo_{0.8}Re_{0.2})_{80}P_{10}B_{10}$	8.71	—	—	—	—	24.2	[9.9]
Thin films							
Nb_3Ge	3.6	222	3.0±0.5	0.4[a]	3.5±0.1[d]	—	[9.41, 42]
$Mo_{68}Si_{32}$	6.7	—	—	—	3.6±0.1[d]	—	[9.41]
$Mo_{80}N_{20}$	8.3	—	—	—	3.6±0.1[d]	—	[9.41, 43, 48]

[a] Estimate derived from specific heat (γ). The $D(0)$ values have been renormalized by the factor $(1+\lambda_M)^{-1}$ determined from McMillan's equations.
[b] Estimate derived from magnetic susceptibility (χ_0).
[c] Determined from nuclear spin relaxation rate (see [9.31]).
[d] Determined from superconductive tunnelling.

superconductivity. Low-temperature specific heat measurements have been used as a means of determining the Debye temperature θ_D, as well as the electronic density of states at the Fermi level $D(\varepsilon_F)$. Such measurements have been carried out on ETM-LTM alloys [9.32–34] and give a value of $D(\varepsilon_F)$ considerably in excess of that predicted by the free electron model. More recent measurements [9.43] on $(Mo_{1-x}Ru_x)_{80}P_{20}$ glasses indicate that $D(\varepsilon_F)$ decreases rapidly with x in this series but is again substantially larger than the prediction of the free electron model throughout the series. These results are not unexpected since $D(\varepsilon_F)$ is expected to contain a significant d band contribution in transition metal alloys. This d band contribution is not well described by a free electron or nearly free electron model. Magnetic susceptibility measurements can also be used to estimate $D(\varepsilon_F)$ [9.29]. In estimating $D(\varepsilon_F)$ from the temperature-independent contribution χ_0 to the paramagnetic susceptibility of $(Mo_{1-x}Ru_x)_{80}P_{20}$ alloys, it was assumed that Pauli paramagnetism was the dominant term in χ_0 [9.29, 44]. Although this is a rather crude approximation, the values of $D(\varepsilon_F)$ obtained appear to agree reasonably well with those obtained from specific heat [9.32]. The data for χ_0 and the estimated density of states are shown for the $(Mo_{1-x}Ru_x)_{80}P_{20}$ series in Fig. 9.3 for comparison with the variation of T_c. The values of T_c and χ_0 are both decreasing functions of x (or e/a) suggesting a close relationship between T_c and $D(\varepsilon_F)$. The inferred smooth variation of $D(\varepsilon_F)$ with (e/a) is in contrast with both theory and observations on crystalline transition metal alloys. In the crystalline case, $D(\varepsilon_F)$ varies rapidly with (e/a) owing to the complex structure of the d-bands [9.45–47]. Ultraviolet photoemission spectroscopy carried out on crystalline and amorphous thin films of Mo have illustrated this difference in band structure directly [9.48]. The data are shown in Fig. 9.4a. The photoemission intensity $I(\varepsilon)$ is shown as a function of electron energy and reflects the d band density of states for energies below ε_F. For comparison, a theoretical calculation of $D(\varepsilon)$ for bcc Mo by *Moruzzi* et al. [9.49] is shown together with the data. The theoretical $D(\varepsilon)$ is appropriately broadened by an amount corresponding to instrumental broadening in the experiment. For $\varepsilon < \varepsilon_F$, the theory predicts three resolved d subbands with ε_F situated near a minimum in $D(\varepsilon)$ above the third band. This type of structure is clearly resolved in the experimental $I(\varepsilon)$ for the bcc Mo film. The amorphous film exhibits a completely different spectrum with $I(\varepsilon)$ varying through a broad smooth maximum. Near ε_F, $I(\varepsilon)$ is larger than in the crystalline case indicating a larger value of $D(\varepsilon_F)$. The amorphous film has $T_c = 8.0$ K while the crystalline film was not superconducting down to 1.5 K. Bulk bcc Mo has $T_c = 0.92$ K [9.50]. Other photoemission data on the metallic glasses $(Mo_{0.5}Ru_{0.5})_{80}P_{20}$ and $(Mo_{0.6}Ru_{0.4})_{80}B_{20}$ have shown a similar broad structureless maximum in $I(\varepsilon)$ for energies $\varepsilon < \varepsilon_F$ [9.48b]. This behavior seems to be characteristic of all amorphous alloys based on Mo and Ru.

From the combined data above, one can conclude that amorphous transition metal alloys have a rather broad featureless d band structure. The multiple structure of d subbands observed in crystalline transition metals is replaced by a single broad maximum for the d band in the amorphous state. A

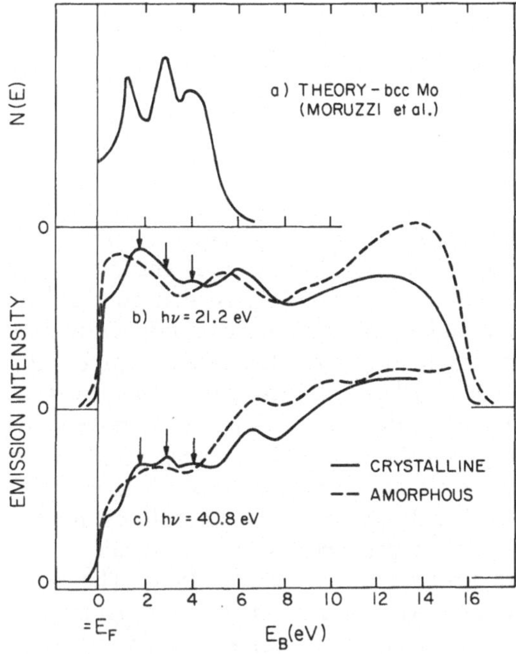

Fig. 9.4. (**a**) Theoretical one-electron density of states for bcc Mo as predicted [9.49]. (**b**) Experimental photoemission spectrum for amorphous and crystalline Mo films for photon energy $h\nu=21.2$ eV. (**c**) The same as (**b**) with $h\nu=40.8$ eV. (Taken from [9.48])

similar behavior has previously been predicted from theoretical calculations of $D(\varepsilon)$ for liquid transition metals [9.51–53].

9.2.3 Analysis in Terms of Microscopic Theory

a) Free Electron Metals

In this section, an attempt will be made to analyze the systematics of superconductivity in amorphous simple metals and alloys thereof. It will be assumed that McMillan's solution to the Eliashberg equations gives a reasonably accurate strong-coupling expression for T_c in terms of well-defined microscopic parameters. McMillan's T_c expression can be questioned when $\lambda > 1$; in such cases it tends to overestimate T_c [9.54, 55]. Also, since *McMillan* used the measured Eliashberg function $\alpha^2(\omega)F(\omega)$ of bcc niobium in arriving at his solution, it follows that this solution may not remain accurate for materials where the phonon spectrum differs markedly from that of niobium. In spite of this, the errors associated with using McMillan's expression will generally be of less consequence than other approximations made. Therefore, the question of solving the Eliashberg equations for an $\alpha^2(\omega)F(\omega)$ appropriate to amorphous metals will not be directly addressed. Instead, attention will be focused on estimating the microscopic parameters which enter an expression such as

McMillan's. *Garland* has proposed a modified expression for T_c of amorphous alloys [9.54] which does not differ greatly from McMillan's.

The parameters of interest here include $D(\varepsilon_F)$, $\langle\omega^2\rangle_M$ as defined by (9.4), and $\langle I^2\rangle$. One defines [9.56]

$$I_\nu(\boldsymbol{K},\boldsymbol{K}')=\int \psi^*_{\boldsymbol{K}}(\boldsymbol{r})[\hat{\varepsilon}_{q,\nu}\cdot\nabla U(\boldsymbol{r})]\psi_{\boldsymbol{K}'}(\boldsymbol{r})\,d\boldsymbol{r}, \tag{9.8}$$

the electronic matrix element of the change in crystal potential $U(\boldsymbol{r})$ per unit displacement of a single atom along the νth direction. Here $\boldsymbol{K}$ and $\boldsymbol{K}'$ refer to the initial and final electron states. The parameter $\langle I^2\rangle$ is then obtained by averaging $I_\nu(\boldsymbol{K},\boldsymbol{K}')$ over all possible $\boldsymbol{K}$ and $\boldsymbol{K}'$ and summing over polarizations ν. For amorphous materials, $\boldsymbol{K}$ and $\boldsymbol{K}'$ are no longer good quantum numbers and should in the strict sense be regarded as labels for electron states. In what follows, the crystal potential $U(\boldsymbol{r})$ will be treated as a weak potential in the spirit of the free electron approximation. In this limit, the plane wave approximation describes the unperturbed electron wave function regardless of whether or not $U(\boldsymbol{r})$ is periodic (i.e., whether we deal with crystalline or amorphous metals). *Ziman* [9.57] has discussed the electronic properties of liquid metals in this limiting case; the discussion here proceeds along similar lines. The weak potential approximation enables one to proceed in evaluating the parameters of (9.3–5) using spherical Fermi surface in $\boldsymbol{K}$ space with $\boldsymbol{K}$ retaining its usual interpretation as a momentum. This approximation may not be as serious as one might expect since amorphous metals are rather isotropic on any spatial scale larger than a few atomic distances. The absence of long-range order means that the wave functions for electrons of given energy will on the average contain a superposition of plane waves with isotropically distributed wave vectors. The spherical Fermi surface is therefore a more appropriate approximation for amorphous metals than their crystalline counterparts.

In order to evaluate $\langle I^2\rangle$, one must now adopt a model for $U(\boldsymbol{r})$ and $\nabla U(\boldsymbol{r})$. Since the present analysis deals with simple metals, the pseudopotential formalism is one natural choice. One can express $U(\boldsymbol{r})$ as a superposition of atomic potentials

$$U(\boldsymbol{r})=\sum_{\boldsymbol{R}_l} U^{\mathrm{a}}(\boldsymbol{r}-\boldsymbol{R}_l), \tag{9.9}$$

where the sum runs over atomic positions $\boldsymbol{R}_l$. Now taking $U^{\mathrm{a}}(\boldsymbol{r}-\boldsymbol{R}_l)$ to be the atomic pseudopotential which can be written in terms of its form factor $U^2_{\boldsymbol{q}}$ as

$$U^{\mathrm{a}}(\boldsymbol{r})=\sum_{\boldsymbol{q}} U^{\mathrm{a}}_{\boldsymbol{q}}\mathrm{e}^{\mathrm{i}\boldsymbol{q}\cdot\boldsymbol{r}}, \tag{9.10}$$

one obtains

$$I_\nu(\boldsymbol{K},\boldsymbol{K}')=-\mathrm{i}\hat{\varepsilon}_{\boldsymbol{K}-\boldsymbol{K}',\nu}\cdot\boldsymbol{q}\,U^{\mathrm{a}}_{\boldsymbol{q}}. \tag{9.11}$$

Table 9.3. Summary of parameters used and results of the model calculation for λ and T_c of simple amorphous metals. Experimental results for several cases are also shown for comparison with corresponding alloys in parentheses

Metal (alloy)	$\langle I^2\rangle^{\text{calc}}$ [$\text{eV}^2/\text{Å}^2$]	$\langle\omega^2\rangle_M^{1/2\,\text{calc}}$ [K]	$D^{\text{f.e.}}(0)$ $\left[\frac{\text{states}}{\text{eV-atom-spin}}\right]$	λ^{calc}	T_c^{calc}[K]	λ^{exp}	T_c^{exp}[K]	Ref.
Pb	8.2	101	0.320	1.1	5.2			
($Pb_{90}Cu_{10}$)						2.0	6.5	[9.2, 4]
Sn	10.8	143	0.299	1.2	8.2			
($Sn_{90}Cu_{10}$)						1.84	6.76	[9.2, 4]
In	6.4	118	0.262	1.1	4.4			
($In_{80}Sb_{20}$)						1.7	5.6	[9.4, 69]
Ga	10.8	184	0.217	0.86	5.9			
(Ga)						1.9–2.2	8.4	[9.1, 4]
Al	11.1	324	0.193	0.66	4.7			
Be	21.0	617	0.106	0.57	4.8			
(Al)							5.84	[9.68]
(Be)						0.5	9.9	[9.22 a]
($Be_{70}Al_{30}$)							6.1	[9.22 c]
Zn	8.2	155	0.160	0.72	3.1			
(Zn)							1.4–1.5	[9.20]
Ca	0.65	111	0.321	0.37	0.03			
($Ca_{70}Al_{30}$)							<1.2	[9.23, 24]
Mg	3.1	195	0.210	0.62	2.3			
($Mg_{70}Zn_{30}$)							~ 0.7	[9.20, 21]

Then the average $\langle I^2\rangle$ can after some algebra be reduced to

$$\langle I^2\rangle = \frac{1}{2k_f^2}\int_0^{2k_f} (U_q^{\text{a}})^2 q^3\, dq\,. \tag{9.12}$$

The polarization factor $|\hat{\varepsilon}_{q,\nu}\cdot q|$ has been replaced by its average value $q^2/3$. Using the *Heine* and *Animalu* [9.58] pseudopotential[1], $\langle I^2\rangle$ was computed for several simple metals of interest. The results are given in Table 9.3. It is worth pointing out that these values are in reasonable agreement with those deduced for several crystalline simple metals by *McMillan* [9.11]. He deduced $\langle I^2\rangle$ from an empirical analysis of tunnelling and low-temperature specific heat data using (2.3–5).

Having estimated $\langle I^2\rangle$, one now needs to determine $D(\varepsilon_F)$ and $\langle\omega^2\rangle_M$ in order to calculate λ and T_c. The above arguments suggest that the nearly free electron model should provide a good description of $D(\varepsilon)$ in amorphous metals. The departure of $D(\varepsilon)$ in crystalline metals from the predictions of the free electron model are due mainly to the influence of Brillouin zone (BZ)

1 Here, we have used tabulated values of V_Q given in [9.67].

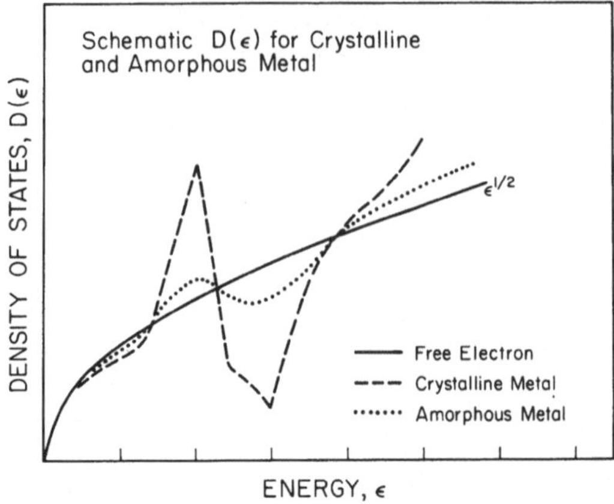

Fig. 9.5. Schematic illustration of the influence of disorder on the electronic density of states function $D(\varepsilon)$ for a nearly free electron metal

boundaries on the size and shape of ther Fermi surface. The structural details of $D(\varepsilon)$ arising from this Fermi surface-BZ interaction will be "smeared out" in the amorphous case. This effect is schematically illustrated in Fig. 9.5. Roughly speaking, the Fermi surface in $\boldsymbol{K}$ space is smeared by an amount $\Delta K \sim \hbar/l_{\mathrm{mfp}}$ where l_{mfp} is the electron mean free path. One result of this smearing should be an overall smoothing out of the function $D(\varepsilon)$ on going from the crystalline to amorphous metal. This in turn should result in $D(\varepsilon)$ being well approximated in amorphous metals by its free electron value, $D^{\mathrm{f.e.}}(\varepsilon)$. In the spirit of this argument one can use $D^{\mathrm{f.e.}}(\varepsilon)$ to estimate $D(\varepsilon_{\mathrm{F}})$.

Finally, we turn to $\langle\omega^2\rangle_M$ defined by (9.4). This parameter is probably the most difficult to calculate since it involves a complicated frequency average involving the weight function $\alpha^2(\omega)F(\omega)$. In the spirit of the previous approximations, we treat the amorphous metal as a homogeneous and isotropic distribution of atoms lacking long-range order. In the usual jellium model, transverse oscillations have vanishing frequency–jellium does not resist shear deformation. *Weaire* et al. [9.59] have computed the elastic moduli of an amorphous metal using an atomic pair potential model together with an atomic distribution given by the Bernal-Finney model [9.60]. Comparing a monatomic amorphous metal to the corresponding crystalline metal, they found the bulk modulus to be essentially unchanged while the shear modulus was reduced by $\sim 30\%$ in the amorphous case. This suggests that the elastic properties of amorphous metals lie somewhere between those of a close-packed crystal and jellium, being closer to the former. Of course, the elastic moduli determine only the long wavelength phonon properties. At short wavelengths (comparable to atomic spacings) the phonon spectrum will be quite complex depending on the details of both atomic scale structure and effective atomic interactions. *Von Heimendahl* and *Thorpe* [9.61] have used computer simulation to determine the spectral function $F(\omega)$ for a large noncrystallographic cluster of atoms. They found an $F(\omega)$ having a broad featureless maximum showing no distinct transverse-longitudinal mode structure. *Bergmann* and *Rainer* [9.15], and more recently *Poon* and *Geballe* [9.62], have actually attempted to calculate $\alpha^2(\omega)$ for amorphous simple metals. In Poon and Geballe's calculation, the Heine-Animalu pseudopotential was used together with an experimental structure factor of the amorphous metal and a simple Debye-type model phonon spectrum containing longitudinal and transverse Debye frequencies. *Poon* and *Geballe* found their computed $\alpha^2(\omega)F(\omega)$ to be in reasonable agreement with results of tunnelling experiments [9.17, 18] on amorphous simple metals. *Bolz* and *Pobell* [9.63] have used Mössbauer techniques to compare $\int_0^\infty [F(\omega)/\omega]\,d\omega$ in crystalline and amorphous tin and found the two values to be nearly identical.

In the present case, one would like to make a simple but reasonable first principles estimate of $\langle\omega^2\rangle$ for amorphous metals. One can employ the following argument based on the nearly free electron concept. The phonon frequency for a particular wave vector $\boldsymbol{q}$ and polarization ν consists of two main contributions [9.64, 65]

$$\omega^2_{\boldsymbol{q},\nu} = \Omega^{\mathrm{coul}^2}_{\boldsymbol{q},\nu} - \omega^{\mathrm{el}^2}_{\boldsymbol{q},\nu}, \tag{9.13}$$

where $\Omega^{\mathrm{coul}}_{\boldsymbol{q},\nu}$ is the coulomb frequency of the bare ions interacting pairwise through the coulomb potential $Z^2e^2/|\boldsymbol{R}|$, $|\boldsymbol{R}|$ being the distance between ions. For a given solid, these frequencies can be computed for each $\boldsymbol{q}$ and ν. A simple sum rule is obeyed by the coulomb frequencies for any array of atoms and can be stated as [9.66]

$$\sum_\nu \Omega^{\mathrm{coul}^2}_{\boldsymbol{q},\nu} = \Omega^2_{\mathrm{p}}, \tag{9.14}$$

where Ω^2_{p} is the ionic plasma frequency

$$\Omega^2_{\mathrm{p}} = \left(\frac{4\pi N e^2 Z^2}{M}\right)^{1/2} \tag{9.15}$$

with N the ionic density (ions/cm^3), M the ionic mass, and Z the valence. In the present calculations, N is taken to be the density of ions in the corresponding crystalline metal. The term $\omega^{\mathrm{el}^2}_{\boldsymbol{q},\nu}$ in (9.13) refers to the screening effect of the electrons. Under the ionic displacements associated with the phonon mode of interest, the electronic charge density is redistributed to screen out the deformation potential of the ions [9.56, 57]. We can approximate the electronic screening effect for free electrons by

$$\omega^2_{\boldsymbol{q},\nu} = \frac{\Omega^{\mathrm{coul}^2}_{\boldsymbol{q},\nu}}{\varepsilon(\boldsymbol{q},0)}, \tag{9.16}$$

where $\varepsilon(\boldsymbol{q},0)$ is the static dielectric function of a free electron gas. Equations (9.14–16) give

$$\sum_{\nu}\omega_{\boldsymbol{q},\nu}^{2}\approx\sum_{\nu}\frac{\Omega_{\mathrm{p}}^{2}}{\varepsilon(\boldsymbol{q},0)}. \tag{9.17}$$

Treating each of the three polarizations on an equal footing gives typical squared frequency

$$\omega_{\boldsymbol{q}}^{2}\approx\frac{\Omega_{\mathrm{p}}^{2}}{3\varepsilon(\boldsymbol{q},0)}. \tag{9.18}$$

If for convenience one uses the Thomas-Fermi screening, then $\varepsilon(\boldsymbol{q},0)=\varepsilon(q)=(q^2+\lambda_{\mathrm{s}}^2)/q^2$, where the Thomas-Fermi screening length is $\lambda_{\mathrm{s}}=(6\pi ne^2/\varepsilon_{\mathrm{F}})^{1/2}$ and n is the electronic density. This then gives an approximation for the phonon dispersion relation. The McMillan average square frequency $\langle\omega^2\rangle_M$ involves the weight function $\alpha^2(\omega)F(\omega)$. The full electron phonon matrix element is given by [9.11, 56]

$$g_{\nu}(\boldsymbol{K},\boldsymbol{K}')=\left(\frac{\hbar}{2MNV\omega_{\boldsymbol{K}-\boldsymbol{K}',\nu}}\right)I_{\nu}(\boldsymbol{K},\boldsymbol{K}') \tag{9.19}$$

so that if we ignore the dependence of $I_{\nu}(\boldsymbol{K},\boldsymbol{K}')$ on phonon wave vector (frequency), we could approximate the frequency dependence of $g^2(\boldsymbol{K},\boldsymbol{K})$ and therefore $\alpha^2(\omega)$ as $\alpha^2(\omega)=\alpha_0/\omega$ where α_0 is a constant. In this approximation, $\langle\omega^2\rangle_M$ becomes

$$\langle\omega^2\rangle_M=(4\pi/3)q_{\mathrm{D}}^3\left[\int\frac{d\omega\,F(\omega)}{\omega^2}\right]^{-1} \tag{9.20}$$

with $\int_0^{q_{\mathrm{D}}}F(\omega)\,d\omega=(4\pi/3)q_{\mathrm{D}}^3$. Now taking ω_q from (9.18) finally gives

$$\langle\omega^2\rangle_M=(\Omega_{\mathrm{p}}^2/3)\left(\frac{1+3\lambda_{\mathrm{s}}^2}{q_{\mathrm{D}}^2}\right)^{-1} \tag{9.21}$$

for the case of amorphous metals. The approximations made are most appropriate for a material lacking long-range order.

One can now proceed to estimate λ using the results of (9.3, 12, 21) together with the free electron value of $D(\varepsilon_{\mathrm{F}})$ for various simple metals. It is important to note that this calculation involves only the pseudopotential (Heine-Animalu potential), the density of atoms N (taken to be that of the corresponding crystalline metal), the valence Z, and the ionic mass M, for each simple metal.

There are no adjustable parameters. The results for $D(\varepsilon_F)$, $\langle I^2 \rangle$, $\langle \omega^2 \rangle_M$, and λ are given in Table 9.3. Now using (9.5), we can approximate $\mu^* = 0.13$ [9.11] and calculate T_c. The results are also given in Table 9.3 and compared to the experimentally observed T_c for both amorphous and crystalline metals. One notes that the predicted values of T_c are in much better agreement with those observed for amorphous as opposed to crystalline metals. The jellium type of theory presented here has involved numerous approximations which clearly break down when, for example, the Fermi surface deviates strongly from a spherical shape. It has already been proposed that such deviations should be much more likely for crystalline metals. The above results confirm this hypothesis. For example, consider the case of divalent metals. The free electron Fermi surface of a crystalline divalent metal has a volume in $\boldsymbol{K}$ space exactly equal to that of the first BZ [9.70]. As such, the degree of intersection of the Fermi surface with the first BZ boundary should be extensive. The linear term γT in low-temperature specific heat data can be used to estimate $D(\varepsilon_F)$ for simple metals if one corrects by the factor $(1+\lambda)^{-1}$ which is due to electron-phonon renormalization [9.11]. *McMillan* evaluated $D(\varepsilon_F)$ for several simple metals and compared his results to those obtained from the free electron model. The results are given in [Ref. 9.11, Table III]. As a group, the divalent metals show the largest discrepancy between $D(\varepsilon_F)$ and $D^{\text{f.e.}}(\varepsilon_F)$, with $D(\varepsilon_F) < D^{\text{f.e.}}(\varepsilon_F)$. This reflects the diminished Fermi surface area due to the Fermi surface-BZ interaction.

In summary, crystalline divalent metals are likely to violate the assumption made in the above model. Indeed, the model considerably underestimates the T_c of crystalline divalent metals. The remarkable fact is that the amorphous metals have T_c's in much better agreement with the model. The above discussion illustrates an important concept. The electronic structure of amorphous simple metals is apparently free electronlike. At least, the free electron picture appears to be much more appropriate for amorphous metals than for their crystalline counterparts.

The model presented above shows how the systematics of T_c in amorphous simple metals can be understood in a simple framework. One can now for example understand the general tendency of T_c to increase with valency for amorphous metals. Both the pseudopotential and therefore $\langle I^2 \rangle$ generally increase with Z as does $D(\varepsilon_F)$. Although Ω_p^2 increases with Z, so does the electronic screening effect [i.e., $\varepsilon(q,0)$] so that $\langle \omega^2 \rangle_M$ does not increase as rapidly as Ω_p^2. The combined Z dependences of $D(\varepsilon_F)$ and $\langle I^2 \rangle$ and $\langle \omega^2 \rangle_M$ lead to a general increase of λ, (9.3), with increasing Z. This is easily demonstrated by plotting the present calculated λ values as a function of Z as shown in Fig. 9.6. It is evident that the calculated λ values systematically increase with increasing valence. The experimental λ values tend to be somewhat larger than the calculated values. This is probably in part due to a systematic error in estimating $\langle \omega^2 \rangle_M$ in the model and also to the use of McMillan's expression for T_c. For both the model and the experimental data, it is nevertheless evident that λ (and thus T_c) tend to increase with Z, indicating the general validity of the

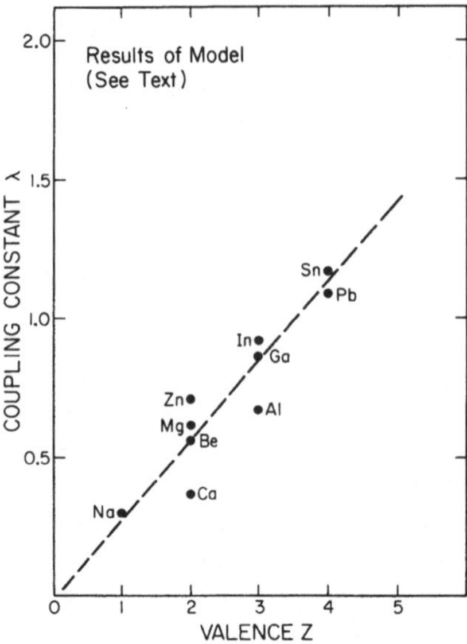

Fig. 9.6. Calculated values of the electron-phonon coupling constant λ plotted as a function of valence Z

model. This is the important conclusion that can be drawn from the above discussion.

b) Transition Metals and Alloys

The electronic structure of transition metals is characterized by the large contribution to $D(\varepsilon)$ arising from partially occupied d bands. Free transition metal atoms have a typical ground state configuration [9.71, 72] $d^{\zeta-1}s$, where ζ is the group number of the element (i.e., the number of valence electrons outside an inert gas core configuration). The atomic d states accommodate up to 10 electrons/atom. The electronic band structure of transition metals has been calculated using a variety of techniques. *Mattheiss* [9.46] has employed the APW method to calculate $D(\varepsilon)$ for several bcc and hcp transition metals. *Friedel* [9.45, 73] was the first investigator to carry out extensive band structure calculations using a tight-binding approximation (TBA) of the d levels. The relatively small spatial extent of atomic d orbitals by comparison with interatomic distances in transition metals renders the TBA description a useful one [9.45]. *Varma* and *Dynes* [9.74] and *Varma* and *Weber* [9.75–77], have carried out extensive calculations of the phonon dispersion curves, the electron-phonon interaction, and superconducting properties of crystalline transition metals using a nonorthogonal tight-binding (NTB) approximation. This amounts to including the direct overlap of atomic orbitals on neighboring sites in addition to the usual TBA calculation. The results of the NTB approximation give phonon spectra possessing all the features of the observed spectra from neutron scattering [9.77]. On the basis of these studies, it is reasonable to adopt the tight-binding description of d electrons.

Recently, *Khanna* and *Cyrot-Lackmann* [9.79, 80], *Gaspard* [9.53], and *Fujiwara* [9.78] have successfully attempted to apply the TBA method to the calculation of $D(\varepsilon)$ for liquid and amorphous transition metals. *Khanna* and *Cyrot-Lackmann* used the fact that $D(\varepsilon)$ can be expanded in a series involving the moments M_p of $D(\varepsilon)$, where

$$M_p = \int D(\varepsilon)(\varepsilon-\varepsilon_0)^p \, d\varepsilon \tag{9.22}$$

with ε_0 the energy of the unperturbed atomic d level. The first moment M_1 gives the shift in the center of gravity of the d band as compared with the atomic d level, while M_2 gives a measure of the total width of the d band. Information concerning the detailed shape of the d band is contained in the higher moments. Calculations for liquid paramagnetic Ni and Co [9.79, 80] show a d band having a width comparable to but slightly less than that of crystalline Ni and Co. The d band density of states $D_d(\varepsilon)$ is similar in shape to that of the close-packed crystalline metals but exhibits far less structural detail. The main feature observed is an overall tendency toward a $d\gamma - d\varepsilon$ splitting of the d band. The experimental ultraviolet photoemission spectra of amorphous and crystalline Mo shown in Fig. 9.4 suggest that a strong splitting does not exist in the d band of amorphous $4d$ transition metals although the 40.8 eV spectrum suggests a slight tendency toward splitting. The magnetic susceptibility data shown in Fig. 9.3 suggest a smoothly decreasing value of $D(\varepsilon_F)$ through the $(Mo_{1-x}Ru_x)_{80}P_{20}$ alloy series. If we use a rigid band model of $D(\varepsilon)$, then this again suggests a broad featureless d band arising from the atomic d states of Mo and Ru. Recent specific heat data [9.32] confirm the inferences made on the basis of susceptibility data for the same alloy series.

One can begin a discussion of superconductivity in amorphous transition metals by assuming a simple model for the d band density of states $D_d(\varepsilon)$. It is a good approximation to neglect the s electron contribution $D_s(\varepsilon)$ since $D_d(\varepsilon)/D_s(\varepsilon) \gg 1$ except at the beginning and end of the transition series where $D_d(\varepsilon)$ is smaller. Furthermore, for the sake of simplicity, one can to lowest order neglect the effect of $s-d$ hybridization. Such an approximation is put on a more concrete basis by recent detailed calculations of the parameter $\langle I^2 \rangle$ by *Varma* et al. [9.76] using the NTB method and including $s-d$ and $p-d$ hybridization. Their results show $\langle I^2 \rangle$ to be governed mainly by $d-d$ scattering.

Following the general NTB analysis of *Varma* and *Dynes* [9.74], and *Barisic* et al. [9.81], one can analyze the parameters which determine λ and T_c for transition metals in the following way *once the function $D_d(\varepsilon)$ is known.* They argued that two approximate relationships exist between the parameters which determine λ. The first relationship, derived by *Barisic* et al. [9.81], can be written as

$$D_d(\varepsilon_F)\langle I^2 \rangle \approx \frac{E_0}{r_0^2}(1 \mp S), \tag{9.23}$$

where E_0 is the d band contribution to the cohesive energy of the transition metal given by

$$E_0 = -\int 2(\varepsilon - \varepsilon_0) D_d(\varepsilon)\, d\varepsilon \tag{9.24}$$

and ε_0 is the energy of the free atomic d level. The parameter r_0 is a characteristic length for exponential decay of the radial part of the atomic d orbital. Finally, S is the overlap integral for atomic d orbitals located on neighboring sites. The factor $(1 \mp S)$ is due to *Varma* and *Dynes* and takes into account the nonorthogonality of atomic d functions on different sites. The minus (plus) sign is appropriate to ε_F in the lower (upper) half of the d band described by $D_d(\varepsilon)$. *Varma* and *Dynes* [9.74] derived the second approximate relationship

$$\frac{\langle I^2 \rangle}{M\langle \omega^2 \rangle_M} \approx W(1 \mp S), \tag{9.25}$$

where $\langle \omega^2 \rangle_M$ has been defined previously and W is related directly to the width of the d band. The important point here is that these approximate relations might be expected to work better for amorphous metals than for their crystalline counterparts. Again, as in the case of simple metals, the lack of long-range order would tend to wash out structural details in $D_d(\varepsilon)$ and also in $I_\nu(\boldsymbol{K}, \boldsymbol{K})$, giving an averaged behavior for these parameters. This is precisely the spirit in which the above relations (9.23, 25) were derived. Now (9.25) implies that

$$\lambda = \frac{D(\varepsilon_F)\langle I^2 \rangle}{M\langle \omega^2 \rangle_M} \approx D(\varepsilon_F)\, W(1 \mp S). \tag{9.26}$$

To the extent that the bandwidth is constant for a given series of transition metals, one finds that $\lambda \propto D(\varepsilon_F)$ within the upper and lower (antibonding and bonding) halves of the d band. In other words, $D_d(\varepsilon)$ is the principal parameter governing superconductivity. One can proceed to test this hypothesis for amorphous transition metals and alloys.

Using McMillan's equation (9.5) together with experimentally determined values of θ_D (from specific heat data), one can take $\langle \omega^2 \rangle_M^{1/2} \approx (\theta_D/2)$ and determine λ [9.11]. Again, as in the discussion of simple metals, one takes $\mu^* = 0.13$. Having determined λ, one then can evaluate the ratio $\delta = [\lambda / D(\varepsilon_F)] \approx W(1 \mp S)$. There are then two methods of testing the Varma-Dynes approach. First one can compare the values of δ obtained for amorphous and corresponding crystalline transition metals. If the Varma-Dynes model applies, then these values should be nearly equal since the width of the d band should not appreciably change on going to the amorphous state (from the crystalline state) [9.74]. *Varma* and *Dynes* have argued that this should be the case for progressively disordered bcc Nb and A-15 Nb_3Ge. Table 9.4 gives the

Table 9.4. Determination of the Varma-Dynes parameter δ for several amorphous and corresponding crystalline transition metals and alloys. λ, $D(0)$, and δ were determined from specific heat data using McMillan's formulae [9.11]

Metal (alloy)	$\gamma\left[\frac{\text{mJ}}{\text{mole-K}^2}\right]$	θ_D[K]	λ[a]	$D(0)$[b] $\left[\frac{\text{states}}{\text{eV-atom-spin}}\right]$	$\delta=\frac{\lambda}{D(0)}$ $\left[\frac{\text{eV-atom-spin}}{\text{states}}\right]$	Ref.
Amorphous						
$Zr_{70}Pd_{30}$	4.7	180	0.61	0.62	0.98	[9.38]
$Nb_{75}Ge_{25}$	3.0	222	0.63	0.39	1.58	[9.42]
$(Mo_{0.6}Ru_{0.4})_{80}P_{20}$	4.1	265	0.71	0.51	1.41	[9.32]
$(Mo_{0.4}Ru_{0.6})_{80}P_{20}$	3.6	267	0.65	0.46	1.40	[9.32]
Crystalline						
Zr(hcp)	2.78	290	0.41	0.42	0.98	[9.11]
Nb(bcc)	7.8	277	0.82	0.91	0.91	[9.11]
Mo(bcc)	1.83	460	0.41	0.28	1.51	[9.11]
$Mo_{61}Ru_{39}$(σ phase)	4.10	418	0.64	0.53	1.21	[c]

[a] From McMillan's equation [9.11].
[b] Renormalized density of states $D(0)=(3\gamma)2\pi^2k_B{}^2\,(1+\lambda)$.
[c] Specific heat data from [9.82].

relevant parameters for several amorphous alloys for which specific heat data are available. The value of $D(\varepsilon_F)$ used for determining δ was obtained from the coefficient γ of the low-temperature specific heat. The value is then renormalized by the factor $(1+\lambda)^{-1}$ to give the bare band structure density of states.

The metallic glasses $Zr_{70}Pd_{30}$, $Nb_{75}Ge_{25}$, and $(Mo_{1-x}Ru_x)_{80}P_{20}$ are compared with the pure crystalline metals Zr, Nb, and Mo, respectively. For the case of the $(Mo_{1-x}Ru_x)_{80}P_{20}$ alloys, a comparison is also made with the crystalline σ phase $Mo_{61}Ru_{39}$. With the exception of the case of Nb_3Ge, the values of δ for the amorphous alloys are in very good agreement with those for the corresponding crystalline alloys. It would appear that amorphous Nb_3Ge has a δ value characteristic of the upper half of the *d* band while bcc Nb has a value characteristic of the lower half of the *d* band. This could be caused by electron transfer effects which result in a net increase in *d* band occupation at the Nb sites [9.30].

The second comparison with Varma-Dynes model involves the variation of δ with *d* band occupancy. In Table 9.4, one observes a clear "jump" in δ on going from the Zr-based amorphous alloy to the Nb, Mo, and Ru-based alloys. Within the latter group of alloys, δ is nearly constant. This again, agrees with the Varma-Dynes model which attributes the jump to the nonorthogonality factor $(1\mp S)$. In summary, the nonorthogonal tight-binding model seems to provide a good account of the basic systematics of superconductivity in amorphous transition metals.

9.3 Ginzburg-Landau Superconductors in the High κ Limit

9.3.1 The Upper and Lower Critical Fields

The behavior of a superconductor in the presence of a transport current or externally applied magnetic field can be described using the general Ginzburg-Landau (GL) theory of phase transitions [9.83]. Although this theory in its original form is phenomenological, it has been put on a concrete microscopic basis by Abrikosov and Gorkov (AG) [9.84, 85]. The theory is based on an expansion of the free energy of the superconductor measured with respect to the free energy of the same metal in the normal state. *Tinkham* [9.86] has given a particularly readable discssion of the theory. Two natural characteristic lengths can be defined from the GL differential equations. These are $\xi(T)$, the coherence length, and $\lambda(T)$, the penetration depth. They are, respectively, the characteristic lengths for spatial variation of the order parameter ψ and vector potential $\boldsymbol{A}$. The ratio of these lengths $\kappa = [\lambda(T)/\xi(T)]$ is called the Ginzburg-Landau parameter. Superconductors are distinguished as being type I or type II according to whether $\kappa < 1/\sqrt{2}$ or $\kappa > 1/\sqrt{2}$, respectively. The AG theory allows $\xi(T)$, $\lambda(T)$, and $\kappa(T)$ to be expressed in terms of microscopic parameters which are material dependent. In the limit of small electron mean free path $l_e \ll \xi(T)$, $l_e \ll \lambda(T)$, and T near T_c, one obtains [9.84] the "dirty limit" expressions

$$\xi_d(T) = 0.85[\xi_0(0)l_e]^{1/2}\left(\frac{T_c}{T_c - T}\right)^{1/2} \tag{9.27}$$

and

$$\lambda_d(T) = 0.615\,\lambda_L(0)\left(\frac{\xi_0(0)}{l_e}\right)^{1/2}\left(\frac{T_c}{T_c - T}\right)^{1/2}, \tag{9.28}$$

where $\xi_0(0) = \hbar v_F/\pi\Delta(0)$ is the BCS coherence length for a "clean" material expressed in terms of the Fermi velocity v_F and the zero temperature energy gap $\Delta(0)$. The London penetration depth $\lambda_L(0)$ is given by [9.84]

$$\lambda_L(0) = \frac{3c^2}{16\pi e^2 v_F D(\varepsilon_F)}. \tag{9.29}$$

Amorphous metals have an electron mean free path l_e which is of the order of interatomic spacings [9.87] so that the above conditions on l_e are easily satisfied. From (9.27, 28), one finds

$$\kappa = 0.725\frac{\lambda_L(0)}{l_e}. \tag{9.30}$$

In general $\lambda_L(0)$ is of the order of 1000 Å so that κ should typically be of order 10–100 for amorphous superconductors. Estimates based on experimental data give $\kappa \sim 40$–100 [9.88]. Such values of κ are the largest observed for any superconducting materials. Amorphous superconductors are in general extreme type II materials.

Using the AG theory [9.91], *Helfand* and *Werthamer* [9.89], and *Maki* and *Tsuzuki* [9.90] have derived exact expressions for the upper and lower critical field $H_{c2}(T)$, and $H_{c1}(T)$ for type II superconductors in this limit. By definition, $H_{c2}(T)$ is the largest field (at temperature T) for which a superconducting mixed state exists, while H_{c1} is the field at which the material transforms from a homogeneous superconducting state to a mixed state. *Maki* [9.90, 92] expressed his result for H_{c2} in terms of a parameter $\kappa_1(T)$, where $\kappa_1(T)=\kappa$ for $T \to T_c$ and $\kappa_1(T) \to 1.2\kappa$ for $T \to 0$. The upper critical field is

$$H_{c2} = \sqrt{2}\,\kappa_1(T) H_c(T), \tag{9.31}$$

where $H_c(T)$ is the thermodynamic critical field. Similarly, $H_{c1}(T)$ can be written [9.92]

$$H_{c1} = \frac{H_c(T)}{2\kappa_3(T)} \log \kappa_3(T), \tag{9.32}$$

where $\kappa_3(T)=\kappa$ for $T \to T_c$. *Maki* has computed the temperature variation of $\kappa_3(T)$.

It is obvious that the large κ values of amorphous superconductors will lead to large values of H_{c2} and small values of H_{c1} by comparison with $H_c(T)$. Experimentally, it is found the $H_{c2}(T)$ is large and linear in T over a broad range of T [9.9, 88, 93]. This is illustrated in Fig. 9.7 for a typical metallic glass. Also shown is the theoretical prediction of *Maki* which clearly does not describe the data at low temperatures. Maki's theory has been modified [9.94, 95] to include the effects of Pauli paramagnetism and spin-orbit scattering. The former correction alone gives the results in the second theoretical curve shown in Fig. 9.7. The spin-orbit correction, when included, will give a curve lying between the two curves shown. In any case, the theory cannot account for the linear variation of the data and thus requires modification. *Rainer* and *Bergmann* have included strong-coupling effects in the theory and found improved agreement with data on simple metals [9.96]. Amorphous transition metals are not strong-coupling superconductors since $\lambda < 1$ (see Table 9.4). Thus it seems that the behavior of $H_{c2}(T)$ is not understood.

The lower critical field H_{c1} has also been measured for several metallic glasses [9.88] and as expected found to be small by comparison with H_c and H_{c2}. A set of data is shown in Fig. 9.8 and compared with Maki's prediction [(9.32)]. Again, the low-temperature data do not appear to be well described by the theory.

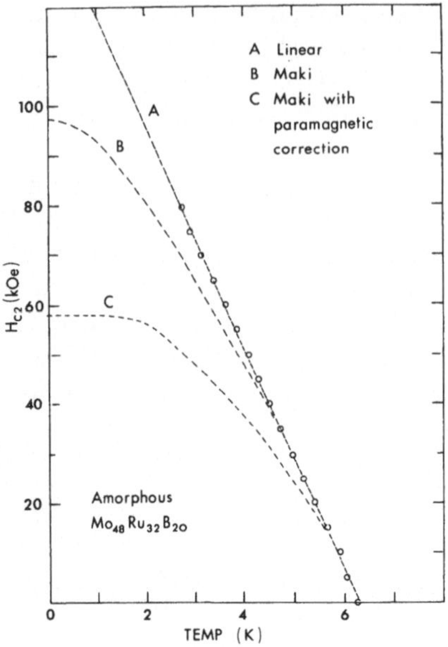

Fig. 9.7. Upper critical field $H_{c2}(T)$ for the metallic glass $(Mo_{0.6}Ru_{0.4})_{80}B_{20}$. Also shown are the predictions of the Maki theory with no corrections and with a paramagnetic correction. The paramagnetic correction parameter was taken to be $\alpha = 1.37$ [9.95]

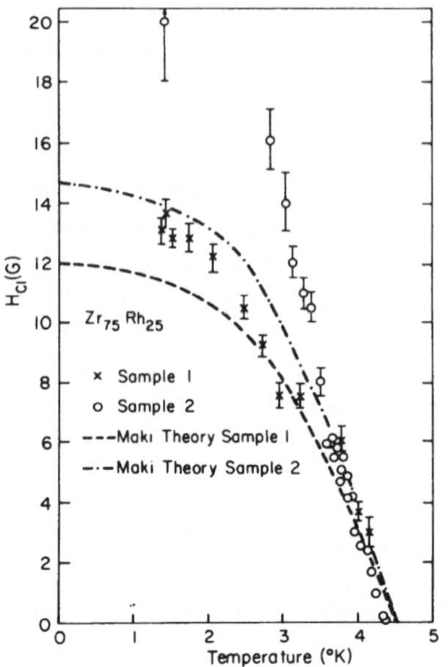

Fig. 9.8. Lower critical field $H_{c1}(T)$ for two samples of amorphous $Zr_{75}Rh_{25}$. Data are compared with the predictions of the Maki theory

The breakdown of the Maki and Werthamer theories for $H_{c2}(T)$ may possibly be related to the relaxation time approximation which treats the motion of electrons as being governed by a diffusion process. When l_e is extremely small (or the order of interatomic spacings) one is in a strong scattering limit. The relaxation time approximation is suspect in this limit (see, for example, [Ref. 9.56, pp. 212–219]).

9.3.2 Critical Phenomena

The transition from the normal state to the superconducting state occurs over a very narrow temperature interval for a homogeneous crystalline superconductor. The influence of critical fluctuations is restricted to a temperature interval of order 10^{-6} K in this case. In amorphous superconductors, the small coherence length ξ permits rather localized fluctuations of the order parameter ψ involving a relatively smaller region of material than in the crystalline case. This in turn means that such fluctuations involve a smaller volume free energy. As a result, the effects of fluctuations are observable over a much broader temperature range in amorphous superconductors. *Aslamasov* and *Larkin* (AL) [9.97] first derived explicit expressions for the temperature dependence of such critical fluctuations. In particular, they calculated the electrical conductivity

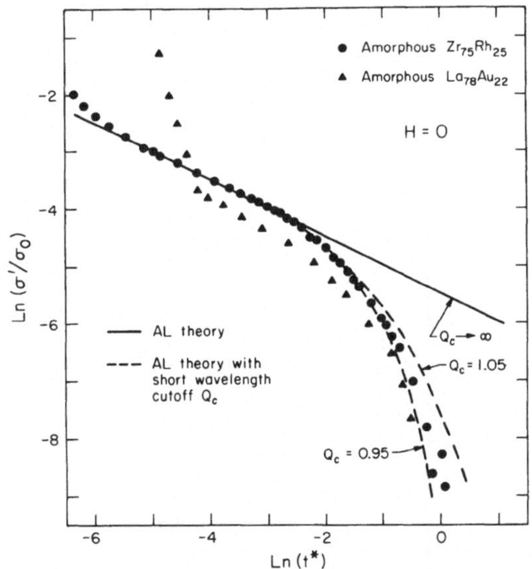

Fig. 9.9. The dependence of the excess conductivity $\Delta\sigma$ on reduced temperature $t=[T-T_c)/T_c]$ for several bulk amorphous superconductors. The results are compared with σ_{AL}, the excess conductivity predicted by *Azlamasov* and *Larkin*. (See [9.101] for further details)

due to fluctuations into the superconducting state for $T>T_c$. *Maki* [9.98], *Thompson* [9.99], and others [9.100] have extended the theoretical work of AL. *Johnson* et al. [9.101, 102] have carried out detailed measurements of the fluctuation conductivity of bulk amorphous superconductors and compared their results to both the AL and extended theories. An example of the data compared with the AL theory is shown in Fig. 9.9. The AL theory describes the fluctuation conductivity $\Delta\sigma$ near T_c, but fails to account for the temperature dependence far from T_c. Recently, *Ami* and *Maki* [9.103] have extended the theory to show that the data far from T_c can also be accounted for as well as the dependence of $\Delta\sigma$ on an externally applied field.

9.3.3 Homogeneity, Critical Current Density, and Flux Pinning

In Sect. 9.3.1, the critical fields H_{c1} and H_{c2} of amorphous superconductors were discussed. For $H_{c1}<H<H_{c2}$, a type II superconductor is in a mixed state consisting of a regular array of superconducting and normal domains. For amorphous superconductors $\lambda \gg \xi$, so that variations of the microscopic magnetic field in space are slow compared with variations in the order parameter ψ. The mixed state can be described by a domain or flux lattice structure over which ψ varies rapidly over the domain dimension while $\boldsymbol{H}$, the microscopic field, varies little. When a transport current flows normal to the applied field direction, the flux lattice experiences a Lorentz force. In the absence of any other forces, the flux lattice flows viscously [9.84]. This flow involves energy dissipation so that the sample is, strictly speaking, no longer superconducting. Under these circumstances, the critical current density $J_c=0$

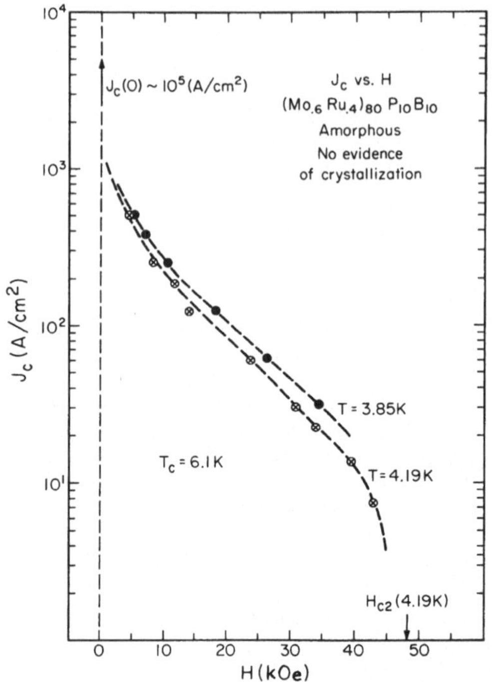

Fig. 9.10. Critical current density as a function of applied magnetic field for a homogeneous metallic glass

for $H > H_{c1}$. The situation described applies to a homogeneous type II superconductor.

For an inhomogeneous material, there are additional interactions between the flux lattice and the material inhomogeneities. Under favorable conditions, these pinning interactions resist or prevent flux lattice motion and $J_c > 0$ for $H > H_{c1}$. To determine the temperature and field dependence of $J_c = J_c(H, T)$, one must therefore understand the nature and effect of material inhomogeneities on the flux lattice. A single-phase amorphous alloy is spatially rather homogeneous. This can be deduced, for example, from the fact that the transition to the superconducting state is very sharp [9.101, 102]. One can use the term "local T_c" to refer to the T_c of a local region of the sample with characteristic dimension d. One type of inhomogeneity could then be described by the standard deviation Δ_d of the local T_c throughout the various local regions of the sample. Broadening of the transition of the sample as a whole by an amount of order Δ_d will then occur if and only if

$$d \gtrsim \xi(T_c \pm \Delta_d) = \xi(0)(\Delta_d/T_c)^{1/2} . \tag{9.33}$$

We can conclude from the studies of critical phenomena [9.101, 102] that $\Delta_d \sim 10^{-2} - 10^{-3}$ K when $d \sim \xi(0)$. The fractional variation is $\Delta_{\xi(0)}/T_{c0} \sim 10^{-3}$ where T_{c0} is the mean T_c of the sample as a whole. Amorphous superconductors are thus quite homogeneous on the spatial scale ξ. For H near H_{c2}, this

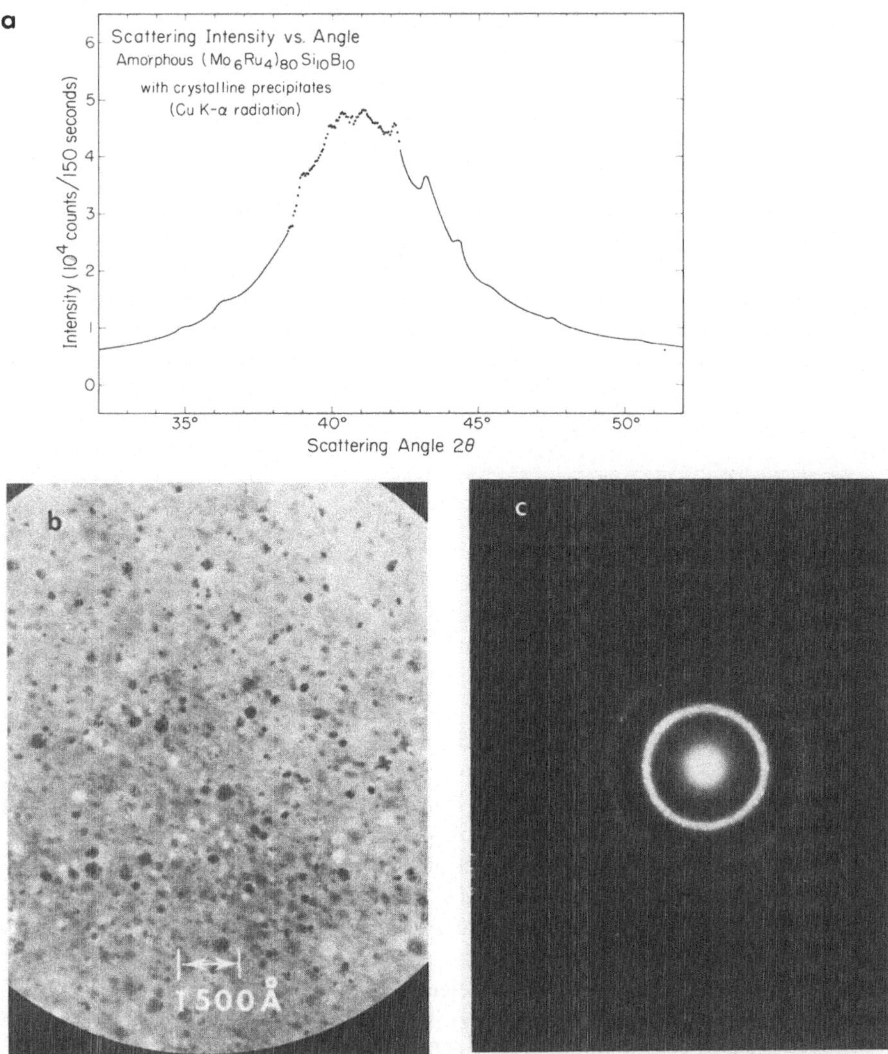

Fig. 9.11. (**a**) X-ray diffraction pattern showing small peaks superimposed on a broad maxima for the metallic glass $(Mo_{0.6}Ru_{0.4})_{80}Si_{10}B_{10}$. The sample contains 3–5% (by volume) of crystalline σ phase. (**b**) Electron micrograph showing size and morphology of σ phase particles. (**c**) Electron diffraction pattern of the same region showing diffuse halos together with Laue spots

is precisely the spatial scale of the flux lattice since [9.84]

$$H_{c2}(T)=\phi_0/2\pi\xi(T)^2\,, \tag{9.34}$$

where ϕ_0 is the fundamental flux quantum and exactly equal to the magnetic flux in a single cell of the flux lattice. Inhomogeneity on the scale of the flux lattice is therefore small; the flux lattice should not exhibit appreciable pinning

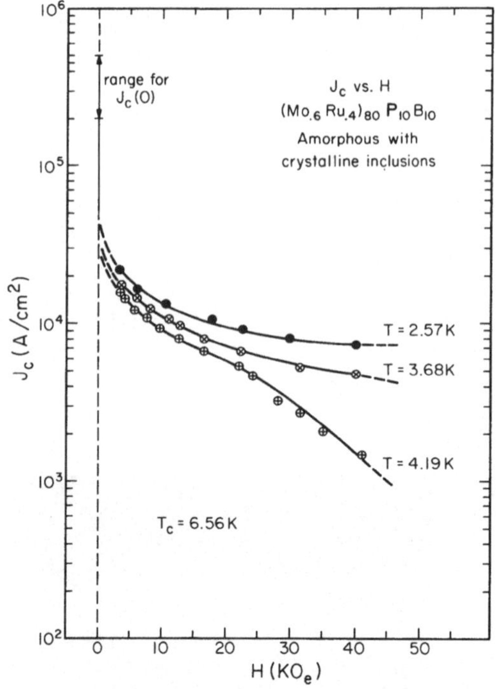

Fig. 9.12. Critical current density as a function of applied magnetic field for the $(Mo_{0.6}Ru_{0.4})_{80}P_{10}B_{10}$ sample with crystalline precipitates (see Fig. 9.11)

interactions. This expectation is confirmed by experiments. For $H > H_{c1}$ and fixed T, one observes a rapid decrease in J_c with increasing H. An example is shown in Fig. 9.10 for a homogeneous amorphous sample of $(Mo_{0.6}Ru_{0.4})_{80}P_{10}B_{10}$ [9.104, 105]. A second sample of similar composition but containing a small volume fraction of finely dispersed crystalline precipitates was studied for comparison [9.104]. The crystalline precipitates have dimension ~100–300 Å and have the σ phase structure. The bulk T_c of the crystalline σ phase is ~7 K while the amorphous matrix has $T_c \sim 5.3$ K. The morphology is illustrated by the electron micrograph in Fig. 9.11. This inhomogeneous two-phase material exhibits strong flux pinning interactions as evidenced by a dramatic increase in $J_c(H, T)$ for $H > H_{c1}$. The results are shown in Fig. 9.12. The details of the pinning mechanism have been discussed by *Clemens* et al. [9.105].

The above example illustrates how the introduction of spatial inhomogeneities influences $J_c(H, T)$. Other forms of inhomogeneity in addition to crystalline precipitates are possible. Phase separation into two amorphous phases [9.106, 107], magnetic impurities [9.108], and shear bands [9.109, 110] are all examples of inhomogeneities that could result in flux pinning interactions. The author has recently studied the effect of concentrated magnetic impurities on amorphous superconductors where the impurities interact to form a magnetically ordered state [9.111]. A compositional range was found in which this ordered state and superconductivity coexist.

9.4 Materials for Applications

9.4.1 High Field Magnets

One of the most problematic practical aspects of crystalline high field superconductors is their tendency toward brittleness and fracture. Progress in the fabrication of high field superconducting magnets and devices has been noticeably hindered by this tendency. Superconducting metallic glasses, on the other hand, frequently exhibit desirable mechanical properties such as high strength [9.112, 113] and some degree of ductility [9.36]. Furthermore, low cost commercial methods for producing continuous wires and ribbons of metallic glass already exist. Together with the large values of H_{c2} observed for metallic glasses, these observations suggest possible applications as superconducting magnet windings.

In addition to a large H_{c2} (typically $H_{c2} > 100$ kG) a practical superconductor must carry a high current density at high fields. A critical current density of $\sim 10^5$ A/cm^2 for $H \sim 100$ kG is typically required. As seen in the previous section, such current densities are possible for amorphous superconductors only if spatial inhomogeneities are introduced. The technological potential of superconducting metallic glasses thus depends on one being able to control the introduction of pinning centers. This can be accomplished by partially crystallizing metallic glasses during controlled thermal treatment, or by cooling the metallic melt at a rate insufficient to completely suppress crystallization. The latter method offers the possibility of fabricating a useful conductor in a single-step casting process.

9.4.2 Effects of Neutron Irradiation

The effect of neutron irradiation on the properties of crystalline superconductors has recently been the subject of much study [9.116, 117]. For example, *A*-15 superconductors typically exhibit a rapid degradation of T_c when exposed to fast neutron fluences of 10^{18}–10^{20} n/cm^2. The T_c of Nb_3Sn is reduced by 8–10 K by a fluence of 10^{19} n/cm^2. This degradation is particularly important for applications of superconducting magnets in fusion technology. The disordered nature of metallic glasses suggests that their electronic properties should not be appreciably influenced by radiation damage. A recent study [9.116] shows that this is indeed the case. For the metallic glass $(Mo_{0.6}Ru_{0.4})_{82}B_{18}$, T_c is actually increased slightly following exposure to fast reactor neutrons to a fluence of 10^{19} n/cm^2. The superconducting transition was found to be sharper in the irradiated glass and the critical field H_{c2} was essentially unchanged. Changes in other properties of the material following the irradiation are summarized in Table 9.5. These observations provide an additional incentive for developing practical amorphous superconductors.

Table 9.5. Effects of fast reactor neitrons (~1 MeV) on the properties of the superconducting metallic glass $(Mo_{0.6}Ru_{0.4})_{82}B_{18}$. A fluence of 10^{19} n/cm^2 was used in irradiating the sample. X-ray data was obtained using Cu K_α radiation. X-ray data are given in terms of scattering vector $\boldsymbol{K}=(4\pi\sin\theta)/\lambda$. See [9.116] for details

Property	Before irradiation	After irradiation	Percent change
X-ray data			
A) Position of first diffraction maxima	2.83 Å^{-1}	2.83 Å^{-1}	No change
B) Width of first diffraction maxima (at half maximum)	0.45 Å^{-1}	0.48 Å^{-1}	7% decrease
Superconductivity			
A) T_c	6.05 ± 0.02 [K]	6.19 ± 0.02 [K]	2% increase
B) Transition width ΔT_c	0.20 ± 0.02 [K]	0.06 ± 0.02 [K]	70% increase
C) $[dH_{c2}/dT]_{T=T_c}$	− 23.6 ± 1.2 [kOe/K]	− 24.4 ± 1.2 [kOe/K]	Not significant
Density	10.37 ± 0.05 [g/cm^3]	10.22 ± 0.05 [g/cm^3]	1.5% decrease
Resistivity	131 ± 12 μΩcm	136 ± 14 μΩcm	Not significant

9.4.3 Composite Materials

Starting with a metallic glass, it is possible to produce a variety of partially (or fully) crystallized materials by thermal treatment. The metallurgical state of these composite or mixed phase materials is generally a nonequilibrium one. For example, crystalline phases with highly refined grain size can be obtained which exhibit a high density of grain boundaries suitable for flux pinning [9.117, 118]. *Tsuei* [9.119] has shown, for example, that *A*-15 Nb_3Ge with a high T_c can be obtained by heat treatment of amorphous Nb_3Ge films. More recently, *Tenhover* [9.120] has shown that metallic glasses of the form $(Hf_{1-x}Zr_x)_{60}V_{40}$ can be partially (or completely) crystallized to give the high T_c *C*-15 crystalline superconductor $V_2(Hf_{1-x}Zr_x)$. In these cases, the crystalline superconducting phase of interest is a brittle phase with poor mechanical properties. The amorphous matrix from which it is obtained has more desirable mechanical properties. By first fabricating a magnet or device using the amorphous material and then using a thermal treatment, it may be possible to circumvent some of the practical difficulties associated with processing a brittle material. These and other potential applications of metallic glasses remain for the most part unexplored. Future research will determine whether or not superconducting metallic glasses are to play a major role in technology.

References

9.1 W. Buckel, R. Hilsch: Z. Phys. **138**, 109 (1954)
9.2 W. Buckel, R. Hilsch: Z. Phys. **146**, 27 (1956)

9.3 W.L.McMillan, J.M.Rowell: In *Superconductivity*, Vol. 1, ed. by R.D.Parks (Dekker, New York 1969) p. 449
9.4 G.Bergmann: Phys. Rep. **27**C, 161 (1976)
9.5 M.M.Collver, R.H.Hammond: Phys. Rev. Lett. **30**, 92 (1973)
9.6 P.Duwez: *Progress in Solid State Chemistry*, Vol. 3 (Pergamon Press, Oxford 1966) pp. 377–406
9.7 P.Duwez: In Annu. Rev. Mater. Sci. **6**, 83–117 (1976)
9.8 W.L.Johnson, S.J.Poon, P.Duwez: Phys. Rev. B**11**, 150 (1975)
9.9 W.L.Johnson: In *Rapidly Quenched Metals III*, Proc. 3rd Intern. Conf. on Rapidly Quenched Metals, Vol. 2 (British Metals Society, London 1978) pp. 1–16
9.10 J.Bardeen, L.Cooper, J.Schrieffer: Phys. Rev. **108**, 1175 (1957)
9.11 W.L.McMillan: Phys. Rev. **167**, 331 (1968)
9.12 G.M.Eliashberg: Zh. Eksp. Teor. Fiz. **38**, 966 (1960)
9.13 P.W.Anderson: J. Phys. Chem. Solids **11**, 26 (1959)
9.14 G.Bergmann: Phys. Rev. B**3**, 3797 (1971)
9.15 G.Bergmann, D.Rainer: Z. Phys. **263**, 59 (1973)
9.16 K.Knorr, N.Barth: Solid State Commun. **8**, 1085 (1970)
9.17 J.E.Jackson, C.V.Briscoe, H.Wühl: Physica **55**, 447 (1971)
9.18 C.G.Grangvist, T.Claeson: Z. Phys. B**20**, 165 (1975)
9.19 W.Klement, Jr., R.H.Willens, P.Duwez: Nature **187**, 869 (1960)
9.20 W.L.Johnson: Unpublished results
9.21 A.Calka, M.Madhaua, D.E.Polk, B.C.Giessen, H.Matyja, J.VanderSande: Scr. Metall. **11**, 65 (1977)
9.22 a) C.G.Grandqvist, T.Claeson: Z. Phys. B**20**, 13 (1975)
b) C.G.Grandqvist, T.Claeson: J. Low Temp. Phys. **10**, 735 (1973)
c) C.G.Grandqvist, T.Claeson: Z. Phys. **20**, 241 (1975)
9.23 B.C.Giessen, J.Hong, L.Kabacoff, D.E.Polk, R.Raman, R.St.Amand: In *Rapidly Quenched Metals III*, Proc. 3rd Intern. Conf. on Rapidly Quenched Metals, Vol. 1, ed. by B. Cantor (British Metals Society, London 1978) p. 249
9.24 C.O.Kim, W.L.Johnson: Unpublished results
9.25 A.Comberg, S.Ewert: Z. Phys. **268**, 241 (1974)
9.26 J.Fortman, W.Buckel: Z. Phys. **162**, 93 (1961)
9.27 B.T.Matthias: *Progress in Low Temperature Physics* (Interscience, New York 1957) Vol. 2, p. 138
9.28 M.M.Collver, R.H.Hammond: Solid State Commun. **22**, 55 (1977)
9.29 W.L.Johnson, S.J.Poon, J.Durand, P.Duwez: Phys. Rev. B**18**, 206 (1978)
9.30 W.L.Johnson, A.R.Williams: Phys. Rev. B**20**, 1 (1979)
9.31 D. Aliaga Guerra, J.Durand, W.L.Johnson, P.Panissod: Solid State Commun. **31**, 487 (1979)
9.32 R.Shull, W.L.Johnson: J. Phys. Paris (1980)
9.33 W.H.Shull, D.G.Naugle: Phys. Rev. Lett. **39**, 1580 (1977)
9.34 W.H.Shull, D.G.Naugle, S.J.Poon, W.L.Johnson: Phys. Rev. B**18**, 3263 (1978)
9.35 K.Togano, K.Tachikawa: Phys. Lett. **54**A, 205 (1975)
9.36 R.Hasegawa, L.E.Tanner: Phys. Rev. B**16**, 3925 (1977)
9.37 W.L.Johnson, S.J.Poon: J. Appl. Phys. **46**, 1787 (1975)
9.38 J.E.Graebner, B.Golding, R.J.Schultz, F.S.L.Hsu, H.S.Chen: Phys. Rev. Lett. **39**, 1480 (1977)
9.39 O.Rapp, B.Lindberg, H.S.Chen, K.V.Rao: J. Less Common Met. (in press)
9.40 K.Agyeman, R.Müller, C.C.Tsuei: Phys. Rev. B**19**, 193 (1979)
9.41 a) C.C.Tsuei, W.L.Johnson, R.L.Laibowitz, J.M.Viggiano: Solid State Commun. **24**, 615 (1977)
b) W.L.Johnson, C.C.Tsuei, S.I.Raider, R.B.Laibowitz: J. Appl. Phys. **50**, 4240 (1979)
9.42 C.C.Tsuei, S.Von Molnar, J.M.Coey: Phys. Rev. Lett. **41**, 664 (1978)
9.43 B. Schroeder, W.L.Johnson, C.C.Tsuei, P.Chaudhari: In *Proc. IV Intern. Conf. on the Physics of Noncryst. Solids*, ed. by G.H.Frischat (Trans. Tech. Publications, Switzerland 1977) p. 190

9.44 A.M.Clogston, V.Jaccarino, Y.Yafet: Phys. Rev. **134**, A650 (1964)
F.I.Ajami, R.K.MacCrone: J. Phys. Chem. Solids **36**, 7 (1975)
9.45 J.Friedel: In *Physics of Metals*, ed. by J. M. Ziman (Cambridge University Press, Cambridge 1969) pp. 340–403
9.46 L.F.Mattheiss: Phys. Rev. B**1**, 373 (1970): Phys. Rev. **151**, 450 (1966)
9.47 I.R.Gomersoll, B.L.Gyorffy: Phys. Rev. Lett. **33**, 1286 (1974)
9.48 a) B.Schroeder, W.Grobman, W.L.Johnson, C.C.Tsuei, P.Chaudhari: Solid State Commun. **28**, 631 (1978)
b) A.Amamou, W.L.Johnson: Unpublished results
9.49 V.L.Moruzzi, J.F.Janak, A.R.Williams: *Calculated Electronic Properties of Metals* (Pergamon, Oxford 1978) p. 129
9.50 B.W.Roberts: J. Phys. Chem. Ref. Data **5**, 581 (1976)
9.51 F.Cyrot-Lackmann: Adv. Phys. **16**, 393 (1967)
9.52 F.Cyrot-Lackmann: J. Phys. **35**, C4–109 (1974)
9.53 J.P.Gaspard: AIP Conf. Proc. **31**, 372 (1976)
9.54 J.W.Garland: Unpublished results, quoted in T. T. Chan, J. T. Chen, J.D.Leslie, H.J.Smith: Phys. Rev. Lett. **22**, 526 (1969)
9.55 P.B.Allen: Phys. Rev. B**6**, 2577 (1972)
9.56 J.M.Ziman: *Electrons and Phonons* (Clarendon, Oxford 1960) Chap. V
9.57 J.M.Ziman: *Principles of the Theory of Solids* (Cambridge University Press, Cambridge 1972) Chap. 7
9.58 A.O.E.Animalu, V.Heine: Philos. Mag. **12**, 1249 (1965)
9.59 D.Weaire, M.F.Ashby, J.Logan, M.J.Weins: Acza Metall. **19**, 779 (1971)
9.60 J.L.Finney: Proc. R. Soc. Ser. A**319**, 495 (1970)
9.61 L. von Heimendahl, M.F.Thorpe: J. Phys. F**5**, L87 (1975)
9.62 S.J.Poon, T.Geballe: Phys. Rev. B (to be published)
9.63 J.Bolz, F.Pobell: Z. Phys. B**20**, 95 (1971)
9.64 W.A.Harrison: Phys. Rev. **129**, 2522 (1963)
9.65 L.J.Sham, J.M.Ziman: In *Solid State Physics*, Vol. 15, ed. by F. Seitz, D. Turnbull (Academic Press, New York 1963)
9.66 A.O.E.Animalu: *Intermediate Quantum Theory of Solids* (Prentice-Hall, Englewood Cliffs, NJ 1977) Chap. 4
9.67 W.A.Harrison: *Pseudopotentials in the Theory of Metals* (Benjamin, New York 1966)
9.68 G.V.Minnigerode, J.Rothenberg: Z. Phys. **213**, 397 (1968)
9.69 A.Comberg, S.Ewert, G.Bergmann: Z. Phys. **271**, 317 (1974)
9.70 See for example N.F.Mott, H.Jones, *The Theory of the Properties of Metals and Alloys* (Dover, New York 1958) Chap. V
9.71 L.Brewer: In *Phase Stability in Metals and Alloys*, ed. by P.S.Rudman, J.Stringer, R.I.Jaffee (McGraw-Hill, New York 1967) p. 50
9.72 W.Hume-Rothery: Prog. Mater. Sci. **13**, 229 (1967)
9.73 J.Friedel: J. Phys. **16**, 342, 829 (1955)
9.74 C.M.Varma, R.C.Dynes: *Superconductivity in d- and f-band Metals*, ed. by D.H.Douglass (Plenum, New York 1976)
9.75 C.M.Varma, W.Weber: Phys. Rev. Lett **39**, 1094 (1977)
9.76 C.M.Varma, E.I.Blout, P.Vashishta, W.Weber: Phys. Rev. B**19**, 6130 (1979)
9.77 C.M.Varma, W.Weber: Phys. Rev. B**19**, 6142 (1979)
9.78 T.Fujiwara: J. Phys. F**9**, 2011 (1979)
9.79 S.N.Khanna, F.Cyrot-Lackmann, M.C.Desjonquères: J. Phys. F**9**, 79 (1979)
9.80 S.N.Khanna, F.Cyrot-Lackmann: Philos. Mag. B**38**, 197 (1978)
9.81 S.Barisic, J.Labbé, J.Friedel: Phys. Rev. Lett. **25**, 919 (1970)
9.82 A.K.Sinha: Prog. Mater. Sci. **15**, 79 (1972)
9.83 V.L.Ginzburg, L.D.Landau: Zh. Eksp. Teor. Fiz. **20**, 1064 (1950)
9.84 D.Saint-James, G.Sarma, E.J.Thomas: *Type II Superconductivity* (Pergamon, Oxford 1969) Chap 5

9.85 L.P.Gorkov: Zh. Eksp. Teor. Fiz. **36**, 1918 (1959); [Sov. Phys. JETP **9**, 1364 (1959)]
9.86 M.Tinkham: *Introduction to Superconductivity* (McGraw-Hill, New York 1975) Chap. 4
9.87 H.Güntherodt: In *Metallic Glasses* (American Society for Metals, Metals Park, OH 1978) pp. 247–272
9.88 a) E.Domb, W.L.Johnson: J. Low Temp. Phys. **33**, 29 (1978)
b) R.Koepke, G.Bergmann: Solid State Commun. **19**, 435 (1976)
9.89 E.Helfand, N.R.Werthamer: Phys. Rev. Lett. **13**, 686 (1964)
9.90 K.Maki, T.Tsuzuki: Phys. Rev. **139** A, 868 (1965)
9.91 A.A.Abrikosov, L.P.Gorkov: Zh. Eksp. Teor. Fiz. **42**, 1088 (1962); [Sov. Phys. JETP **15**, 752 (1962)]
9.92 K.Maki: Physics **1**, 21, 127, 201 (1964)
9.93 M.Tenhover, W.L.Johnson, C.C.Tsuei, S.Foner: Phys. Rev. Lett.
9.94 N.R.Werthamer, E.Helfand, P.C.Hohenberg: Phys. Rev. **147**, 295 (1966)
9.95 K.Maki: Phys. Rev. **148**, 362 (1966)
9.96 D.Rainer, G.Bergmann: J. Low Temp. Phys. **14**, 501 (1974)
9.97 L.G.Aslamasov, A.I.Larkin: Phys. Lett. A**26**, 238 (1968)
9.98 K.Maki: Prog. Theor. Phys. **39**, 897 (1968)
9.99 R.S.Thompson: Phys. Rev. B**1**, 327 (1970)
9.100 R.A.Craven, G.A.Thomas, R.D.Parks: Phys. Rev. B**7**, 157 (1973)
9.101 W.L.Johnson, C.C.Tsuei: Phys. Rev. B**13**, 4827 (1976)
9.102 W.L.Johnson, C.C.Tsuei, P.Chaudhari: Phys. Rev. B**17**, 2884 (1978)
9.103 S.Ami, K.Maki: Phys. Rev. B**19**, 1403 (1979)
9.104 W.L.Johnson, B.M.Clemens: *Proc.* 1979 *Fall Meeting of TMS-AIME*, (TMS publication series (Metallurgical Society of AIME, Warrendale, PA 1979)
9.105 B.M.Clemens, W.L.Johnson, J.Bennett: J. Appl. Phys. **51** (in press)
9.106 C-P.P.Chou, D.Turnbull: J. Non-Cryst. Solids **17**, 169 (1975)
9.107 M.A.Marcus: J. Non-Cryst. Solids **30**, 317 (1979)
9.108 C.C.Koch, D.M.Kroeger: Philos. Mag. **30**, 501 (1974)
9.109 H.S.Chen, H.J.Leamy, M.J.O'Brien: Scr. Metall. **7**, 415 (1973)
9.110 S.Takayama, R.Maddin: Acta Metall. **23**, 943 (1975)
9.111 W.L.Johnson: Solid State Commun. **32**, 981 (1979)
9.112 L.Davis: In *Metallic Glasses* (American Society for Metals, Metals Park, OH 1978) Chap. 8
9.113 T.Masumoto, R.Maddin: Mater. Sci. Eng. **19**, 1 (1975)
9.114 A.R.Sweedler, D.G.Schweitzer, G.W.Webb: Phys. Rev. Lett. **33**, 168 (1974)
9.115 A.R.Sweedler, D.A.Cox, S.Moehleck: J. Nucl. Mater. **72**, 50 (1978)
9.116 E.A.Kramer, W.L.Johnson, C.Cline: Appl. Phys. Lett. **35**, 815 (1979)
9.117 J.Bevk: In *Rapidly Quenched Metals III*, Proc. Third Intern. Conf. on Rapidly Quenched Metals, Vol. 2, ed. by B. Cantor (British Metals Society, London 1978) p. 17
9.118 K.Lo, J.Bevk, D.Turnbull: J. Appl. Phys. **48**, 2957 (1977)
9.119 C.C.Tsuei: Appl. Phys. Lett. **33**, 262 (1978)
9.120 M.Tenhover: Appl. Phys. **21**, 279 (1980)

10. Crystallization of Metallic Glasses

U. Köster and U. Herold

With 25 Figures

Relaxation towards an ideal amorphous state is distinct from the heterogeneous process of crystallization which results in a metastable or stable crystalline state of the material. The aim of this chapter is to review our current knowledge on the micromechanisms of crystallization in metallic glasses.

10.1 Background

Metallic glasses whichever way they are prepared are not in configurational equilibrium, but are relaxing slowly by a homogeneous process towards an "ideal" metastable amorphous state of lower energy. The ideal amorphous structure would be related to that of the prepared amorphous solid in much the same way as a perfect crystal is related to the normally available crystalline material, i.e., by the presence of defects in the real material. The concept of such an ideal glass is well known in amorphous semiconductors such as germanium or silicon [10.1, 2]. In metallic glasses relaxation is assumed to occur by annealing-out of defects with so far unknown structure, or free volume, or by changes in both topological and compositional short-range order [10.3–5]. As a consequence of low-temperature annealing which does not cause crystallization, various changes in physical properties (e.g., magnetic anisotropy [10.6], Curie temperature [10.7, 9], electrical resistivity [10.10], superconductivity [10.11], specific heat [10.12], diffusivity [10.13], in some alloys loss of ductility [10.14, 15], etc.) have been observed. These changes are assumed to be due to such relaxation processes.

Since the amorphous state is essentially a metastable one, it inherently possesses the possibility of transforming into a more stable crystalline state. However, the most promising properties of metallic glasses, e.g., the excellent magnetic behavior or the high hardness and strength combined with ductility and high corrosion resistance, have been found to deteriorate drastically during crystallization. Understanding the micromechanisms of crystallization to impede or control crystallization is therefore a prerequisite for most applications, as the stability against crystallization determines their effective working limits. On the other hand, controlled crystallization of metallic glasses can be used for designing very special partially or fully crystallized microstructures that cannot be obtained from the liquid or crystalline states. For example, $(Mo, Ru)_{80}B_{20}$ glasses have been found to be somewhat ductile superconductors with a T_c of about 6 K. Due to a lack of pinning, however, the critical current density is extremely small, making this material useless for application, unless pinning centers are produced by controlled partial crystallization (see Chap. 9).

Besides technological aspects, crystallization behavior is attracting an increasing scientific interest. To compare properties like structure, density of states, specific heat, thermopower, corrosion resistance, etc., it is of utmost importance to use amorphous alloys which crystallize only into one single phase with the same composition. So far, there is an extreme lack in such comparative experiments. Thermal stability [10.16] of the amorphous state is often used in theoretical approaches; however, it is not clear by what parameters "stability" of an "ideal" glass can be measured. As will be shown further on, so-called crystallization temperatures are not very useful, because kinetics of crystallization depend on a variety of parameters: the mode of crystallization, the number of quenched-in nuclei, the activation energy for diffusion, and finally the driving force, i.e., the difference in free energy between the amorphous and possible crystalline phases [10.17].

10.2 Experimental Methods

The amorphous to crystalline transition in metallic glasses has been studied in numerous investigations by observing one or more of the physical properties of the material such as electrical resistivity, saturation magnetization, Mössbauer spectroscopy, etc., which will change discontinuously during the crystallization process. Electrical resistivity is one of the most structure-sensitive parameters of a solid; Fig. 10.1a shows the changes of resistivity during continuous isochronal heating of $Fe_{32}Ni_{36}Cr_{14}P_{12}B_6$ (METGLAS 2826A) [10.18]. Crystallization or other phase transformations are indicated by the steps in the resistivity versus temperature plot. Differential scanning calorimetry (DSC) is the method commonly used to obtain informations as "crystallization temperature" and heat of crystallization on phase changes occurring in metallic glasses as a function of temperature and heating rate. Figure 10.1b shows a DSC plot for the same glass obtained at a heating rate higher than that for the resistivity measurements [10.19]. The peaks indicate very rapid crystallization; the peak temperatures called "crystallization temperatures" are not as sharply defined temperatures as the melting temperature of crystalline material; rather a temperature range exists within which the crystallization rate increases very strongly with the temperature. Since measuring this temperature is very simple, it is a convenient, but – as will be shown later on – suspicious method to compare the "stability" of different amorphous alloys.

In addition to isochronal heating during which crystallization reactions occur usually at a very fast rate, more significant data, e.g., the rate of crystallization, can be obtained from isothermal heat treatment experiments, by following one of the properties mentioned at different, but constant temperatures versus annealing time.

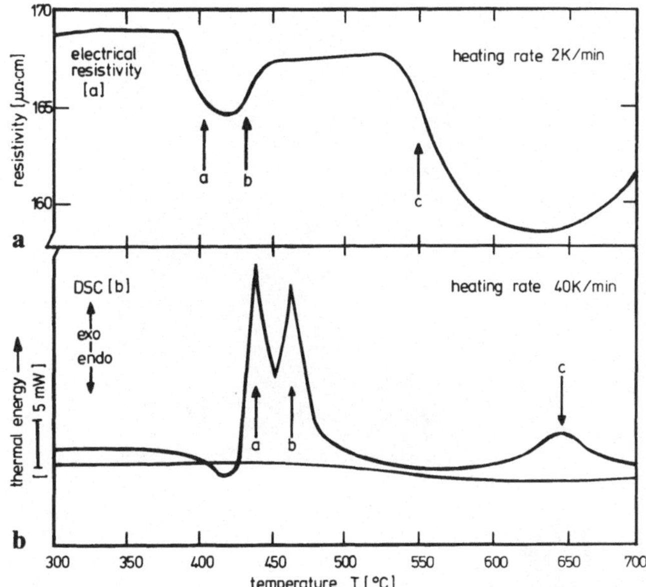

Fig. 10.1a, b. Crystallization of $Fe_{32}Ni_{36}Cr_{14}P_{12}B_6$ (METGLAS 2826A); the arrows indicate different stages of crystallization corresponding to the micrographs shown in Fig. 10.2. (**a**) Electrical resistivity during isochronal heating [10.18]; (**b**) differential scanning calorimetry (DSC) during isochronal heating [10.19]

All the methods mentioned so far are indirect ones, indicating only that some phase changes have occurred. If the crystallization process is complex, a correlation between a discontinuity in a physical property and a particular crystallization reaction is not possible until the structure has been analyzed, for example by means of x-ray diffraction or electron diffraction techniques. Figure 10.2 shows the microstructure of METGLAS 2826A after different stages of crystallization [10.19] indicated by the peaks in the DSC plot or the steps in the resistivity versus temperature plot in Fig. 10.1.

In situ observations of crystallization processes in the heating stage of a light or transmission electron microscope will give some valuable information on the micromechanisms of these reactions. However, one has to point out that it is indeed necessary to examine the microstructure after annealing in the bulk state [10.20]. As shown in the last part of this chapter, due to close surfaces, crystallization can be drastically modified in thin material such as in unthinned, but electron-transparent areas of splat-quenched foils, vapor-deposited films, or specimens thinned prior to any annealing treatment. Transmission electron microscopy allows one to study micromechanisms of crystallization reactions, but only in a very small part of the specimen which may not be representative. Therefore, combining this method with others mentioned above which give information integrated over much larger parts would be a better approach for analyzing crystallization in metallic glasses.

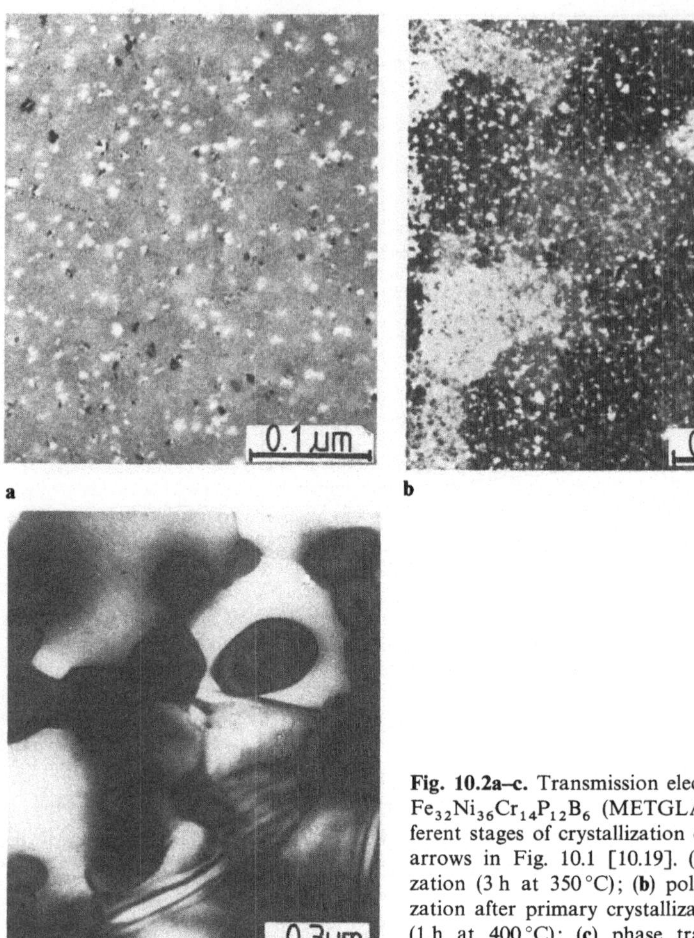

Fig. 10.2a–c. Transmission electron micrographs of $Fe_{32}Ni_{36}Cr_{14}P_{12}B_6$ (METGLAS 2826A) after different stages of crystallization corresponding to the arrows in Fig. 10.1 [10.19]. (**a**) Primary crystallization (3 h at 350 °C); (**b**) polymorphous crystallization after primary crystallization as shown in (**a**) (1 h at 400 °C); (**c**) phase transformation in the crystalline state after crystallization (1 h at 900 °C)

10.3 General Considerations Concerning Thermodynamics and Kinetics of Crystallization

10.3.1 Crystallization Reactions in Amorphous Alloys

Crystallization has been observed to occur generally by nucleation and growth processes. Most metallic glasses, however, do not crystallize by a polymorphous reaction, i.e., crystallization without any concentration changes into just one crystalline phase, but crystallization behavior is complicated by decomposition reactions. In order to gain an overall picture of the reactions which occur during crystallization of an amorphous alloy, a (hypothetical) diagram of

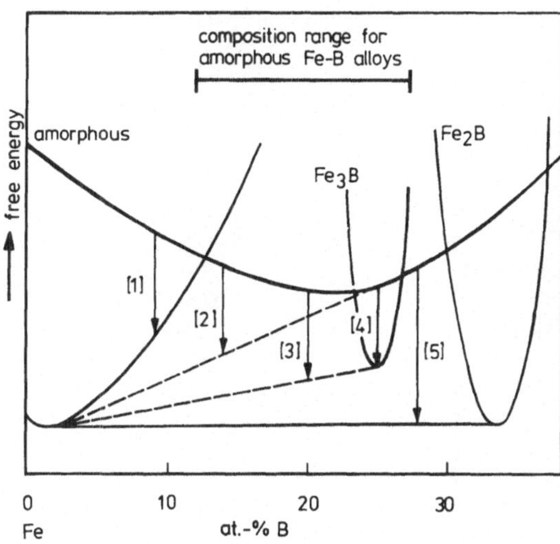

Fig. 10.3. Hypothetical diagram of the free energy for the various phases in Fe–B alloys vs concentration [10.22]. The numbers [1] to [5] correspond to crystallization reactions mentioned in the text

the free energy for the various phases versus concentration has been found to be very useful [10.21, 22]. In Fig. 10.3 such a diagram is shown for Fe–B alloys with the plots of free energy for the stable phases α-iron and Fe_2B, the metastable phase Fe_3B, and the amorphous alloy. The equilibrium coexistence tangent is shown by a solid line; possible metastable equilibria are marked by dashed tangents. Depending on concentration, the transition of the metastable amorphous phase into the crystalline phases can proceed by one of the following reactions:

a) Polymorphous crystallization, i.e., crystallization of the amorphous alloy without any change in concentration into a supersaturated alloy or a metastable or stable crystalline compound. This reaction can occur only in concentration ranges near the pure elements or compounds. As far as supersaturated phases crystallize during this reaction they decompose by subsequent precipitation reactions; a metastable crystalline compound will undergo a phase transformation into the stable equilibrium phases.

As shown in Fig. 10.3 such reactions should be possible in the iron-rich range (reaction 1: polymorphous crystallization of α-iron) and near the composition of Fe_3B (reaction 4: polymorphous crystallization of Fe_3B).

b) Primary crystallization of one of the phases, e.g., α-iron (reaction 2). During this reaction the amorphous phase will be enriched in boron until further crystallization is stopped by reaching the metastable equilibrium, α-iron + amorphous Fe–B; this amorphous matrix can transform later or at higher temperatures by one of the mechanisms described here. The dispersed primary crystallized phase may act as the preferred nucleation site for the following crystallization of the amorphous matrix.

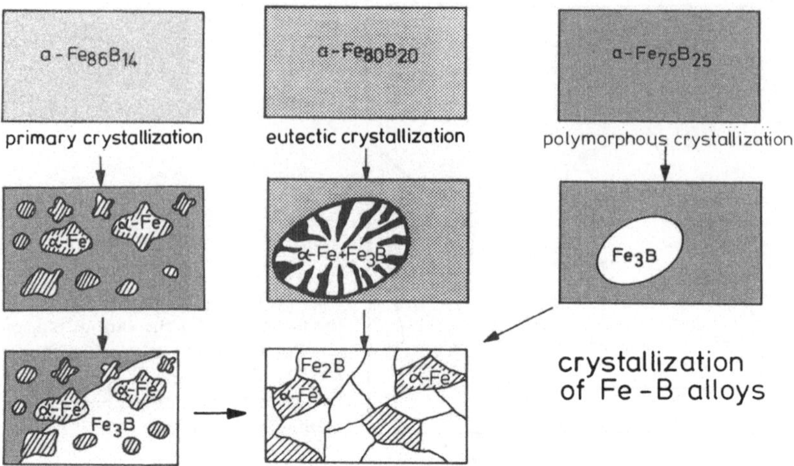

Fig. 10.4. Schematic diagram of typical crystallization reactions in amorphous Fe–B alloys [10.22]

c) Eutectic crystallization, i.e., simultaneous crystallization of two crystalline phases (e.g., reaction 3: α-Fe+Fe_3B, or reaction 5: α-Fe+Fe_2B) by a discontinuous reaction. This reaction has the largest driving force and can occur in the whole concentration range between the two stable phases. There is no concentration difference across the reaction front; however, in the reaction front the two components have to separate into two phases, which is why this reaction usually should take more time compared to the polymorphous reaction, i.e., a reaction without any separation of the components.

These typical crystallization reactions are shown schematically in Fig. 10.4 for three Fe–B metallic glasses. However, these reactions are not limited to Fe–B metallic glasses; they are rather very common and one or the other has been observed in all metallic glasses investigated so far.[1]

Glass forming ability is related to extremely deep eutectics, indicating a strong negative heat of mixing [10.23]; however, there is some evidence for the existence of systems which show at least in small concentration ranges a different behavior, i.e., a positive heat of mixing which may be due to some singularities in the structure of the amorphous state. Therefore, a concentration range should exist in which the free energy of a mixture of two amorphous phases is lower than that of the single amorphous phase. Amorphous materials in this concentration range have not only the possibility of transforming by one of the crystallization reactions mentioned above, but may also decompose into two amorphous phases. *Amorphous phase separation* occurring by nucleation and growth processes or even by spinodal decomposition without any nu-

1 We believe that the sequential nomenclature: *SS→MS-I→MS-II* is not very useful, because the meaning of these topics may change from one metallic glass to the other, and they do not explain anything except crystallization is occurring by a sequence of reactions.

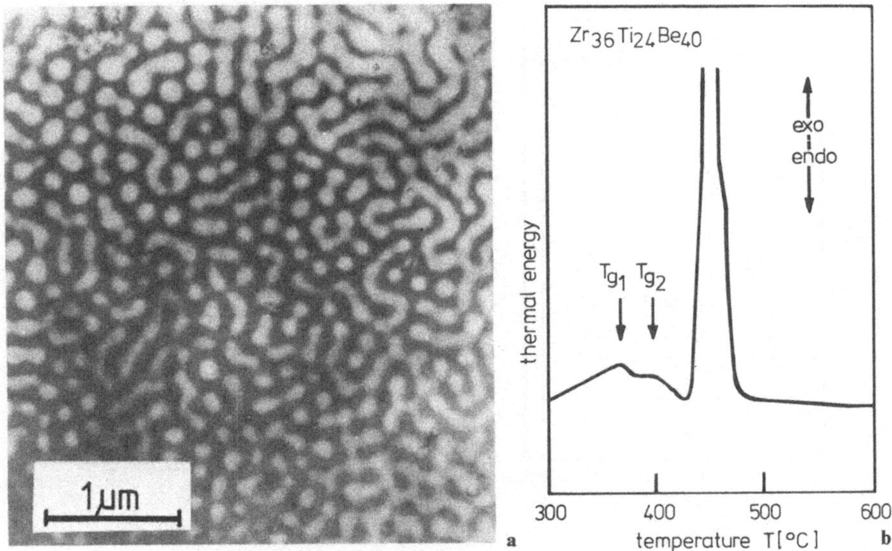

Fig. 10.5a, b. Amorphous phase separation in liquid-quenched $Zr_{36}Ti_{24}Be_{40}$ [10.27]. (**a**) Electron micrograph of the as-quenched glass; (**b**) DSC plot showing two glass transition temperatures T_g typical for the existence of two amorphous phases (20 K/min).

cleation process are well known from oxide glasses [10.24]. Evidence for amorphous phase separation in metallic glasses has been reported in $Pd_{74}Au_8Si_{18}$ [10.25] and $(Pd_{0.5}Ni_{0.5})_{81}P_{19}$ [10.26] alloys investigated by small-angle x-ray scattering and differential scanning calorimetry, respectively. In $Be_{40}Ti_{24}Zr_{36}$ amorphous phase separation has been found in as-quenched specimens (Fig. 10.5) by transmission electron microscopy and DSC [10.27]; the relatively coarse structure in Fig. 10.5a has been reported to be subdivided into much finer scaled regions. The problem of how amorphous phase separation which can occur on a very fine scale affects the subsequent crystallization reactions has not yet been answered.

10.3.2 Growth of Phases During Crystallization

For the following considerations on kinetics of crystallization, we shall assume that a stable nucleus is available and starts growing just at the beginning of isothermal annealing by a *polymorphous reaction* at a constant rate *u*. As shown elsewhere [10.28, 29], the temperature dependence for the rate of crystal growth *u* is given by

$$u = u_0 \cdot \exp\left(\frac{-Q_g}{RT}\right)\left[1 - \exp\left(\frac{-\Delta G}{RT}\right)\right], \tag{10.1}$$

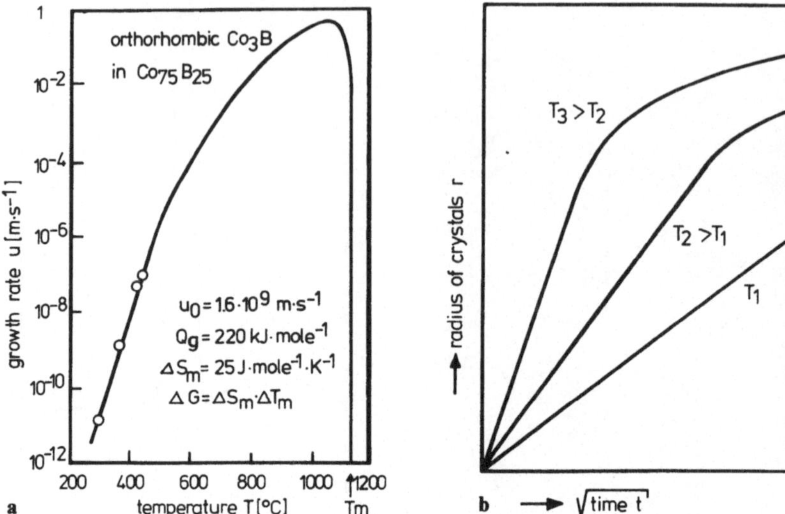

Fig. 10.6a, b. Growth rates for the different crystallization reactions. (**a**) Growth rate vs temperature for polymorphous crystallization; included are data for polymorphous crystallization of orthorhombic Co_3B in amorphous $Co_{75}B_{25}$ (ΔS_m and ΔG are rather arbitrary assumptions; slightly different numbers, however, will not change the plot). (**b**) Schematic diagram of the crystal radii during primary crystallization controlled by volume diffusion

where u_0 is a preexponential factor of the order of 10^3 m/s and Q_g is the activation energy for an atom to leave the amorphous phase, cross the interface, and attach itself to the crystallite. ΔG is the change in free energy per mole due to crystallization. The growth rate versus temperature according to this equation is shown in Fig. 10.6a including growth rates observed for the crystallization of orthorhombic Co_3B in amorphous $Co_{75}B_{25}$ alloys. As one can determine from these data, the preexponential factor is much higher than expected for crystallization processes by the transfer of single atoms from the amorphous phase to the crystallite. Figure 10.7a shows an electron micrograph of partially crystallized $Co_{75}B_{25}$ glasses. As seen from this micrograph for Co_3B crystals, growth rates of crystals depend often strongly on the crystallographic direction. This may be due to the particular growth mechanism which has to produce sufficient steps at the interface necessary for further growth.

For crystallization of metallic glasses, the situation when T is far below the melting temperature T_m is of special interest. In this case, $\Delta G \gg RT$ and $\exp(-\Delta G/RT)$ is very small. Therefore, for large undercooling the growth rate will obey an Arrhenius-type equation

$$u \simeq u_0 \cdot \exp\left(\frac{-Q_g}{RT}\right). \tag{10.2}$$

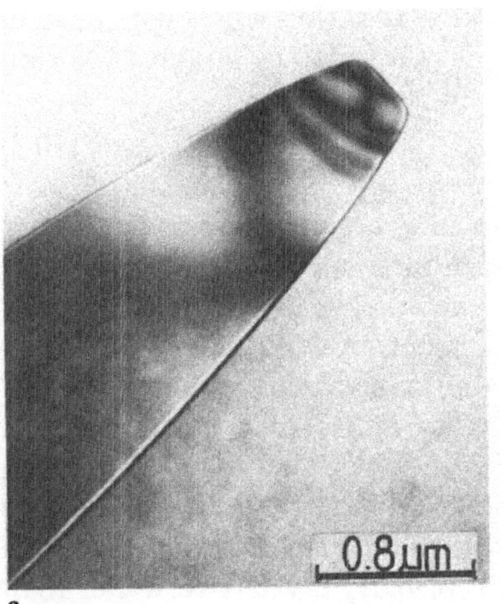

a

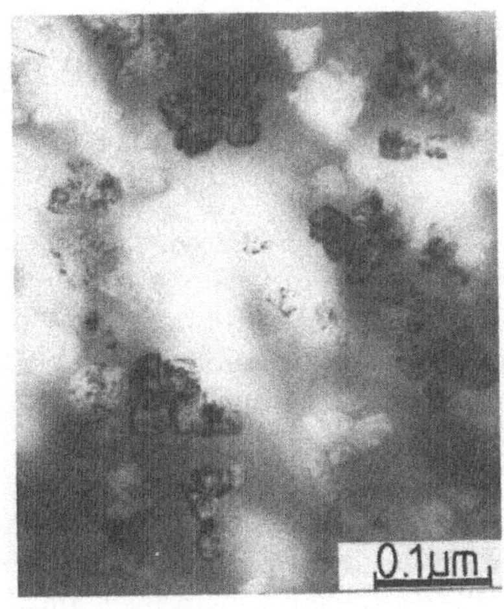

b

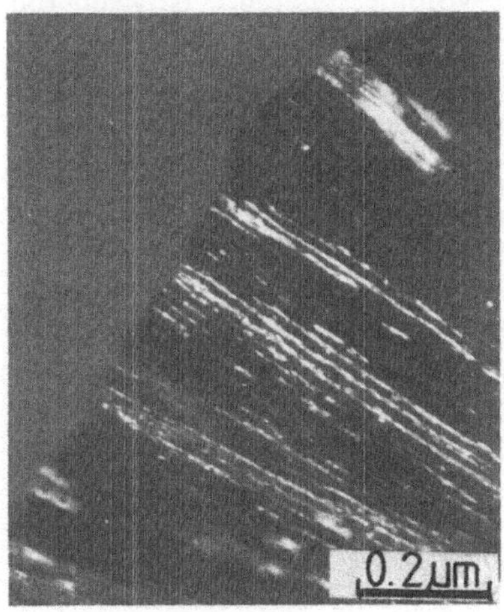

c

Fig. 10.7a–c. Transmission electron micrographs showing the morphology of the different crystallization reactions. (**a**) Polymorphous crystallization of orthorhombic Co_3B in $Co_{75}B_{25}$ glasses (1 h at 375 °C); (**b**) primary crystallization of α-Fe in $Fe_{86}B_{14}$ glasses (30 min at 380 °C); (**c**) eutectic crystallization of (Fe, Ni)+(Fe, Ni)$_3$B in $Fe_{40}Ni_{40}B_{20}$ glasses (30 min at 400 °C)

In cases where the latent heat of transformation is large as during crystallization, transformation of a large mass does not ordinarily occur isothermally. The temperature will rise at the crystallization front and the crystallization is further accelerated [10.30]. Probably this will happen often in crystallization of amorphous solids at relatively high growth rates. From amorphous semiconductors some cases of such "explosive" crystallization are known, e.g., in amorphous antimony [10.31].

During *primary crystallization* a concentration gradient is produced ahead of the crystallization front. Since long-range diffusion is involved in this reaction, the growth rate decreases with time. In a number of metallic glasses [*10.22, 32, 33*] which exhibit primary crystallization the radii r of spherical crystals have been found to follow a parabolic relationship with the annealing time t (see Fig. 10.6b), indicating that growth is controlled by volume diffusion

$$r=\alpha\sqrt{D\cdot t}\,, \tag{10.3}$$

where D is the volume diffusion coefficient and α is a dimensionless parameter evaluated from the compositions at the particle interface and the composition of the sample [10.34]. After reaching the metastable equilibrium between primary crystals and the amorphous matrix, further growth occurs only due to Ostwald ripening at a much slower rate; the crystallized volume does not change any longer until the amorphous matrix starts to crystallize into a different phase. As shown in Fig. 10.7b for Fe–B alloys, after longer annealing times the crystals may change their morphology from a spherical into a more star-shaped or dendritic one, which may be due to the anisotropy of interfacial energy or the particular growth mechanism. If primary crystals are softer than the amorphous matrix as in the case of α-iron crystals, they have been found to contain a high density of defects due to the volume difference between the amorphous and the crystalline states.

When crystallization occurs by a *eutectic reaction*, the growth rate u for a eutectic cell can be described by an equation very similar to (10.1). However, in contrast to polymorphous crystallization, during eutectic crystallization long-range diffusion is necessary for the redistribution of the components into the two phases and the growth rate u is expected to decrease with increasing interlamellar spacing λ. If diffusion occurs in the amorphous phase ahead of the crystallization front, $u\propto\lambda^{-1}$; if diffusion occurs along the amorphous/crystalline interface, $u\propto\lambda^{-2}$ [10.29, 35]. Therefore, assuming a driving force $\Delta G\gg RT$, i.e. for large undercooling, the following equations can be set up:

for volume diffusion:

$$u\propto\frac{4\cdot D}{\lambda}, \tag{10.4a}$$

for interfacial diffusion (δ is the thickness of the crystallization front):

$$u\propto\frac{8\cdot\delta\cdot D}{\lambda^2}. \tag{10.4b}$$

Figure 10.7c shows a micrograph of the crystallization front in a $Fe_{40}Ni_{40}B_{20}$ glass, which is very typical for eutectic crystallization. In $Pd_{78.1}Cu_{5.5}Si_{16.4}$ [10.36], Metglas 2826 [10.37], and $Fe_{78}Mo_2B_{20}$, the only known metallic glasses where interlamellar spacing λ versus temperature has been investigated so far, interlamellar spacing seems to increase as the annealing temperature is lowered (e.g., in $Fe_{78}Mo_2B_{20}$ from 28 nm at 470 °C to 50 nm at 390 °C). This is in opposition to the observations of eutectoid decomposition of crystalline alloys; however, it may occur due to the fact that investigations of metallic glasses are commonly carried out in the low-temperature range, where the growth rate increases with the temperature. Most studies on eutectoid decomposition are done only just below the eutectoid temperature. A similar behavior of interlamellar spacing versus temperature is known from the eutectic crystallization of amorphous $Al_{50}Ge_{50}$ alloys or the eutectoid decomposition of the metastable AlGe phase.

10.3.3 Nucleation of Crystallization

Based on the concepts of the classic Becker-Volmer nucleation theory, the following expression is generally used to predict homogeneous nucleation rates [10.28, 29]:

$$I_{st}=I_0\cdot\exp\left(\frac{-N\Delta G_c}{RT}\right)\exp\left(\frac{-Q_n}{RT}\right), \tag{10.5}$$

where I_{st} equals the steady-state nucleation rate and I_0 generally varies from 10^{30} to 10^{35} nuclei/sm^3 depending on the specific theory used. N is the Loschmidt number, Q_n the activation energy for the transport of an atom across the interface of an embryo, and ΔG_c is the free energy required to form a nucleus of critical size r_c

$$r_c=\frac{2\sigma}{\Delta G}, \tag{10.6}$$

where ΔG is the change in chemical free energy per mole and can be assumed to be proportional to the amount of undercooling. For spherical nuclei with an isotropic interfacial energy σ and in the absence of strain energy this critical energy ΔG_c is equal to

$$\Delta G_c=\frac{16\pi\sigma^3V_m^2}{3\Delta G^2}, \tag{10.7}$$

where V_m is the molar volume. The importance of σ and ΔG cannot be overemphasized since they appear as cubed and squared terms, respectively, in the exponent, and therefore will have a pronounced effect on the nucleation rates. Thus, the crystal nucleus forming first may not be the phase with the lowest free energy, but an alternative metastable phase which has a lower ΔG_c. This is possible, if the interfacial energy of the metastable phase is lower than that of the stable one. At very large undercooling, ΔG_c will be very small and the Q_n terms dominates in (10.5). In this case

$$I_{st}\simeq I_0\exp\left(\frac{-Q_n}{RT}\right); \tag{10.8}$$

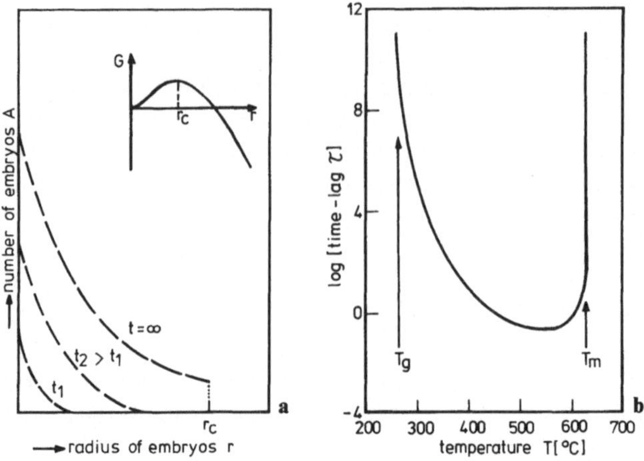

Fig. 10.8a, b. Transient nucleation of crystallization. (**a**) Schematic diagram of the distribution of embryos after different annealing times; $t=\infty$ reflects the situation after reaching the steady state. (**b**) Time lag τ for transient nucleation in $NaPO_3$ glasses [10.41]

that means that as the growth rate in the low-temperature range the nucleation rate can be described by an Arrhenius-type equation.

This behavior is assumed to be complicated by two points. Firstly, the critical radius r_c of a nucleus which decreases with the annealing temperature should not become smaller than the unit cell of the particular phase [10.38]. Therefore, at large undercooling, nucleation is expected to be much slower than given by the classic theory. An atomistic theory of nucleation which includes this long-known aspect has not yet been established in detail. However, there is evidence that thermal stability of an amorphous alloy can be reduced drastically by the occurrence of a phase with a relatively simple unit cell and the same composition as the metallic glass. This may be for example the reason for the difference in glass forming ability between Fe–B and Fe–C alloys due to the occurrence of the ε-phase in Fe–C [10.39]. Secondly, and we believe this point to be even more important, in nucleation theory usually it is assumed that a steady-state concentration of clusters or embryos exists at all times. However, at the very beginning there must be a finite period during which this steady-state distribution of clusters is being established (see Fig. 10.8a) [10.40]. Such transient or time-dependent nucleation has been predicted for crystallization of amorphous materials with high viscosity and has been reported for some oxide glasses [10.41] and amorphous silicon [10.42]. An approximate expression for the transient nucleation rate $I(t)$ is given by the following equation [10.43]:

$$I(t)=I_{st}\left[1+2\sum_{n=1}^{\infty}(-1)^n \exp\left(-n^2\frac{t}{\tau}\right)\right] \tag{10.9}$$

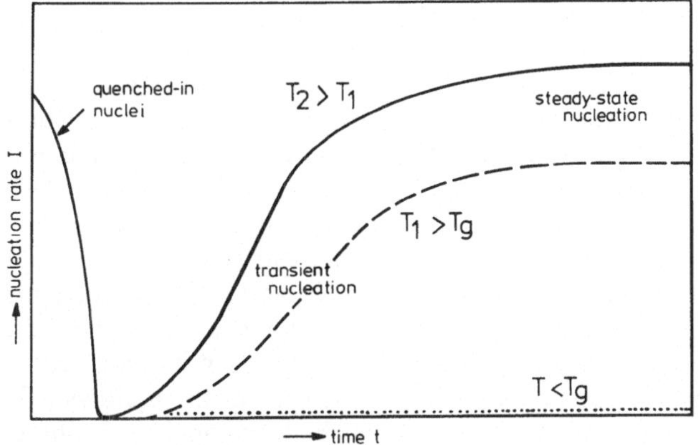

Fig. 10.9. Schematic diagram of the apparent nucleation rate vs time during transient nucleation superimposed by some quenched-in nuclei

with I_{st} being the steady-state nucleation rate given by (10.5) and τ the time lag. The time lag is given by

$$\tau \propto \left(\frac{T_m}{\Delta T}\right)\eta, \tag{10.10}$$

where T_m is the absolute temperature of melting and η the melt viscosity at undercooling $\Delta T = T_m - T$. As shown in Fig. 10.8b for a $NaPO_3$ glass [10.41], the time lag can become quite significant near the glass transition temperature T_g. Therefore, nucleation rates will be lowered drastically near T_g, and below T_g no homogeneous nucleation can be expected.

Homogeneous nucleation, however, is just one possibility of nucleation, very often superimposed by heterogeneous nucleation, athermal nucleation, or even quenched-in nuclei. Surfaces or interfaces may catalyze nucleation as the crystalline new phase replaces a portion of the surface, thus reducing the total surface energy in (10.7). But since the energy for an amorphous/crystalline interface is relatively small compared to crystalline/crystalline interfaces, heterogeneous nucleation at surfaces or interfaces is not as common as in crystalline materials (see Sect. 10.6). Quenched-in nuclei or embryos which were too small at higher temperatures but large enough to be nuclei at typical annealing temperatures for crystallization are much more important. These quenched-in nuclei or fluctuations may not always possess the right structure and there may be some need for transformation taking some time until they become effective nuclei. On the other hand, they can even dissolve during annealing et elevated temperatures.

Figure 10.9 is a schematic diagram of nucleation rate versus annealing time for different temperatures. Below T_g homogeneous nucleation will take too much time and nucleation due to quenched-in fluctuations or nuclei is responsible for crystallization. The microstructure of metallic glasses crystallized this way will contain nearly equal-sized crystals, since all of them start growing at the beginning of the annealing treatment. Above T_g, in addition homogeneous nucleation will be observed starting with an increasing nucleation rate during the transient period until the steady state is reached. This will occur much earlier as the temperature for crystallization increases.

This superimposing nucleation behavior is not restricted to isothermal annealing or only to the polymorphous crystallization reaction; it will occur for eutectic and primary crystallization as well as during isochronal heating. Nucleation for eutectic or primary crystallization has not been investigated so far in any detail. However, there is some evidence at least in Fe–B alloys that eutectic crystallization starts with the formation of the tetragonal Fe_3B phase, thus enriching the surrounding in iron which will crystallize with a strong orientation relationship to the Fe_3B.

10.3.4 Kinetics of Crystallization in Metallic Glasses

Time-temperature transformation diagrams (TTT diagrams) are known to be very useful for understanding kinetics of crystallization during isothermal or isochronal heat treatments. Taking into account the nucleation and growth behavior discussed above, one can draw schematically a TTT diagram for the start of crystallization, i.e., about 1 to 5% crystallized volume fraction, combining nucleation and growth rates as shown in Fig. 10.10: Below T_g the kinetics of crystallization will depend on the number of quenched-in nuclei N. This is in good agreement with experiments on the influence of quenching rate as will be discussed in more detail in Sect. 10.6. The activation energy estimated

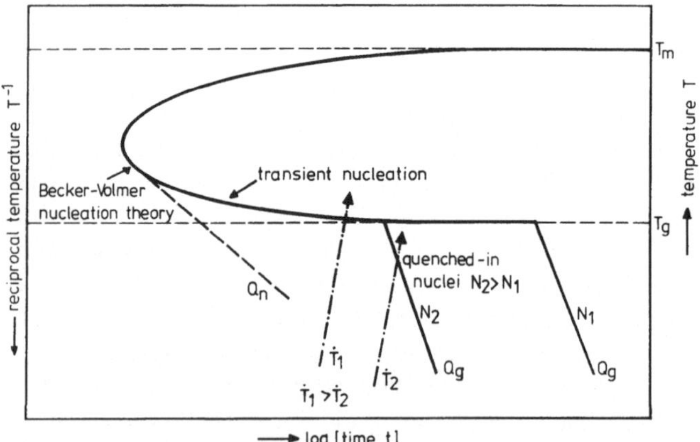

Fig. 10.10. Schematic time-temperature transformation diagram for the start of crystallization (crystallized fraction $X \propto I \cdot u^3 \cdot t^4 \simeq 0.01$). The dashed arrows indicate isochronal annealing treatments with different heating rates $\dot{T}$

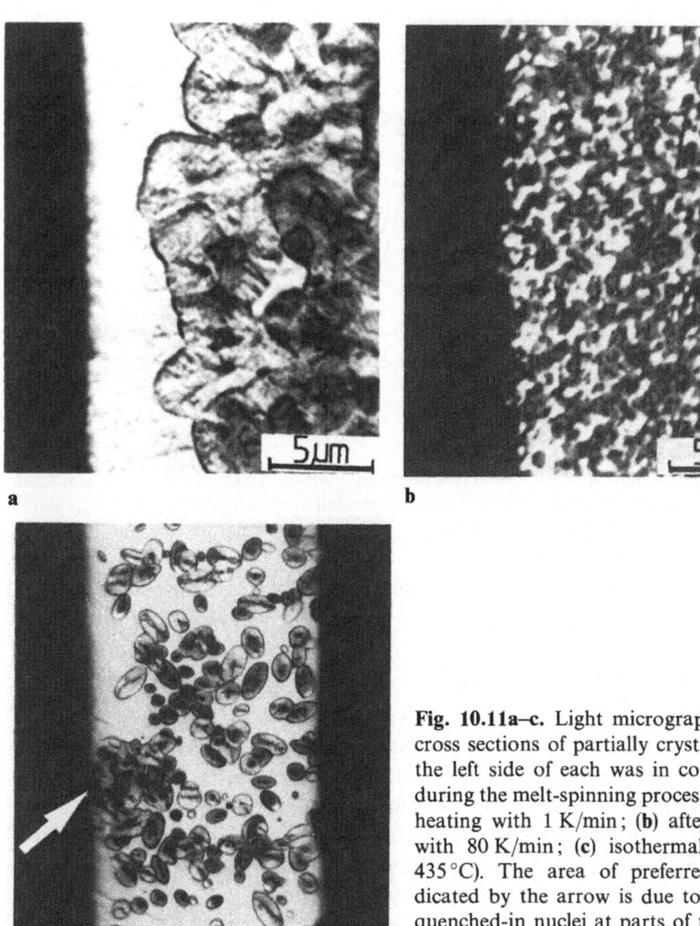

Fig. 10.11a–c. Light micrographs showing parts of cross sections of partially crystallized $Fe_{78}Mo_2B_{20}$; the left side of each was in contact with the wheel during the melt-spinning process. (**a**) After isochronal heating with 1 K/min; (**b**) after isochronal heating with 80 K/min; (**c**) isothermally annealed (4 h at 435 °C). The area of preferred crystallization indicated by the arrow is due to a higher number of quenched-in nuclei at parts of the ribbon where the cooling rate was lowered by gas bubbles during the casting process

from such a plot will be only about that for growth of the crystallites. Above T_g crystallization will depend on the time lag τ bringing about a very large acceleration of crystallization in a relatively narrow temperature range.

With such a diagram crystallization during isothermal or isochronal annealing in $Fe_{78}Mo_2B_{20}$ alloys has been explained. Figure 10.11 shows light micrographs of cross sections of partially crystallized $Fe_{78}Mo_2B_{20}$; the left boundary of each of them was in contact with the wheel during the melt-spinning process. Due to the higher cooling rate we expect a much lower density of quenched-in nuclei on this side, which is confirmed as shown in Fig. 10.11a, where no eutectic cells have been formed near that side of the ribbon. At a much higher heating rate, however, one will reach the temperature range of homogeneous nucleation before the whole specimen has crystallized and there

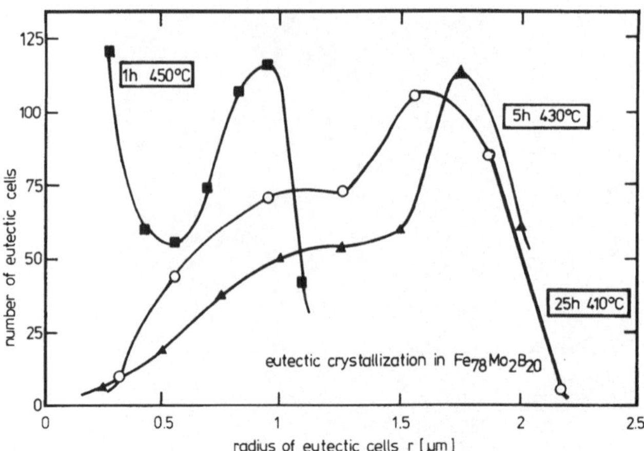

Fig. 10.12. Distribution of the apparent size of eutetic cells (small semi-axis of the elliptical sections resulting from a plane intersecting the eutectic cells) in partially crystallized $Fe_{78}Mo_2B_{20}$ glasses after isothermal annealing at different temperature

should not be any difference due to the cooling rate as is demonstrated in Fig. 10.11b. This interpretation is confirmed by isothermal crystallization experiments (Figs. 10.11c, 12). Whereas annealing at 410 or 430 °C results in the formation of large, nearly equal-sized eutectic cells, the microstructure after annealing at 450 °C contains in addition a large number of smaller cells due to the beginning of homogeneous nucleation. At higher temperatures homogeneous nucleation becomes more pronounced and the microstructure consists of an increasing number of very small eutectic cells.

From this, one can imagine, how difficult it may be to analyze kinetics of crystallization even with only one crystallization mode, but without knowing which nucleation regime is responsible in the temperature range investigated. Therefore, we believe that it is indeed imperative to combine studies of both microstructure and kinetics for a better understanding of the crystallization mechanism of metallic glasses, especially if different crystallization reactions compete with each other. Which of the possible reactions mentioned above will occur depends not only on the driving force but also on the activation energy for the transport of atoms across the interface for long-range diffusion and on the thermal history of the metallic glass, i.e., the number of quenched-in nuclei. Furthermore, different surfaces or strain energies, not discussed here, will influence the crystallization behavior. Due to this very complex behavior, the occurrence of a distinct crystallization reaction has to be discussed in the particular case.

10.4 Diffusion in Metallic Glasses

We have observed that crystallization of metallic glasses occurs by nucleation and growth processes which are governed by diffusion. Therefore, understand-

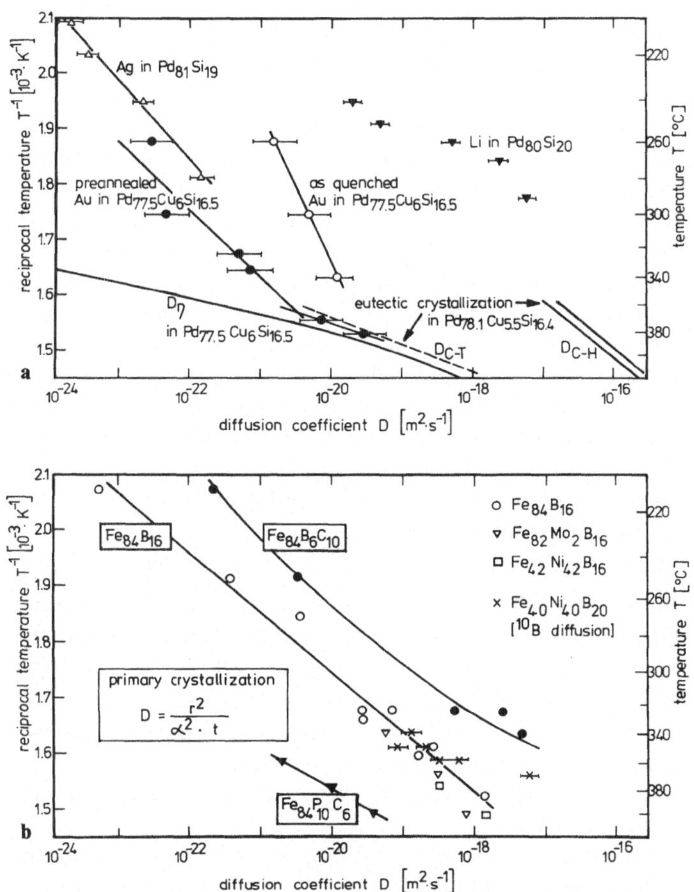

Fig. 10.13a, b. Diffusion coefficients in metallic glasses. (**a**) Diffusion coefficients D in Pd–Si based glasses: Li in $Pd_{80}Si_{20}$ [10.43]; Ag in $Pd_{81}Si_{19}$ [10.44]; D_η and Au in $Pd_{77.5}Cu_6Si_{16.5}$ [10.13]; D estimated with Chen-Turnbull (C–T) or Cahn-Hagel (C–H) analysis from eutectic crystallization data in $Pd_{78.1}Cu_{5.5}Si_{16.4}$ [10.36]. (**b**) Diffusion coefficients in Fe based metallic glasses estimated from primary crystallization data [10.47]; included are data from direct measurements of ^{10}B diffusion in $Fe_{40}Ni_{40}B_{20}$ [10.46]

ing diffusion in metallic glasses is of utmost importance for influencing their thermal stability, e.g., reducing the diffusivity by adding small amounts of other elements. But so far, only very few diffusion data have been published, and these were concentrated until recently only on metal diffusion in Pd–Si based glasses.

In Fig. 10.13a the metal diffusion coefficients measured in Pd–Si based glasses are summarized. The diffusivity below T_g has been found to depend strongly on the thermal history. The initial diffusivity of gold in the as-quenched metallic glass is orders of magnitude larger than D_η evaluated from viscosity data in the relaxed state; however, it approaches D_η as the structure relaxes towards the ideal amorphous structure, e.g., after a thermal treatment near T_g [10.13]. Above T_g measured diffusion coefficients agree fairly well with the diffusivity D_η.

Standard direct diffusion measurements in metallic glasses are impeded by generally very slow diffusivities and because annealing times are limited by the need to avoid any crystallization. However, it has been proposed that data from crystallization reactions for estimating diffusion coefficients can be used. This was done first for $Pd_{78.1}Cu_{5.5}Si_{16.5}$ [10.36] using the eutectic reaction; but as shown in Fig. 10.13a, depending on the analysis [Chen-Turnbull (C–T) or Cahn-Hagel (C–H)] used the estimated diffusivities differ by more than two orders of magnitude. On the other hand, as mentioned above, eutectic crystallization may be controlled by interface diffusion which is as in crystalline material expected to be much faster than volume diffusion, even if these amorphous/crystalline interfaces have been found to be less defective than their crystalline/crystalline counterparts [10.45]. A diffusivity of $3 \times 10^{-21}\,m^2/s$ at 385 °C estimated from amorphous phase separation in $Pd_{74}Au_8Si_{18}$ is in good agreement with the data in Fig. 10.13a [10.25].

For primary crystallization in early stages growth has been found to obey a $\sqrt{time}$ law, indicating that growth is controlled by volume diffusion. Assuming a concentration-independent diffusion coefficient, one can calculate from the radii r of spherical particles versus $\sqrt{time}$ the diffusion coefficient using (10.3). Figure 10.13b shows diffusion coefficients in some Fe–B based glasses calculated by this method which are assumed to reflect metalloid diffusion. A direct measurement of metalloid diffusion has been reported very recently for boron diffusion in $Fe_{40}Ni_{40}B_{20}$ [10.46]. These data shown for comparison in Fig. 10.13b are in very good agreement.

With increasing boron content progressive filling of the holes in the Bernal structure of the metallic glasses should lead to a reduction of available empty holes to sustain diffusion. But diffusivity has not been observed to drop with increasing boron content, at least in the range of boron contents between 14 and 20 at.%. All these data indicate that diffusion of boron does not occur by an interstitial mechanism but as in crystalline α-iron by a substitutional one, i.e., iron and boron diffusion seems to occur by the same mechanism [10.47].

In $Cu_{56}Zr_{44}$ alloys Zr diffusion is assumed to be rate determining and has been estimated from isothermal growth rates during primary crystallization to be $5.3 \times 10^{-19}\,m^2/s$ at 440 °C [10.33].

10.5 Micromechanism of Crystallization in Some Metallic Glasses

10.5.1 General Considerations

To date, most of the literature concerning crystallization of metallic glasses has been centered on alloys composed of metal and metalloid atoms. On a chemical

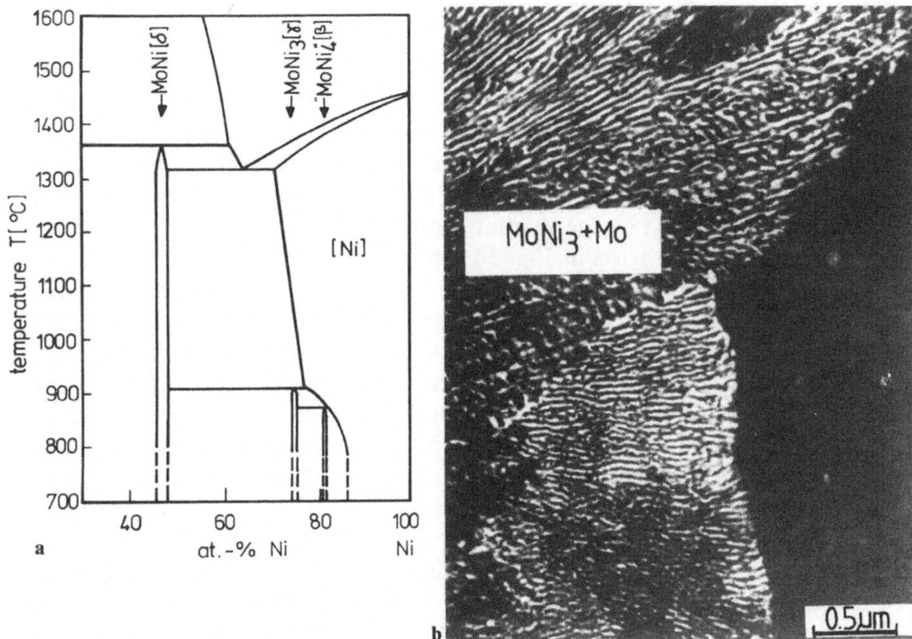

Fig. 10.14a, b. Crystallization of amorphous Mo–Ni alloys. (**a**) Part of the Mo–Ni phase diagram [10.48]; (**b**) electron micrograph of partially crystallized $Mo_{50}Ni_{50}$ sputter-deposited films (48 h at 700 °C) [10.49]

basis, metallic glasses can be divided into two groups: transition metal-metalloid glasses such as Fe–B, Co–B, or Pd–Si, and metallic glasses such as Cu–Zr or Mg–Zn which contain no metalloid atoms. There are only a few crystallization studies so far concerning this second group of metallic glasses.

Due to the high melting temperatures most amorphous transition metal alloys are prepared by coevaporation or sputtering. Figure 10.14a shows the relevant part of the Mo–Ni phase diagram [10.48] in which amorphous alloys have been formed by sputtering [10.49]. Figure 10.14b shows the microstructure of $Mo_{50}Ni_{50}$ after partial crystallization by a eutectic reaction into Mo and $MoNi_2$; ahead of the crystallization front there is some evidence for primary crystallization of another phase [10.49]. During isothermal annealing at higher temperatures these two crystallization modes compete with each other, leading to a more heterogeneous structure. Transformation into MoNi (δ phase) has been found to occur at temperatures of about 900 to 1000 °C. Crystallization of such amorphous films seems to occur by the very same reactions as in metal-metalloid glasses.

With the help of the general principles discussed above crystallization behavior of some particular metallic glasses will be discussed in more detail.

10.5.2 Fe–B Metallic Glasses

Metallic glasses stabilized by boron, Fe–B in particular, have been intensively studied by numerous authors over the last few years [10.22, 50–61], and binary Fe–B alloys have become the model system of metal-metalloid glasses, especially for the crystallization behavior. Figure 10.15a shows the relevant part of the Fe–B phase diagram where the glass forming range from 12 to 27 at.% boron is indicated by the shading; the structures of stable and metastable crystalline phases are listed in Fig. 10.15b.

As shown schematically in Fig. 10.4, depending on the alloy composition three distinct crystallization modes have been observed: primary crystallization of α-iron in the more iron-rich alloys up to about 20 at.% boron, but in alloys with more than 16 at.% only concurrently with the eutectic crystallization. Eutectic crystallization itself into α-iron and a tetragonal Fe_3B phase occurs in the concentration range from 17 to 24 at.% boron. The concentration range for polymorphous crystallization of Fe_3B has been found to be very narrow, about 1 at.%. In alloys with boron concentration of 26 at.% or more eutectic crystallization into the stable phases α-iron and Fe_2B has been observed.

Figure 10.16 shows a crystallization diagram of Fe–B alloys with crystallization temperatures T_{cryst} versus boron concentration using the differential scanning calorimetry data of several authors. Crystallization in the high boron alloys is due to eutectic or polymorphous crystallization, respectively. The low-temperature branch T^1_{cryst} in the low boron alloys is generally assumed to indicate primary crystallization of α-Fe. The increase in T^1_{cryst} with the boron content reflects a strong decrease in the nucleation rate for α-Fe due to the

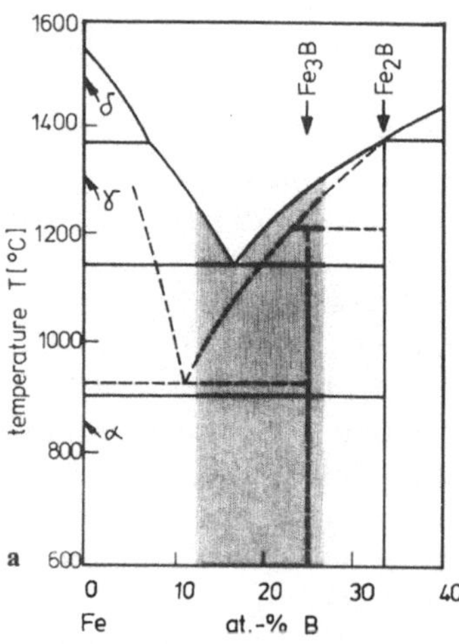

b

α-Fe	bcc	a = 0.287nm
γ-Fe	fcc	a = 0.359nm
Fe_3B	tetragonal Ti_3P type	a = 0.864nm c = 0.428nm
Fe_3B	orthorhombic Fe_3C type	a = 0.543nm b = 0.666nm c = 0.445nm
Fe_2B	tetragonal $CuAl_2$ type	a = 0.511nm c = 0.425nm

Fig. 10.15a, b. Iron-boron system. (**a**) Part of the phase diagram; the glass forming range is indicated by the shading [10.50, 56, 62]. (**b**) Crystal structure of stable and metastable phases in iron-rich Fe–B alloys [10.63]

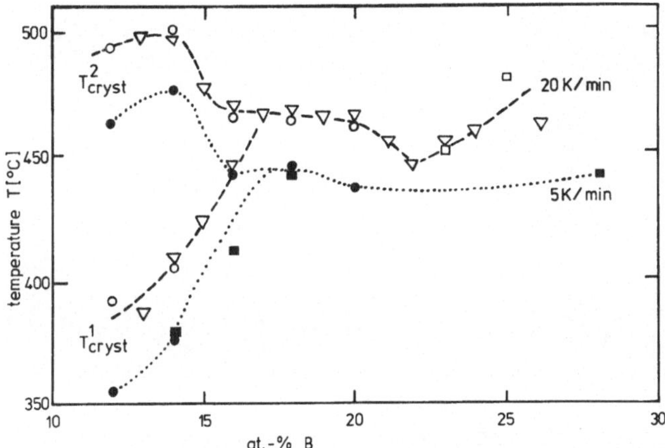

Fig. 10.16. Crystallization temperatures T_{cryst} (peak in the crystallization exotherm) of Fe–B metallic glasses; the DSC data are taken from [10.51] ▽; [10.53] ■; [10.60] □; [10.61] ○, ●

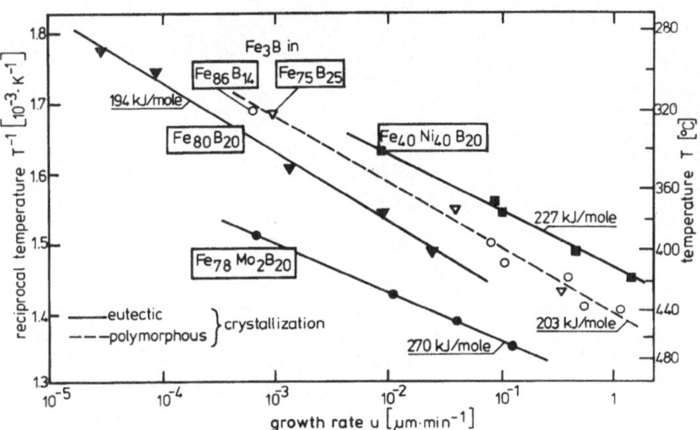

Fig. 10.17. Growth rates for polymorphous and eutectic crystallization (large semi-axis in the case of anisotropic growth) in iron-based metallic glasses

decreasing driving force for this reaction (given by the length of the corresponding arrow in Fig. 10.3). As mentioned in Sect. 10.4, growth of these crystals is controlled by volume diffusion. Diffusion coefficients calculated from these growth data have been found to be independent on the boron concentration. The second crystallization reaction at T^2_{cryst} occurs by polymorphous crystallization of the amorphous matrix into Fe_3B. As shown in Fig. 10.17 growth rates for polymorphous crystallization of Fe_3B are equal in $Fe_{75}B_{25}$ metallic glasses and in the matrix of $Fe_{86}B_{14}$ after finished primary crystallization, at least in the temperature range investigated between 300 and 450 °C. But the number of nucleation sites was found to be some orders of magnitude smaller for the polymorphous crystallization of the amorphous matrix in $Fe_{86}B_{14}$ (compare

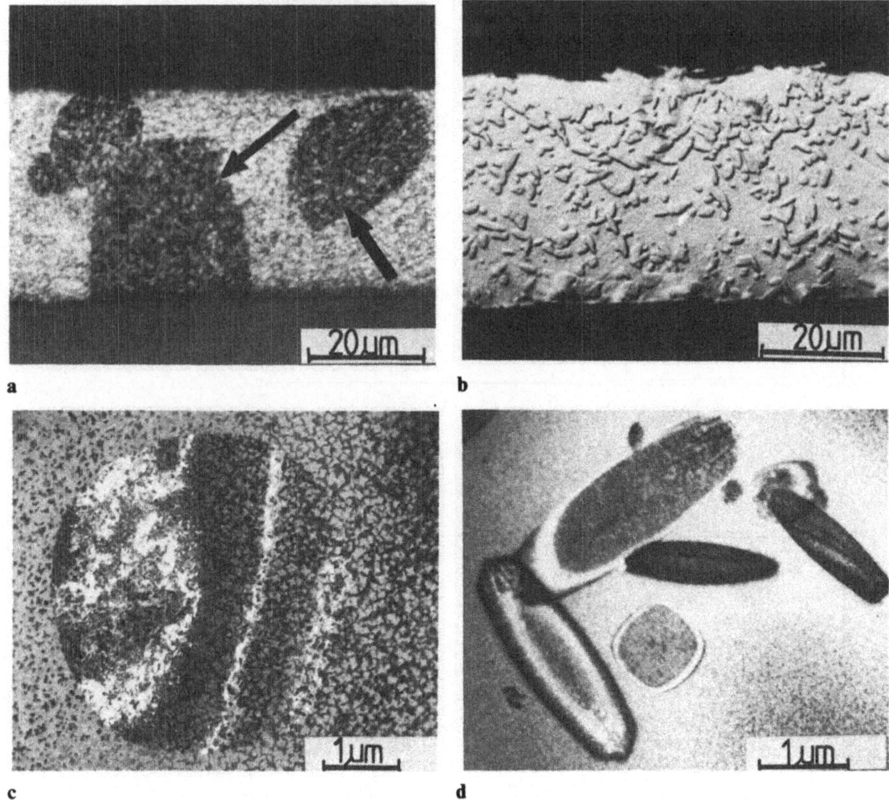

Fig. 10.18a–d. Microscopic investigation of crystallization in Fe–B metallic glasses. (**a**) Polymorphous crystallization of Fe_3B in $Fe_{86}B_{14}$ after primary crystallization of α-Fe is indicated by arrows; 10 h at 375 °C (LM). (**b**) Polymorphous crystallization of Fe_3B in $Fe_{75}B_{25}$ glasses; 1 h at 375 °C (LM). (**c**) Reaction as in (**a**); $Fe_{86}B_{14}$ [5 h at 375 °C (TEM)]. (**d**) Reaction as in (**b**); $Fe_{75}B_{25}$ [5 min at 430 °C (TEM)]

Fig. 10.18a with b). This is consistent with a T^2_{cryst} for the second crystallization reaction shifted to higher temperatures than for the polymorphous crystallization in $Fe_{75}B_{25}$; high crystallization temperatures as measured by DSC can reflect two totally different reasons: very slow growth rates or lack of nucleation sites. Lack of nucleation sites could be explained, assuming that the number of quenched-in nuclei of the tetragonal phase Fe_3B increases with the boron content or is reduced by the diffusion processes necessary for the primary crystallization. The growth rates for the eutectic reaction (Fig. 10.17) are slower than those for the polymorphous one with shorter diffusion distances and (assuming an equal number of nucleation sites) T^2_{cryst} could be shifted to higher values for the eutectic reaction.

It is well known that ternary amorphous alloys are more stable against crystallization than the basic binary ones. For $Fe_{78}Mo_2B_{20}$ and $Fe_{40}Co_{40}B_{20}$

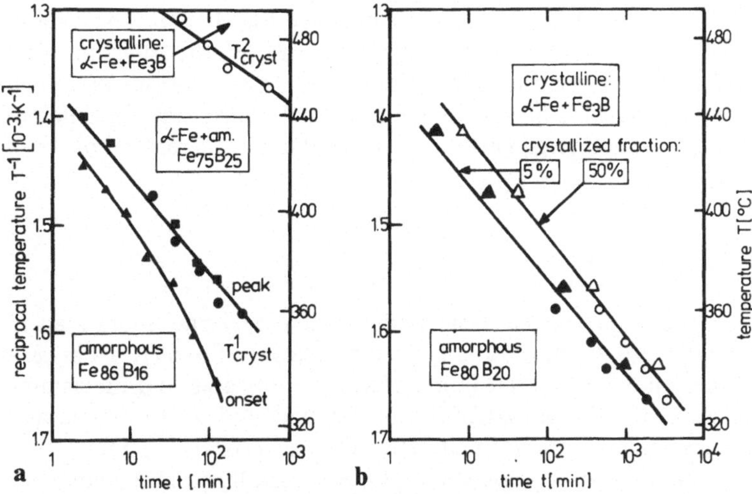

Fig. 10.19a, b. Time-temperature transformation diagrams for crystallization of Fe–B metallic glasses. (**a**) For isochronal annealing of $Fe_{86}B_{14}$; data are taken from [10.61] ○, ●; [10.53] ▲■; apparent activation energies are 215 kJ/mole for T^1_{cryst} (peak) and 280 kJ/mole for T^2_{cryst}. (**b**) For isothermal annealing of $Fe_{80}B_{20}$; data are taken from [10.58] ▲△; [10.57] ○, ●; activation energies are about 220 kJ/mole

this higher thermal stability has been found to be due to retarded nucleation as well as to decreased growth rates. The stabilizing effect of third elements is generally observed in alloys exhibiting eutectic or polymorphous crystallization. The significant larger activation energy for growth as well as the observed strong increase in the preexponential factors for ternary alloys, e.g., 2.3×10^9 m/s for growth in $Fe_{40}Ni_{40}B_{20}$ or 6.7×10^9 m/s in $Fe_{80}P_{14}B_6$ compared to 9×10^5 m/s in $Fe_{80}B_{20}$ indicate correlated jumps of groups of atoms during the crystallization process. In addition, ternary alloys very often exhibit a lower eutectic temperature than the basic binary systems, thus reducing the critical quenching rate for glass formation or reducing the number of quenched-in nuclei during equal quenching rates.

No such general stabilization by third elements has been found in alloys exhibiting primary crystallization. On replacement of iron in $Fe_{84}B_{16}$ by molybdenum or nickel no significant influence of the diffusivity has been observed. On the other hand, replacement of boron by carbon in $Fe_{84}B_{16}$ increases the diffusion coefficient by about one order of magnitude, thus decreasing thermal stability; but in addition, the number of primary crystals is reduced drastically, which would increase stability. Superimposing these two effects leads in Fe–B–C alloys to the observed reduced thermal stability of the ternary alloys.

Using known crystallization data [10.53, 57, 61] one can draw a TTT diagram for isochronal annealing as shown in Fig. 10.19a for $Fe_{86}B_{14}$ alloys;

an isothermal TTT diagram for $Fe_{80}B_{20}$ is shown in Fig. 10.19b. The data of other authors [10.54] show very similar behavior; however, the annealing time for a particular state of crystallization is shifted to slightly shorter or longer times which may be due to differences in the actual concentration, different amounts of contamination (e.g., carbon or oxygen), and different numbers of quenched-in nuclei.

10.5.3 Pd–Si Metallic Glasses

Crystallization of Pd–Si glasses has been studied extensively since more than 15 years [10.64], but there are still quite large discrepancies [10.65]; we believe this is due to the following points: Our knowledge on stable and metastable crystalline phases in the Pd–Si system is still very incomplete; the Pd_5Si and Pd_4Si or the Pd_9Si_2 phases are not generally accepted [10.66, 67]. Due to the relatively large number of complicated phases crystallization behavior is expected to be more complex than in Fe–B alloys; some inaccuracy in the exact composition of the alloy should have a much larger influence. As mentioned above, crystallization in edges of splat-formed foils thin enough to transmit electrons may not give a representative picture of crystallization. But quite a lot of older work with Pd–Si glasses was done using this technique.

Figure 10.20 shows the Pd-rich part of the Pd–Si phase diagram. Metallic glasses can be prepared by rapidly quenching in the concentration range from 16 to 22 at.% Si [10.68]. The following crystallization reactions have been identified [10.69, 70]: Whereas in $Pd_{83}Si_{17}$ crystallization starts depending on temperature with the formation of a palladium-rich fcc phase or a palladium-rich silicide (Pd_9Si_2) followed by the growth of complex silicides in the remaining amorphous matrix, crystallization of $Pd_{80}Si_{20}$ occurs by a eutectic reaction. The eutectic cells consist of Pd_3Si and a more palladium-rich silicide.

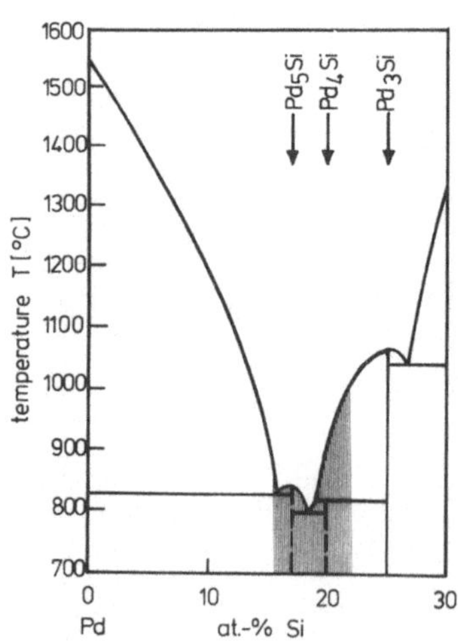

Fig. 10.20. Part of the Pd–Si phase diagram [10.67]; the glass forming range is indicated by the shading [10.68]

The crystallization in $Pd_{78}Si_{22}$ seems to start with dendriticlike crystallization of Pd_3Si, but has not been investigated in detail. There is a strong evidence that the temperature from which the melt was quenched has an influence on crystallization; with decreasing melt temperature crystallization is shifted to lower temperatures which may be due to a larger number of quenched-in nuclei.

In addition, there are some crystallization studies on ternary Pd–Si glasses with Cu [10.36], Ag [10.71], or Au [10.26]. In $Pd_{67}Fe_{13}Si_{20}$ [10.72] some evidence for amorphous phase separation has been observed using Mössbauer spectroscopy and crystallization is reported to start with the formation of Pd_3Fe crystallites in the iron-rich regions.

10.5.4 Cu–Zr Metallic Glasses

Cu–Zr alloys are so far the best-known glasses composed predominantly of transition metals, which can be prepared by quenching from the melt. Crystallization reactions in metallic glasses with compositions ranging from $Cu_{70}Zr_{30}$ to $Cu_{30}Zr_{70}$ [10.73] have been investigated by means of calorimetry, x-ray diffraction, and transmission electron microscopy. The phase diagram Cu–Zr [10.73] is shown in Fig. 10.21; however, there is still uncertainty about the existence of some phases. The exact crystal structures of most Cu–Zr phases have not been determined. The "stability" indicated by the glass transition temperatures T_g increases with the copper content as shown in Fig. 10.21, too [10.74]. Amorphous $Cu_{30}Zr_{70}$ cannot be formed using the melt-spinning technique, but only by splat-quenching [10.75].

So far, the following four alloys have been studied in more detail. The most striking feature in crystallization of these glasses is the significantly different behavior below and above the glass transition temperature T_g.

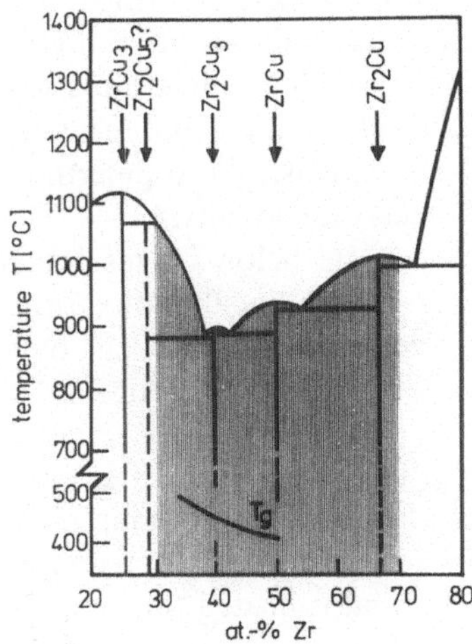

Fig. 10.21. Part of the Cu–Zr phase diagram [10.62]; the glass forming range is indicated by the shading [10.73]

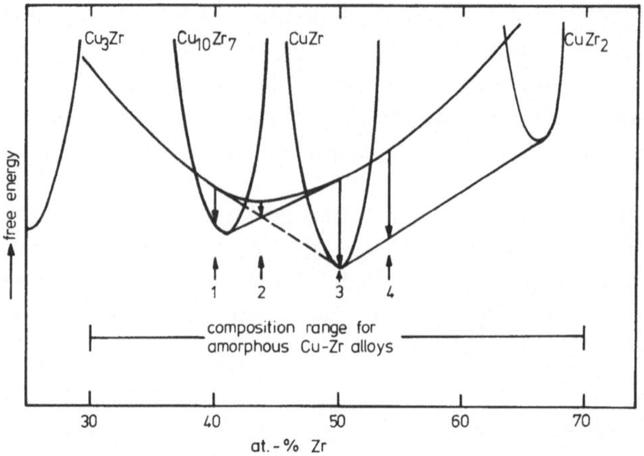

Fig. 10.22. Hypothetical diagram of the free energy for the various phases in Cu–Zr alloys vs concentration. The numbers *1* to *4* correspond to the four alloys mentioned in Sect. 10.5.4

1) $Cu_{60}Zr_{40}$ [10.76]: Calorimetric studies indicate that during isochronal heating crystallization occurs by a single exothermic reaction into (probably) $Cu_{10}Zr_7$. Samples previously isothermally annealed for some time below T_g exhibit two exothermic reactions, whereof the first disappears progressively with increasing preannealing times. After such annealing treatment below T_g very small crystals with fcc structure have been found by electron microscopy which are assumed to be Cu or a copper-rich phase leaving a more stable amorphous matrix.
2) $Cu_{56}Zr_{44}$ [10.33]: After isothermal annealing just below T_g primary crystallization of $Cu_{10}Zr_7$ controlled by long-range diffusion has been observed. After longer annealing times the amorphous matrix enriched in zirconium crystallizes into CuZr. At temperatures above T_g only isochronal experiments have been performed, indicating that CuZr is formed first leaving a more stable copper-rich glass which crystallizes at a higher temperature into $Cu_{10}Zr_7$. But so far, is has not been determined precisely that such a copper-rich glass co-exists with CuZr following the first reaction. The two reactions occur too fast and overlap too extensively to determine the microstructural changes associated which each individually.
3) $Cu_{50}Zr_{50}$ [10.77]: Isothermal annealing about 14 °C below T_g results in crystallization of CuZr after a relatively long incubation time. Crystallization during isochronal heating has been observed to occur by a two-step reaction process above T_g via a metastable so far unknown crystalline phase.
4) $Cu_{46}Zr_{54}$ [10.77, 78]: Eutectic-like crystallization with dendritic morphology of CuZr and $CuZr_2$ has been found after annealing about 15 °C below T_g, whereas above T_g, $CuZr_2$ is crystallized during the first reaction, while CuZr has not been identified until the material is heated through the second exothermic reaction.

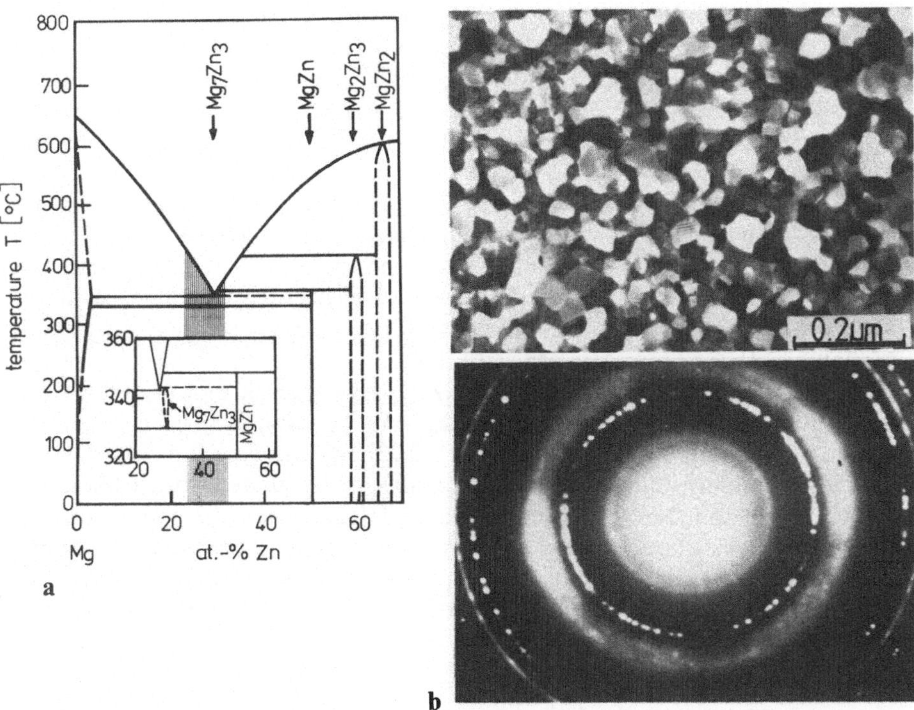

Fig. 10.23a, b. Crystallization of Mg–Zn metallic glasses. (**a**) Part of the Mg–Zn phase diagram [10.81]; the glass forming range is indicated by the shading. (**b**) Electron micrograph and diffraction pattern of $Mg_{72}Zn_{28}$ after crystallization into $Mg+Mg_7Zn_3$ (30 min at 100 °C)

All these reactions are understandable qualitatively using a hypothetical free energy versus zirconium-content diagram as shown in Fig. 10.22. However, we are still far off an understanding for example why reaction paths below and above T_g are so different; and there is quite a large number of unsolved problems concerning the nucleation processes in these glasses.

10.5.5 Mg–Zn Metallic Glasses

Amorphous Mg–Zn alloys which are model glasses for those nontransition metals which form glasses on liquid quenching have been produced by splat-quenching in the range of compositions from 24 to 32 at.% Zn [10.79]; production of amorphous ribbons by melt-spinning is limited to the immediate region of $Mg_{70}Zn_{30}$ [10.80]. In Fig. 10.23a the relevant section of the equilibrium phase diagram is shown [10.81]. The intermediate phase Mg_7Zn_3 occurs over a narrow range of compositions, very near to the relatively deep eutectic. The structures of MgZn and Mg_7Zn_3 are still unknown; Mg_7Zn_3 has been found to possess a large, slightly distorted cubic unit cell ($a \simeq 1.4$ nm).

In amorphous $Mg_{70}Zn_{30}$ the principal feature of thermal analysis (80 K/s) has been reported to be an initial doublet peak (107 °C/128 °C) associated with the polymorphous crystallization of very fine grained Mg_7Zn_3 which is highly defective and a recrystallization process to more perfect grains; a much smaller peak occurs at a higher temperature (224 °C) due to the precipitation of some magnesium [10.80].

In $Mg_{74}Zn_{26}$ the activation energy for polymorphous crystallization into a very fine grained supersaturated Mg_7Zn_3 phase is about 160 kJ/mole [10.80]; the activation energy for the subsequent precipitation of magnesium and formation of more perfect Mg_7Zn_3 grains has been reported to be about 120 kJ/mole [10.80], which is equal to that for eutectoid decomposition of bulk crystalline Mg_7Zn_3 into Mg+MgZn [10.82].

After isothermal annealing of $Mg_{72}Zn_{28}$ splats for 30 min at 100 °C, Mg crystals and extremely fine grained Mg_7Zn_3 crystals have been found by electron microscopy as shown in Fig. 10.23b; the magnesium crystals are strongly textured with the (001) planes parallel to the surface of the ribbon. No partial crystalline state has been found so far after any thermal treatment, indicating a lack of quenched-in nuclei and a very high nucleation rate above T_g. Further annealing at higher temperatures leads to decomposition of Mg_7Zn_3 into Mg+MgZn until these two phases react by forming Mg_7Zn_3 again at about 330 °C.

10.6 Influences of External Factors on Crystallization

10.6.1 Influence of Surfaces

Surfaces are supposed to have a strong influence on crystallization which has been neglected for a long time. In crystalline alloys surfaces act as preferred nucleation sites for phase transformation due to the reduction of the total surface energy and accelerated surface diffusion. In the case of metallic glasses there is a strong evidence that this is not the dominating influence on crystallization, but selective oxidation and segregation create concentration gradients at or near surfaces which strongly influence thermal stability [10.20, 83, 84]. In amorphous $Fe_{40}Ni_{40}B_{20}$, for example, the nickel concentration has been assumed to increase near the surface due to selective oxidation, i.e., preferred oxidation of iron, thus reducing locally the crystallization temperature. As shown in Fig. 10.24a crystallization has been observed to start at the surface of such a ribbon. In amorphous $Fe_{80}P_{20}$ preferred eutectic crystallization of α-Fe+Fe_3P at internal cracks induced by cold rolling prior to the annealing is supposed to be due to phosphorus segregation to these oxygen-free fresh surfaces (Fig. 10.24b). In thin foils, the very close surfaces can change the crystallization behavior drastically by a locally significant rise of the free energy due to the surface energy term. In $Fe_{80}B_{20}$ for example such a rise of free energy

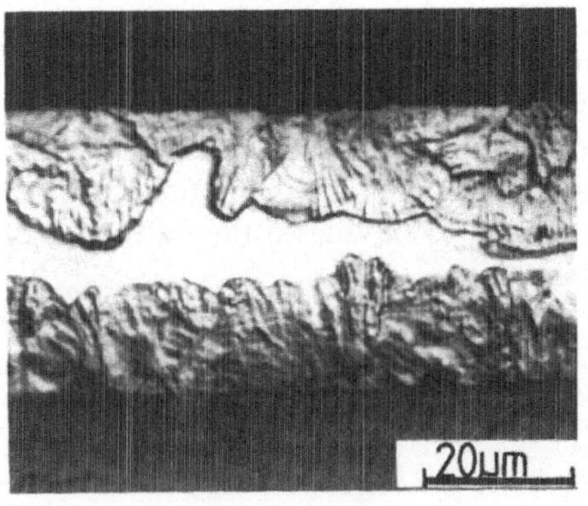

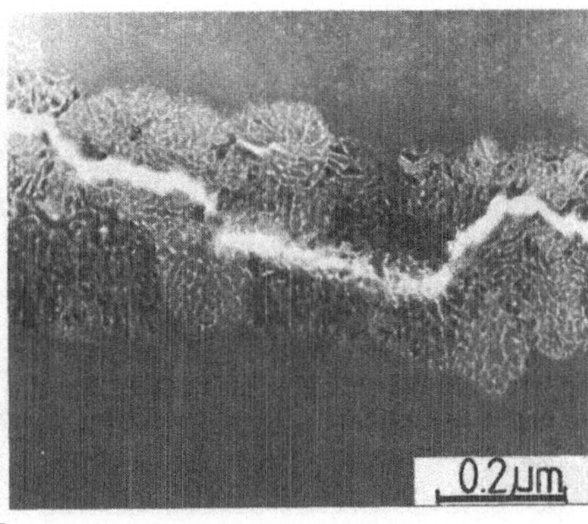

a b

Fig. 10.24a, b. Influence of surfaces on crystallization. (**a**) Preferred crystallization at the surface of $Fe_{40}Ni_{40}B_{20}$ metallic glasses; 1 h at 400 °C (LM). (**b**) Preferred crystallization at internal cracks in $Fe_{80}P_{20}$ metallic glasses; 5 min at 440 °C (TEM)

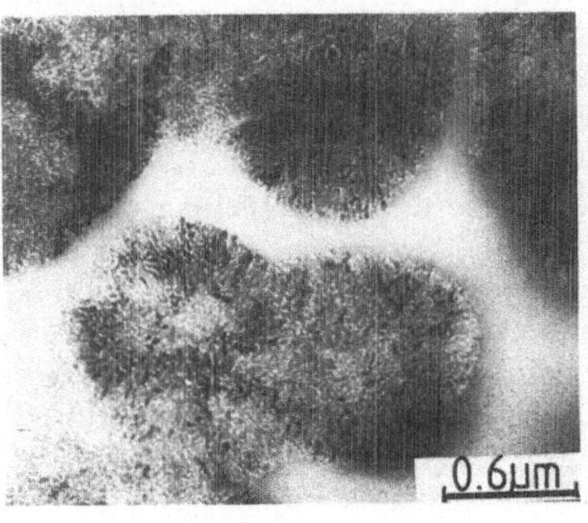

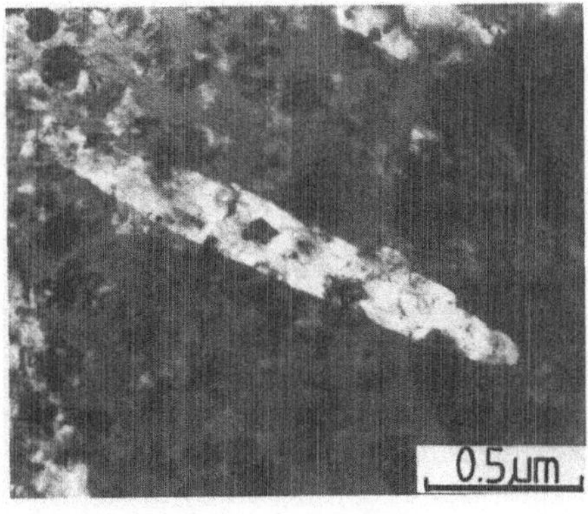

a b

Fig. 10.25a, b. Dimensional effect on crystallization. (**a**) Partially crystallized $Fe_{80}B_{20}$ annealed in the bulk state; 1 h at 380 °C (TEM); (**b**) partially crystallized $Fe_{80}B_{20}$ annealed after thinning by electro-polishing; 5 min at 400 °C (TEM)

could be reduced by iron segregation to the surface; iron enrichment at the surface or only the locally higher free energy both should increase the driving force for α-iron primary crystallization at the surface. Figure 10.25b shows a thinned $Fe_{80}B_{20}$ foil after in situ annealing in a hot stage of an electron microscope. Instead of eutectic crystallization (Fig. 10.25a) in ribbons annealed in the bulk state one observes primary crystallization of α-iron followed by eutectic or nearly polymorphous crystallization of Fe_3B. And in addition to the crystallization of tetragonal Fe_3B some orthorhombic Fe_3B has been found which is assumed to occur in the thin foil due to the easier relaxation of stresses induced by the density differences between the amorphous and crystalline phases.

So far we do not have any direct measurements of such oxidation or segregation processes. But all these results show very clearly that crystallization observed during in situ hot stage electron microscopy of specimens thin enough to be transparent for electrons cannot be assumed to be representative for bulk amorphous ribbons. Much of the discrepancy between different studies on the micromechanisms of crystallization is assumed to be due to the influences of surfaces during such in situ investigations.

10.6.2 Influence of the Quenching Rate and Temperature

Since complete structural relaxation of metallic glasses is generally assumed to occur prior to crystallization, quenching rate, irradiation or plastic deformation, all influencing structural and chemical disorder, should have no influence on crystallization behavior. To the authors' knowledge, there are no systematic studies of the influence of quenching rate and temperature on crystallization. Higher quenching rates are known to increase structural and chemical disorder; therefore, one would expect an acceleration of crystallization due to the higher driving force, but the contrary is observed: As shown in Fig. 10.11c that side of a ribbon which was in contact with the wheel during the casting process, i.e., which was quenched at a higher rate, does not show any crystallization. As mentioned in Sect. 10.3, during quenching even by surpassing any detectable crystallization some fluctuations or even nuclei may be quenched in. Higher quenching rates, therefore, will reduce size and number of such quenched-in nuclei, thus retarding crystallization which is in accordance with our observations. Only near some pockets which are due to gas bubbles during the casting process and therefore areas of locally reduced cooling rates can one observe a high density of crystallites. Further on, this assumption is confirmed by observations that for example $Fe_{40}Ni_{40}B_{20}$ glasses which form very easily, i.e., are produced with a relatively high quenching rate, do not crystallize from any quenched-in nuclei, but as shown in Fig. 10.24a from the surface. Reducing the quenching temperature as mentioned for Pd–Si alloys seems to increase the number of quenched-in nuclei, thus reducing the "crystallization temperature" [10.69].

10.6.3 Influence of Irradiation

Irradiation by neutrons have been found to stabilize $Pd_{80}Si_{20}$ glasses as indicated by raising the crystallization temperature [10.85]. This behavior could be interpreted by amorphization of some quenched-in nuclei during the irradiation process. Both retardation (in $Fe_{80}P_{13}C_7$ glasses [10.86]) and enhancement (in $Fe_{75}B_{25}$ glasses [10.87]) of crystallization by electron irradiation have been reported; these studies, however, were done with in situ hot stage electron microscopy of thinned ribbons.

10.6.4 Influence of Plastic Deformation

On the influence of plastic deformation, there are again some controversial reports. Most authors believe that "defects" introduced by plastic deformation will anneal out during the relaxation processes prior to any crystallization and they do not expect any influence. After deformation in compression in $Fe_{83}P_{10}C_7$ glasses, however, an acceleration of crystallization as indicated by electrical resistivity measurements has been reported [10.88], whereas amorphous $Fe_{80}P_{13}C_7$ alloys deformed by cold rolling by about 30% reduction in thickness have been claimed to show the opposite [10.89]. This discrepancy may result from the different deformation modes used or from the expected differences in the crystallization mode due to the particular composition. Internal cracks, very often produced during severe deformation of metallic glasses, are preferred nucleation sites for crystallization in Fe–P alloys and probably in Fe–P–C alloys, too.

Microscopic investigations do not show any influence of cold rolling on crystallization in such Fe–P–C alloys or other metallic glasses, except in Co–Fe–B alloys, where surface nucleation is accelerated due to the creation of new surfaces during the very localized deformation.

10.6.5 Influence of Hydrostatic Pressure

One might expect that an increased pressure during annealing would accelerate a phase change which is accompanied by an increase in density, but the reduced atomic mobility at high pressure in fact very often retards the reaction. When pressure is applied to some amorphous elements like arsenic (18 kbar [10.90]) or germanium (60 kbar [10.91]), the amorphous state is transformed into a crystalline state even at room temperature. In amorphous metals, however, high pressure has been shown to retard crystallization. By means of electrical resistivity measurements the "crystallization temperature" of $Pd_{75}Ag_5Si_{20}$ is increased by 1.4 K/kbar under hydrostatic pressure up to 6 kbar [10.92]. This result has been confirmed by other investigations where pressure up to 100 kbar has been found to retard crystallization in amorphous $Pd_{80}Si_{20}$ and to change the crystallization reaction: crystallization of a fcc phase (almost pure palladium) is followed by the crystallization of a so far unknown bct phase [10.93]. In $Fe_{80}B_{20}$ alloys onset of crystallization has been observed up to pressures of 25 kbar to be retarded by about 1.5 K/kbar [10.94].

10.6.6 Influence of Uniaxial Tension

An acceleration of crystallization observed in some metallic glasses with the application of uniaxial stress does not contradict the results mentioned above, since uniaxial tension unlike hydrostatic pressure is known to produce appreciable shear components within the material which provide a driving force for atoms to move. In $Pd_{80}Si_{20}$ the crystallization of a fcc solid solution at low annealing temperatures has been found to be accelerated by the application of tensile stress [10.95]. In $Pd_{82}Si_{18}$, however, this accelerated crystallization has not been observed [10.96]. Very recently, in $Fe_{40}Ni_{40}P_{14}B_6$ glasses the activation energy for crystallization during deformation calculated from thermal analysis data has been reported to be appreciably lower than that in unstressed material by an amount which depends on the applied stress [10.97]. Unfortunately, at the moment there is no information available on whether nucleation or growth, or even the mode of crystallization is influenced by hydrostatic pressure or uniaxial tension.

10.7 Conclusions

In metallic glasses, depending on the concentration of a particular alloy, the following modes of crystallization have been found to occur by nucleation and growth processes:

a) Polymorphous crystallization into a single crystalline phase with the same composition as the metallic glass, e.g., $Fe_{75}B_{25} \rightarrow Fe_3B$. Depending on the annealing temperature, steady-state or transient homogeneous nucleation has been observed as well as just growth of quenched-in nuclei at relatively low temperatures. Near surfaces or in thinned foils oxidation and segregation effects lead to drastic changes of the observed crystallization reactions, for example to primary crystallization of α-iron instead of the polymorphous crystallization.

b) Eutectic crystallization produces a microstructure consisting of two-phase "spherulites". Interlamellar spacings which have been found to increase with decreasing annealing temperature can be used together with the growth rate for calculating diffusion coefficients in metallic glasses.

c) By primary crystallization, the homogeneous amorphous alloy transforms into a metastable equilibrium of crystals in an amorphous matrix which in turn can crystallize by one of the other reactions: polymorphous or eutectic crystallization. Growth of primary crystals has been found to be controlled by volume diffusion as indicated by the parabolic relationship between crystal radii and annealing time.

For amorphous phase separation prior to any crystallization some evidence has been found only in ternary glasses, e.g., in Ti–Zr–Be alloys. All these reactions can be understood qualitatively by using a hypothetical diagram of free energy of the different phases versus concentration. Whereas on a qualitative basis our knowledge on crystallization may be reasonable, in details or on a more quantitative basis we are far from understanding crystallization, especially nucleation phenomena or diffusion reactions.

Studies so far seem to support the suggestion that the lack of crystalline phases, especially with small unit cells and relative similar concentrations rather than the stabilization of the amorphous state, is responsible for the stability of metallic glasses. Thermal stability of metallic glasses has been found to be determined by the fastest mode of crystallization available in a particular alloy, the driving force, the kind and number of quenched-in nuclei, the time lag for homogeneous nucleation, and the diffusivity. Usually used so-called crystallization temperatures can give only an integrated view on the influence of all these parameters on thermal stability.

Acknowledgement. The authors would like to thank Professor E. Hornbogen for his interest in this work and his stimulating discussions. The financial help for the experimental work by the "Deutsche Forschungsgemeinschaft" (DFG) is gratefully acknowledged.

References

10.1 H.Ehrenreich, D.Turnbull: Comments Solid State Phys. BIII **3**, 75–81 (1970)
10.2 D.E.Polk: J. Non-Cryst. Solids **5**, 365–376 (1971)
10.3 F.Spaepen: J. Non-Cryst. Solids **31**, 207–221 (1978)
10.4 R.W.Cahn: In *C. R. du 21e Colloque de Métallurgie* (C. E. N., Saclay 1979)
10.5 T.Egami: Mater. Res. Bull. **13**, 557–562 (1978)
10.6 H.H.Liebermann, C.D.Graham, Jr., P.J.Flanders: IEEE Trans. MAG-**13**, 1541–1543 (1977)
10.7 H.S.Chen: J. Appl. Phys. **49**, 4595–4597 (1978)
10.8 B.Cantor (ed.): *Proceedings of the Third International Conference on Rapidly Quenched Metals, Brighton* (Metals Society, London 1978)
10.9 A.L.Greer, J.A.Leake: In Ref. 10.8, Vol. I, pp. 299–306
10.10 M.A.Marcus: Acta Metall. **27**, 879–891 (1979)
10.11 A.J.Drehmann, W.L.Johnson: Phys. Status Solidi (a) **52**, 499–507 (1979)
10.12 J.Krause, T.C.Long, D.G.Onn, P.F.Gleeson, T.Egami: In *Extended Abstracts*, MRS Annual Meeting 1979, Symposium K, Cambridge, MA (Materials Research Society, University Park, PA 1979)
10.13 H.S.Chen, L.C.Kimmerling, J.M.Poate, W.L.Brown: Appl. Phys. Lett. **32**, 461–463 (1978)
10.14 J.L.Walter, D.G.Legrand, F.E.Luborsky: Mater. Sci. Eng. **29**, 161–167 (1977)
10.15 H.S.Chen: Mater. Sci. Eng. **26**, 79–82 (1976)
10.16 For example, J.Hafner, L. von Heimendahl: Phys. Rev. Lett. **42**, 386–389 (1979)
10.17 E.Hornbogen, I.Schmidt: In *Liquid and Amorphous Metals*, ed. by E.Lüscher, H.Coufal (Sijthoff and Nordhoff, Alphen 1980) pp. 353–380
10.18 M.Sostarich, W.Golenia: Solid State Commun. **31**, 339–342 (1979)
10.19 M. von Heimendahl, G.Maussner: In Ref. 10.8, Vol. I, pp. 424–427
10.20 U.Köster: Krist. Tech. **14**, 1369–1376 (1979)
10.21 U.Köster, P.Weiß: J. Non-Cryst. Solids **17**, 359–368 (1975)
10.22 U.Herold, U.Köster: In Ref. 10.8, Vol. I, pp. 281–290
10.23 M.Marcus, D.Turnbull: Mater. Sci. Eng. **23**, 211–214 (1976)
10.24 J.W.Cahn, R.J.Charles: Phys. Chem. Glasses **6**, 181 (1965)
10.25 C.-P.Chou, D.Turnbull: J. Non-Cryst. Solids **17**, 169–188 (1975)
10.26 H.S. Chen: Mater. Sci. Eng. **23**, 151–154 (1976)
10.27 L.Tanner, R.Ray: Scr. Metall. **14**, 657–662 (1980)

10.28 H.Rawson: *Inorganic Glass-Forming Systems* (Academic Press, London 1967)
10.29 M.E.Fine: *Phase Transformation in Condensed Systems* (MacMillan, New York 1964)
10.30 U.Köster, P.S.Ho: Unpublished results
10.31 A.Götzberger: Z. Phys. **142**, 182–200 (1955)
10.32 G. Kuglstatter: Diploma Thesis, Erlangen (1979)
10.33 R.L.Freed, J.B.VanderSande: Acta Metall. **28**, 103–121 (1980)
10.34 H.B.Aaron, D.Fainstein, G.R.Kotler: J. Appl. Phys. **41**, 4404–4410 (1970)
10.35 E.Hornbogen: Metall. Trans. **10**A, 947–972 (1979)
10.36 P.G.Boswell, G.A.Chadwick: Scr. Metall. **10**, 509–513 (1976)
10.37 S.Ranganathan, T.S.Tivari, M. v. Heimendahl: Private communication
10.38 D.Turnbull: Contemp. Phys. **10**, 473–488 (1969)
10.39 E.Hornbogen, I.Schmidt: In Ref. 10.8, Vol. I, pp. 261–264
10.40 D.Turnbull: Trans. AIME **175**, 774–783 (1948)
10.41 I.Gutzow, S.Toschev: In *Advances in Nucleation and Crystallization of Glasses*, ed. by L.L.Hench (American Ceramic Society, Columbus, OH 1971), pp. 10–23
10.42 U.Köster: Phys. Status Solidi (a) **48**, 313–321 (1978)
10.43 C.Birac, D.Lesueur: Phys. Status Solidi (a) **36**, 247–251 (1976)
10.44 D.Gupta, K.N.Tu, K.W.Asai: Phys. Rev. Lett. **35**, 796–799 (1975)
10.45 F.Spaepen: Acta Metall. **23**, 729–743 (1975)
10.46 R.W.Cahn, J.E.Evetts, J.Patterson, R.E.Somekh, C.K.Jackson: J. Mater. Sci. **15**, 702–710 (1980)
10.47 U.Köster, U.Herold, H.-G.Hillenbrand, J.Denis: J.Mater. Sci. **15**, 2125–2128 (1980)
10.48 F.A.Shunk: *Constitution of Binary Alloys* (McGraw-Hill, New York 1969) Second supplement
10.49 J.L.Brimhall, R.Wang, H.E.Kissinger: In *Extended Abstracts*, AIME Fall Meeting 1979, Milwaukee (The Metallurgical Society of AIME, Warrendale, PA 1979)
10.50 M.Takahashi, M.Koshimura: Jpn. J. Appl. Phys. **16**, 1711–1712 (1977)
10.51 R.Ray, R.Hasegawa, C.-P.Chou, L.A.Davis: Scr. Metall. **11**, 973–978 (1977)
10.52 U.Köster, U.Herold: Scr. Metall. **12**, 75–77 (1978)
10.53 F.E.Luborsky, H.H.Liebermann: Appl. Phys. Lett. **33**, 233–234 (1978)
10.54 J.L.Walter, S.F.Bartram, R.R.Russell: Metall. Trans. **9**A, 803–814 (1978)
10.55 T.Kemény, I.Vincze, B.Fogarassy, S.Arajs: In Ref. 10.8, Vol. I, p. 291–298
10.56 U.Herold, U.Köster: Z. Metallkde. **69**, 326–332 (1978)
10.57 A.S.Schaafsma, H.Snijders, F. van der Woude: In Ref. 10.8, Vol. I, pp. 428–430
10.58 J.Tóth: Rpt. KFKI-1978-33, Central Research Institute for Physics, Budapest (1978)
10.59 R.Hasegawa, R.Ray: J. Appl. Phys. **49**, 4174–4179 (1978)
10.60 E.Coleman: Tech. Mem. TM 78-1521-42, Bell Labs, Murray Hill (1978)
10.61 M.Matsuura: Solid State Commun. **30**, 231–233 (1979)
10.62 M.Hansen: Constitution of Binary Alloys (McGraw-Hill, New York 1958)
10.63 H.Franke, U.Herold, U.Köster, M.Rosenberg: In Ref. 10.8, Vol. I, pp. 155–162
10.64 P.Duwez, R.H.Willems, R.C.Crewdson: J. Appl. Phys. **36**, 2267–2269 (1965)
10.65 M.Scott: In Ref. 10.8, Vol. I, pp. 198–213
10.66 A.Nyland: Acta Chem. Scand. **20**, 2381–2386 (1966)
10.67 E.Röschel, C.-J.Raub: Z. Metallkde. **62**, 840–842 (1971)
10.68 B.G.Lewis, H.A.Davies: Mater. Sci. Eng. **23**, 179–182 (1976)
10.69 P.Duhaj, V.Sládek, P.Mrafko: J. Non-Cryst. Solids **13**, 341–354 (1973–74)
10.70 P.Duhaj, D.Barančok, A.Ondrejka: J. Non-Cryst. Solids **21**, 411–428 (1976)
10.71 N.Funakoshi, T.Kanamori, T.Manabe: Jpn. J. Appl. Phys. **16**, 515–516 (1977)
10.72 P.Duhaj, J.Sitek, M.Prejsa, P.Butvin: Phys. Status Solidi (a) **35**, 223–233 (1976)
10.73 R.Ray: PhD Thesis, MIT, Cambridge, MA (1969)
10.74 A.J.Kerns, D.E.Polk, R.Ray, B.C.Giessen: Mater. Sci. Eng. **38**, 49–53 (1979)
10.75 H.S.Chen, J.T.Krause: Scr. Metall. **11**, 761–764 (1977)
10.76 J.M.Vitek, J.B.VanderSande, N.J.Grant: Acta Metall. **23**, 165–176 (1975)
10.77 R.L.Freed, J.B.VanderSande: J. Non-Cryst. Solids **27**, 9–28 (1978)
10.78 R.L.Freed, J.B.VanderSande: Metall. Trans. **10**A, 1621–1630 (1979)

10.79 P.G.Boswell: Mater. Sci. Eng. **34**, 1–5 (1978)
10.80 A.Calka, M.Madhara, D.E.Polk, B.C.Giessen: Scr. Metall. **11**, 65–70 (1977)
10.81 K.P.Anderko, E.J.Klimek, D.W.Levison, W.Rostokker: Trans. ASM **49**, 778–793 (1957)
10.82 W.J.Kitchingman, I.M.Vesey: J. Inst. Met. **98**, 52–54 (1970)
10.83 U.Herold, U.Köster: In *Extended Abstracts*, MRS Annual Meeting 1979, Symposium K, Cambridge, MA (Materials Research Society, University Park, PA 1979)
10.84 U.Herold, U.Köster, A.G.Dirks: J. Magn. Magn. Mater. **19**, 152–156 (1980)
10.85 H.Kayano, T.Masumoto, S.Tomizawa, S.Yajama: Sci. Rep. Res. Inst. Tohoku Univ. A**26**, 240–245 (1977)
10.86 M.Doi, M.Yoshida, N.Nonoyama, T.Imura, T.Masumoto, Y.Yashiro: Mater. Sci. Eng. **23**, 169–172 (1976)
10.87 M.Kiritani, T.Yoshiie, F.E.Fujita: In Ref. 10.8, Vol. II, p. 308–317
10.88 H.Leda: Przerobka Plastyczno **18**, 223–226 (1979)
10.89 F.E.Fujita, T.Masumoto, M.Kitaguchi, M.Ura: Jpn. J. Appl. Phys. **16**, 1731–1738 (1977)
10.90 C.T.Wu, H.L.Luo: J. Non-Cryst. Solids **18**, 21–28 (1975)
10.91 O.Shimomura, S.Minomura, N.Sakai: Philos. Mag. **29**, 547–558 (1974)
10.92 W.C.Emmens, J.Vrijen, S.Radelaar: J. Non-Cryst. Solids **18**, 299–302 (1975)
10.93 H.Iwasaki, T.Masumoto: J. Mater. Sci. **13**, 2171–2176 (1978)
10.94 M.Cedergren, G.Bäckström: J. Non-Cryst. Solids **30**, 69–76 (1978)
10.95 R.Maddin, T.Masumoto: Mater. Sci. Eng. **9**, 153–162 (1972)
10.96 A.I.Taub, F.Spaepen: Scr. Metall. **13**, 195–198 (1979)
10.97 J.Patterson, D.R.H.Jones: Scr. Metall. **13**, 947–949 (1979)

Notes Added in Proof for Chapter 6

J. Hafner

An independent investigation of the atomic structure of glassy $Mg_{0.7}Zn_{0.3}$ using time-of-flight pulsed neutron diffraction has been performed by *Mizoguchi* et al. [6.136]. Qualitatively, these experiments confirm the results of *Rudin* [6.96], the agreement with the theoretical model being even slightly better. *Higashi* et al. [6.137] determined the crystal structure of $Mg_{51}Zn_{20}$, a phase previously designated as "Mg_7Zn_3" [6.23]. The structure is described as an arrangement of icosahedral coordination polyhedra. If we compare the average partial coordination numbers of this crystalline phase and of the amorphous model, we find that the number of Zn–Zn neighbors is lower in the crystal than in the model (1.7 versus 3.2), whereas the number of Mg–Mg neighbors is increased (9.4 versus 8.7), the mixed and the total coordination numbers being nearly identical. This correlates with the fact that the Mg–Mg pair potential is more attractive than the Zn–Zn potential (cf. Fig. 6.7). Further investigations along this line are presently being done.

An investigation on tunneling states in metallic glasses (including $Mg_{0.7}Zn_{0.3}$), which is similar in spirit to that of *Smith* [6.128] on amorphous semiconductors, has been performed by *Banville* and *Harris* [6.138]. Since only one-atom movements are considered, both models share the same limitations. A more promising line of approach seems to consist in the study of defects and of their dynamical properties, as proposed by *Egami* et al. [6.139].

Ishii and *Fujiwara* [6.140] have calculated the vibrational properties (vibrational density of states (DOS) and dynamical structure factor) of amorphous $Fe_{100-x}P_x$ ($x=24.3$ and 15.1) alloys using the recursion method. *Yamamoto* et al. [6.141] used both the equation-of-motion and the recursion techniques to study the vibrational states of amorphous iron. *Kobayashi* and *Takeuchi* [6.142] investigated the dynamical properties of amorphous Cu–Zr alloys using the recursion method, but the agreement of the theoretical DOS with the experimental DOS of *Suck* et al. [6.143] is only fair. Neutron inelastic scattering experiments have been performed on glassy $Ca_{0.7}Mg_{0.3}$ and $Mg_{0.7}Zn_{0.3}$ alloys by *Suck* et al. [6.144, 145]. For the $Mg_{0.7}Zn_{0.3}$ glass the dispersion of short wavelength collective density excitations could be measured for the first time, but their interpretation in terms of *von Heimendahl*'s [6.24] calculations is difficult.

A more complete investigation (both experimentally and theoretically) of the electronic transport properties of amorphous $Mg_{0.7}Zn_{0.3}$ has been presented by *Hafner* et al. [6.146], for the experimental aspect see also *Mizoguchi* et al. [6.136].

References

6.136 T. Mizoguchi, N. Shiotani, U. Mizutani, T. Kudo, S. Yamada: "Structure and Physical Properties of a Simple Metal-Metal Amorphous Alloy $Mg_{0.7}Zn_{0.3}$", J. Phys. Paris **41**, C8-183 (1980) [Proc. 4th Int. Conf. on Liquid and Amorphous Metals, Grenoble 1980, ed. by F. Cyrot-Lackmann, P. Desré]

6.137 I. Higashi, N. Shiotani, M. Uda, T. Mizoguchi, H. Katoh: "The Crystal Structure of $Mg_{51}Zn_{20}$", preprint, to appear in J. Solid State Chem. **36** (1981)

6.138 M. Banville, R. Harris: "Tunneling States in Metallic Glasses: A Structural Model", Phys. Rev. Lett. **44**, 1136 (1980)

6.139 T. Egami, K. Maeda, V. Vitek: Structural Defects in Amorphous Solids, Philos. Mag. A**41**, 883 (1980)

6.140 Y. Ishii, T. Fujiwara: "Vibrational Spectrum of Amorphous Metallic Alloy $Fe_{100-x}P_x$", J. Phys. F**10**, 2125 (1980)

6.141 R. Yamamoto, K. Haga, T. Mihara, M. Doyama: "The Vibrational States in a Realistic Model of Amorphous Iron", J. Phys. F**10**, 1389 (1980)

6.142 S. Kobayashi, S. Takeuchi: "Vibrational States in a Model Amorphous $Cu_{57}Zr_{43}$", J. Phys. C**13**, L969 (1980)

6.143 J. B. Suck, H. Rudin, H. J. Güntherodt, H. Beck, J. Daubert, W. Gläser: "Dynamical Structure Factor and Frequency Distribution of the Metallic Glass $Cu_{46}Zr_{54}$ at Room Temperature", J. Phys. C**13**, L167 (1980)

6.144 J. B. Suck, H. Rudin, H. J. Güntherodt, D. Tomanek, H. Beck, C. Morkel, W. Gläser: "Frequency Distribution and Dynamic Structure Factor of a Metallic Glass", J. Phys. Paris **41**, C8-175 (1980) [Proc. 4th Int. Conf. on Liquid and Amorphous Metals, ed. by F. Cyrot-Lackmann, P. Desré]

6.145 J. B. Suck, H. Rudin, H. J. Güntherodt, H. Beck: "Short Wavelength Transverse Collective Modes in a Metallic Glass", J. Phys. C (in print)

6.146 J. Hafner, E. Gratz, H. J. Güntherodt: "The Electrical Resistivity of Molten and Glassy Mg–Zn Alloys", J. Phys. Paris **41**, C8-512 (1980) [Proc 4th Int. Conf. on Liquid and Amorphous Metals, Grenoble 1980, ed. by F. Cyrot-Lackmann, P. Desré]

Subject Index

B. K. Agarwal

X-Ray Spectroscopy

1979. 188 figures, 31 tables. XIII, 418 pages
(Springer Series in Optical Sciences, Volume 15)
ISBN 3-540-09268-4

Contents:
Continuous X-Rays. – Characteristic X-Rays. – Interaction of X-Rays with Matter. – Secondary Spectra and Satellites. – Scattering of X-Rays. – Chemical Shifts and Fine Structure. – Soft X-Ray Spectroscopy. – Experimental Methods. – Appendices. – Wavelength Tables. – References. – Author Index. – Subject Index.

Amorphous Semiconductors

Editor: M. H. Brodsky
1979. 181 figures, 5 tables. XVI, 337 pages
(Topics in Applied Physics, Volume 36)
ISBN 3-540-09496-2

Contents:
M. H. Brodsky: Introduction. – *B. Kramer, D. Weaire:* Theory of Electronic States in Amorphous Semiconductors. – *E. A. Davis:* States in the Gap and Defects in Amorphous Semiconductors. – *G. A. N. Connell:* Optical Properties of Amorphous Semiconductors. – *P. Nagels:* Electronic Transport in Amorphous Semiconductors. – *R. Fischer:* Luminescence in Amorphous Semiconductors. – *I. Solomon:* Spin Effects in Amorphous Semiconductors. – *G. Lucovsky, T. M. Hayes:* Short-Range Order in Amorphous Semiconductors. – *P. G. LeComber, W. E. Spear:* Doped Amorphous Semiconductors. – *D. E. Carlson, C. R. Wronski:* Amorphous Silicon Solar Cells.

H. Bilz, W. Kress

Phonon Dispersion Relations in Insulators

1979. 162 figures in 271 separate illustrations.
VIII, 241 pages
(Springer Series in Solid-State Sciences, Volume 10)
ISBN 3-540-09399-0

Contents:
Summary of Theory of Phonons: Introduction. Phonon Dispersion Relations and Phonon Models. – Phonon Atlas of Dispersion Curves and Densities of States: Rare-Gas Crystals. Alkali Halides (Rock Salt Structure). Metal Oxides (Rock Salt Structure). Transition Metal Compounds (Rock Salt Structure). Other Cubic Crystals (Rock Salt Structure). Cesium Chloride Structure Crystals. Diamond Structure Crystals. Zinc-Blende Structure Crystals. Wurtzite Structure Crystals. Fluorite Structure Crystals. Rutile Structure Crystals. ABO_3 and ABX_3 Crystals. Layered Structure Crystals. Other Low-Symmetry Crystals. Molecular Crystals. Mixed Crystals. Organic Crystals. – References. – Subject Index.

Dynamics of Solids and Liquids by Neutron Scattering

Editors: S. W. Lovesey, T. Springer
1977. 156 figures, 15 tables. XI, 379 pages
(Topics in Current Physics, Volume 3)
ISBN 3-540-08156-9

Contents:
S. W. Lovesey: Introduction. – *H. G. Smith, N. Wakabayashi:* Phonons. – *B. Dorner, R. Comès:* Phonons and Structural Phase Transformations. – *J. W. White:* Dynamics of Molecular Crystals, Polymers, and Absorbed Species. – *T. Springer:* Molecular Rotations and Diffusion in Solids, in Particular Hydrogen in Metals. – *R. D. Mountain:* Collective Modes in Classical Monoatomic Liquids. – *S. W. Lovesey, J. M. Loveluck:* Magnetic Scattering.

E. N. Economou

Green's Functions in Quantum Physics

1979. 49 figures, 2 tables. IX, 251 pages
(Springer Series in Solid-State Sciences, Volume 7)
ISBN 3-540-09154-8

Contents:
Green's Functions in Mathematical Physics. – Green's Functions in One-Body Quantum Problems. – Green's Functions in Many-Body Sytems. – Appendices.

Excitons

Editor: K. Cho
1979. 118 figures, 8 tables. XI, 274 pages
(Topics in Current Physics, Volume 14)
ISBN 3-540-09567-5

Contents:
K. Cho: Introduction. – *K. Cho:* Internal Structure of Excitons. – *P. J. Dean, D. C. Herbert:* Bound Excitons in Semiconductors. – *B. Fischer, J. Lagois:* Surface Exciton Polaritons. – *P. Y. Yu:* Study of Excitons and Exciton-Phonon Interactions by Resonant Raman and Brillouin Spectroscopies.

Springer-Verlag
Berlin
Heidelberg
New York

O. Madelung

Introduction to Solid-State Theory

Translated from the German by B. C. Taylor
1978. 144 figures. XI, 486 pages
(Springer Series in Solid-State Sciences, Volume 2)
ISBN 3-540-08516-5

Contents:
Fundamentals. – The One-Electron Approximation. – Elementary Exitations. – Electron-Phonon Interaction: Transport Phenomena. – Electron-Electron Interaction by Exchange of Virtual Phonons: Superconductivity. – Interaction with Photons: Optics. – Phonon-Phonon Interaction: Thermal Properties. – Local Description of Solid-State Properties. – Localized States. – Disorder. – Appendix: The Occupation Number Representation.

Neutron Diffraction

Editor: H. Dachs
1978. 138 figures, 32 tables. XIII, 357 pages
(Topics in Current Physics, Volume 6)
ISBN 3-540-08710-9

Contents:
H. Dachs: Principles of Neutron Diffraction. – *J. B. Hayter:* Polarized Neutrons. – *P. Coppens:* Combining X-Ray and Neutron Diffraction: The Study of Charge Density Distributions in Solids. – *W. Prandl:* The Determination of Magnetic Structures. – *W. Schmatz:* Disordered Structures. – *P.-A. Lindgård:* Phase-Transitions and Critical Phenomena. – *G. Zaccaï:* Application of Neutron Diffraction to Biological Problems. – *P. Chieux:* Liquid Structure Investigation by Neutron Scattering. – *H. Rauch, D. Petraschek:* Dynamical Neutron Diffraction and Its Application.

H. Raether

Excitation of Plasmons and Interband Transitions by Electrons

1980. 121 figures, 17 tables. VIII, 196 pages
(Springer Tracts in Modern Physics, Volume 88)
ISBN 3-540-09677-9

Contents:
Introduction. – Volume Plasmons. – The Dielectric Function and the Loss Function of Bound Electrons. – Excitation of Volume Plasmons. – The Energy Loss Spectrum of Electrons and the Loss Function. – Experimental Results. – The Loss Width. – The Wave Vector Dependency of the Energy of the Volume Plasmon. – Core Excitations. – Application to Microanalysis. – Energy Losses by Excitation of Cerenkov Radiation and Guided Light Modes. – Surface Excitations. – Different Electron Energy Loss Spectrometers. – Notes Added in Proof. – References. – Subject Index.

C. P. Slichter

Principles of Magnetic Resonance

2nd corrected printing of the 2nd revised and expanded edition.
1980. 115 figures, 9 tables. XII, 397 pages
(Springer Series in Solid-State Sciences, Volume 1)
ISBN 3-540-08476-2

Contents:
Elements of Resonance. – Basic Theory. – Magnetic Dipolar Broadening of Rigid Lattices. – Magnetic Interactions of Nuclei with Electrons. – Spin-Lattice Relaxation and Motional Narrowing of Resonance Lines. – Spin Temperature in Magnetism and in Magnetic Resonance. – Double Resonance. – Advanced Concepts in Pulsed Magnetic Resonance. – Electric Quadrupole Effects. – Electron Spin Resonance. – Summary. – Problems. – Appendices. – Selected Bibliography. – References. – Author Index. – Subject Index.

Springer-Verlag
Berlin
Heidelberg
New York